Tiger Bravo's War

Currahee Press LLC
1639 Bradley Park Drive
Suite 500/366, Columbus, Georgia 31904

First Edition Paperback 2017
Copyright © 2017 by Richard St John

ISBN 978-0-9988542-1-2

Certificate of registration from Copyright Office reads:
Registration Number: TXu 2-022-580
Effective Date: July 25, 2016

Cover photo: Rick St John

Design: Toelke Asssociates
www.toelkeassociates.com

Tiger Bravo's War

Rick St John

Currahee Press

This book is dedicated to

the men of Tiger Bravo who did not return home,

and to the families who love them, still.

Contents

Preface

Tiger Bravo's War chronicles the exploits of the men of B Company, 2nd Battalion (Airborne), 506th Infantry—an airborne rifle company in the 3rd Brigade, 101st Airborne Division, during one year of combat operations in the Vietnam War. The year was 1968, the bloodiest year of a decade-long war, marked by strife at home and deadly combat on the battlefield.

On one level, this book is a day-to-day accounting of the operations of B Company, called Tiger Bravo by the soldiers who served in it. At the same time, this book is a history of the war itself in 1968, attempting to place Tiger Bravo's tactical missions in the context of a wider operational scenario and political backdrop—though such a perspective was rarely available to the troops on the ground. This is also a book about everyday life in a war zone and the strange, often harsh, sometimes beautiful, tropical environment in which the war was fought. Through its pages, perhaps those who know little about the Vietnam War, or the country of South Vietnam, will gain some understanding of a war fought halfway around the world.

Last, it is a soldier's tale of the young men of Tiger Bravo, both the old soldiers of today, telling war stories no one wants to hear, and those forever young soldiers, lost so many years ago, who no one will ever forget. I am one of those young men of Tiger Bravo. I joined the company in December 1967 as executive officer, and then served as its commander through many of the events recounted in this book. Because I lived through it, I can speak directly to the bond that connects soldiers—the indescribable, enduring relationship that blossoms wherever men and women fight and die for each other. A bond that was born of friendship and mutual respect, given wings by shared hardships and loss of innocence in a strange land, and hardened in the same blast furnace of combat known to generations of infantrymen. A bond that is as real today as it was yesterday, and will be for all tomorrows.

I will try to be objective in telling Tiger Bravo's story. To be anything less would break the trust placed in me by the Tiger Bravo veterans and families to tell this amazing story just as it happened, both the good and the bad. It will not be an easy task, but I will do my best.

—Rick St John

Glossary

AK-47: Basic automatic assault rifle carried by enemy soldiers. Often referred to as simply as "AK."

ARVN: (Army of the Republic of Vietnam): The South Vietnamese Army.

CA: (combat assault): Heliborne attack into enemy territory by troop-carrying helicopters, normally preceded by artillery and helicopter gunship preparation of the landing zone.

Chieu Hoi: A VC or NVA who had surrendered and volunteered to work with an American combat unit.

CO: (commanding officer): Officer in command of a company, battalion, or brigade. Sometimes called "the Old Man."

Cordon and search: Typical infantry company mission where the company surrounds a village, usually at night, and then, accompanied by the South Vietnamese National Police, searches for the enemy.

CP: (command post): At company level, anywhere the company commander and his radios were located.

Currahee: Motto of the 506th Airborne Infantry Regiment. Cherokee for "Stand Alone" and name of the mountain in the regiment's training center during World War II. Used as a common greeting between members of the regiment.

FO: (forward observer): Individual who calls for and adjusts artillery, mortars, and other supporting fire.

FSB: (fire support base): A temporary base of operations surrounded by bunkers and barbed wire, normally containing one or more batteries of artillery, mortars, a headquarters element, and one or two infantry companies for security.

1SG: (first sergeant): Top noncommissioned officer in the company. Reported directly to the company commander.

KIA: (killed in action): Any soldier killed from hostile action. Does not include death due to accident or disease.

Line unit: Any combat unit such as an infantry company. To be "on the line" was to serve in a combat unit versus a combat support or service support unit.

LP/OP: (listening post/observation post): A small team, usually three to five soldiers, posted for early warning on likely avenues of approach into fire support bases and night defensive positions. Called an OP in daylight hours and an LP at night, when observation is extremely limited and soldiers must rely on "listening" for enemy movement.

LZ: (landing zone): Relatively level area in enemy territory where helicopters offload soldiers during a CA.

M-16: Basic automatic rifle carried by American and South Vietnamese soldiers.

MIA: (missing in action): Any soldier unaccounted for during or after a battle.

NDP: (night defensive position): An overnight, temporary company defensive position. Normally, a 360-degree perimeter of foxholes protected by claymore mines and trip flares.

NCO: (noncommissioned officer): All sergeants, from sergeant (E5) to sergeant major (E9).

NVA: (North Vietnamese Army): Any individual or unit from North Vietnam fighting in South Vietnam against Allied forces.

OPCON: (operational control): When a unit is controlled by a higher headquarters other than its parent unit during combat operations. For example, when Tiger Bravo was "opcon to" the 1/506th Battalion, it responded to orders from the 1/506th Battalion commander during a battle.

PZ: (pickup zone): Relatively level area in enemy territory where helicopters pick up soldiers to conduct a CA.

RPG: (rocket-propelled grenade launcher): A shoulder-fired weapon carried by VC and NVA infantrymen that shoots a rocket tipped with a conical-shaped grenade.

RTO: (radio telephone operator): Soldier who carries a radio for an officer or noncommissioned officer in a leadership position.

S & D: (search and destroy): A typical infantry company mission, sometimes called an RIF (reconnaissance in force) or a sweep, where the company moves along a designated route or in a designated area looking for an enemy whose exact location is not known.

TOC: (tactical operations center): Location, at battalion level or higher, where the commander and his staff monitor and control operations. Battalion TOCs are normally inside an FSB.

VC: (Viet Cong): Any individual or unit, indigenous to South Vietnam, fighting in South Vietnam against Allied forces.

WIA: (wounded in action). Any soldier wounded by hostile action. Does not include loss due to accident or disease.

XO: (executive officer): Second in command of a company, battalion, or brigade.

Prologue

It was Saturday, December 2, 1967, when a small group of young paratroopers from B Company, 2nd Battalion (Airborne), 506th Infantry, stole a six-foot-tall tiger and successfully smuggled it on board the US Air Force C-141 Starlifter carrying them to the war zone in Vietnam. They were well trained, silent, swift, tactically dispersed, and garbed in olive drab jungle fatigues that were more suited for the hot and humid climate of Southeast Asia than the thirty-degree cold of a Tennessee winter morning.[1] They moved with just a hint of youthful exuberance, suggestive of the high school pranks not far in their past. It was all over in minutes. The tiger never saw them coming, and it didn't make a sound.

This was not the result of a daring and dangerous capture in the wild; the tiger was stolen from an Esso gas station during a rest break while the company was enroute to Campbell Army Airfield from nearby Fort Campbell, Kentucky, home of the 101st Airborne Division — the legendary 'Screaming Eagles'. The tiger was not flesh and blood, of course; rather, it was a large banner, used in the iconic Esso gasoline "put a tiger in your tank" advertising campaign of the 1960s. Nevertheless, it was a tiger, and it symbolized B Company, which was known throughout the battalion as "Tiger Bravo."

B Company, along with the other companies in the battalion (A, C, D, E and Headquarters Company), was on the road to join the remainder of the 101st Airborne Division in Operation Eagle Thrust—code name for their deployment to the Republic of Vietnam. Tiger Bravo numbered 160 assigned and attached officers, noncommissioned officers (NCOs), and enlisted men.[2] The company roster was a slice of American youth at the time. They came from small towns with names only they had heard of: Binger, Oklahoma; Red Lion, Pennsylvania; Ewa Beach, Hawaii; Corder, Missouri; Piedmont, Alabama. Others came from neighborhoods in iconic cities such as Los Angeles, Detroit, Chicago, and Philadelphia. One spent his youth in South America, returning to the United States in time for high school because of a revolution in Venezuela. Some were as comfortable in the woods and mountains with a hunting rifle in their hands as others were in catching a bus on a busy street. They were college students and high school dropouts; one walked away just a few days short of graduation to enlist. Some were high school athletes who traded cleats for boots and cheerleaders' coy glances for NCOs' stern stares.

Most were young, many still in their teens. One enlisted man celebrated his twentieth birthday on the plane to Vietnam. Their ancestry could be traced back to waves of immigrants from Europe and the United Kingdom, the Hawaiian Islands, Japan, Mexico, Puerto Rico, Central and South America, and a host of other countries. Included in the mix were descendants of African slaves and Native American warriors who had fought against the army only a century before.

These young paratroopers were no different from other innocents arriving in the war zone for the first time. They found themselves in a strange corner of the world, fighting a war that would reverberate in their lives, and in the consciousness of their nation, for decades after.[3] They followed orders, broke a few regulations, drank too much, trained incredibly hard, and played even harder. They thought themselves to be invincible, yet unabashedly said what everyone knew could be the last good-bye to their families. About heading into combat, they displayed equal parts bravado, curiosity, excitement, anxiety, naïveté, and anticipation. Their mental image of combat stemmed more from Hollywood's interpretation than the harsh, ugly reality. They had less fear of dying than they had of being the one to freeze up or "chicken out" under fire. They thought of themselves as paratroopers, trained to kill; yet their families saw only boys who had learned to jump out of airplanes and shoot a rifle. To a man they preferred going to war as a unit, with comrades they knew and trusted, rather than as an individual replacement lost in the army's personnel system. Instinctively they understood what would later come out in a postwar study that "units that trained and deployed as units generally performed far better_than those in which personnel were assigned as individuals."[4]

As many closely-knit groups do, they had nicknames for each other. Bill Plemons was a decade older than his squad mates; he became "Pop." Dave Spencer was small, weighing no more than 145 pounds. Too small, said a buddy, to be a big bird of prey like a "screaming eagle," so he became "Sparrow." Spencer had originally enlisted in the navy, but mistakenly reported to an army reception center in Texas. When a transfer to the navy bogged down in a bureaucratic shuffle, he opted to stay in the army, put in his three years, and then reenlist in the navy.[5]

Two Hawaiians, one big and one small, were in the same platoon; they became "Big Pineapple" and "Little Pineapple." All medics received the same nickname; "Doc." Sergeant First Class (SFC) Andy Matosky was "Mad Dog," although no one was crazy enough to use it in his presence. Mike Tarpley was walking as pointman when he discovered a massive rice cache and the battalion commander started calling him "Bird Dog." Chris Backman answered to "Babe," for no reason other than that's what popped out of the first sergeant's mouth the first time the two met.[6]

The majority were volunteers to the army, but all had volunteered to be a paratrooper. Some walked into recruiting offices seeking a better life or to live out a boyhood dream. A few volunteered to see what war was all about. Still others wanted to follow fathers and uncles who had been in World War II. But all had a fierce pride in being airborne and a Screaming Eagle, which to their minds made them the best of the best—the elite of the army.

The company was organized into a headquarters section, three rifle platoons, and a mortar platoon. Its chain of command was a mix of a few experienced Vietnam veterans, such as the commanding officer, Captain Freddie D. Rankin (Special Forces), from Little Rock, Arkansas; the first sergeant, Sergeant First Class Herman L. Trent (173rd Airborne Brigade), from Hazard, Kentucky, plus new, untried officers and NCOs. Of the lieutenants, only the executive officer First Lieutenant (1LT) Don Goates, had been a commissioned officer for more than a year.

PREPARING FOR WAR

The deployment had been ordered months earlier. On a typical hot, humid, no-breeze kind of day at Fort Campbell in early August 1967, the 101st Airborne Division received an alert for deployment to "an undisclosed destination in US Army Pacific," with a scheduled arrival of February 1968.[7] Although the message was classified "secret," word flew around the 105,068-acre installation[8] and quickly reached B Company at Camp Natural Bridge in West Point, NY. Along with other companies in the 2nd Battalion (Airborne), 506th Infantry Regiment, Tiger Bravo was supporting West Point's Cadet Summer Training Program for its yearlings (second-year cadets). Few in the division

were surprised at the news, but the reception varied widely, from testoster-
one-charged excitement swirling in the barracks to dread mixed with resig-
nation lying heavily in the family quarters on Fort Campbell and in homes
scattered across the border communities of Hopkinsville, Kentucky, and
Clarksville, Tennessee. Even though for security reasons the destination was
not named, everyone knew it was to the Republic of South Vietnam—the only
war on that side of the world worthy of the commitment of one of this nation's
irreplaceable, strategic reserve airborne divisions.

Initial plans called for surface transportation to take the bulk of the division straight across the Pacific Ocean from ports on the West Coast in late January and early February 1968, giving the division ample time for a nine-week
training schedule in preparation for the combat that lay ahead.[9] However, a
series of changes to the original dates moved up the departure to early December 1967. Soldiers were cautioned not to share the date with their families.
Most, however, found a way to bend but not break the security rules. In a letter
home, one young soldier wrote, "I finally got the official word on when we go
over & also when the leaves start. The departure date is classified but I can tell
you that I think I will be able to see the Bob Hope Christmas Show live this
year." Like many of his age, he was about to achieve his dream to "become a
paratrooper and get into the war."[10]

While perceived by many to be a fully capable, tough, and ready strike
force, as its reputation implied, the 101st Airborne Division in the summer of
1967 was hardly more than a cadre-level nucleus. Its capabilities had been severely depleted from constantly feeding replacements to its 1st Brigade, which
had been in the war zone for more than two years, and to other airborne units
actually fighting the war.[11] In addition to being chronically understrength, the
division struggled to sustain viable unit training for its battalions and brigades,
and to maintain its equipment in a high state of readiness. It was clear to all
involved in the planning that the division had a long, arduous path ahead to
be ready for combat.

Planners soon found that to fill the two remaining infantry brigades at
Fort Campbell to a minimum combat-ready strength level of 75 percent would
require more than forty-five hundred replacements.[12] To exacerbate the personnel problem even further, many currently assigned paratroopers were

deferred from being deployed because they were only sons, their enlistment would be up before or shortly after deployment, or they had recently returned from Vietnam. The 3rd Brigade, which counted the 2/506 as one of its maneuver battalions, had eight hundred of these nondeployable troopers. As quickly as the nondeployables were outprocessed, replacements arrived from the 82nd Airborne Division and scattered units from across the United States and Europe. During the last two weeks of September 1967 approximately five thousand replacements were inprocessed and assigned to divisional units.[13]

Not all of the replacements held the much-needed infantry military occupational specialty (MOS), nor were all of them airborne qualified. The dearth of available airborne infantry officers was so acute that the army's personnel center in Virginia assigned approximately two hundred airborne-qualified armor officers to fill airborne infantry officer slots.[14] There was also an army-wide shortage of airborne-qualified physicians. When a nonairborne-qualified medical doctor arrived to be the 2/506 Battalion's surgeon, it was viewed as a potential "morale problem," so several medics in the headquarters company were instructed to work the doctor into shape before he was sent to the arduous Airborne School at Fort Benning, Georgia. One of those medics recalled, "After a few weeks of PT and running, the doctor was shipped off to Benning, but was sent back to Campbell a week later. We then doubled down on PT and running in preparation for another try. When he was sent back a second time without jump wings, the Battalion Commander pronounced him to be 'honorary airborne,' because the division's deployment to Nam was imminent and we were out of time."[15]

Just when it seemed that the personnel system was making progress in finding replacements and building up unit strength, the totally unexpected happened—an emergency requisition of five hundred replacements to be transferred to the 173rd Airborne Brigade, which was already in Vietnam. All units in the 101st shared in the pain. In mid-november the 173rd had taken part in the Battle of Dak To, which was the longest and most violent battle in the Central Highlands since the Battle of Ia Drang two years before. Enemy casualties numbered in the thousands, with an estimated 1,400 killed. American units had suffered, too. Approximately a fifth of the 173rd were casualties, with 174 killed, 642 wounded, and 17 missing in action.[16]

In just a few days, the replacements were selected, issued orders, rushed through the Preparation for Overseas Movement (POM) process, and taken to the airfield to be loaded on USAF transports headed to South Vietnam. Not all 173rd Brigade replacements were volunteers. Most complied, but a select few had to be escorted planeside by an armed officer with orders to "make sure everyone from the battalion is on the plane, and don't leave the airfield until wheels up." Luckily, the departure went off without incident.

To replenish an understrength division for deployment to a war zone would be daunting under any circumstances; but there was another, equally disruptive force working against the orderly, measured steps needed for the division to be ready. Because of a bureaucratic mandate, each infantry battalion's structure changed overnight from four companies to six.[17] Lieutenant Colonel (LTC) David E. Grange Jr., the newly assigned battalion commander of the 2/506 remembered, "We were faced with the problem of building two new companies. We had to build another rifle company (D Company), and a new combat support company (E Company), train them as companies, and then train them as part of the battalion. We just didn't have very much time."[18]

Tough, charismatic, and professional, Grange was a decorated, three-war combat veteran. In World War II he served as a paratrooper with the 517th Airborne Infantry Regiment fighting in Europe. In Korea, he led a platoon with the 187th Airborne Infantry Regiment, and on his first tour in Vietnam he was an adviser to the South Vietnamese Army (ARVN) Airborne. He was a stickler for bayonet training and physical fitness, personally leading long, grueling runs every morning. One morning the order came to fall out for the run in full field gear, with steel pot and weapon. Dan Bernard, who was the company commander's radio operator, remembered thinking, "Who is the idiot who is ordering this thing? Then when we got out there, who was at the head of the battalion but Grange with his rucksack, steel pot, and weapon. I wasn't about to fall out." Bernard went on to say, "If that old man could do it, then so could I."[19]

As the weeks progressed toward deployment, the logistical woes mounted as well. The most serious of these directed the exchange of all M-16 rifles, the basic infantryman's weapon, for the newer M-16A1 version. While enhancing the combat posture of the division, it resulted in a massive supply/exchange effort and required a duplication of training.[20] Most units had already conducted M-16 zero ranges but now needed to rezero their new weapons, not a small feat when every minute was already planned for and ranges were running day and night. For the soldiers at the end of a massive logistical system in overdrive, the drawing of new equipment, weapons, and jungle uniforms never seemed to end. "Our new rifles came this week. They have a few changes over the old M-16s," wrote one soldier. "Our jungle boots also came in and we tried them on yesterday. Last week we were issued green towels, socks, underwear, and handkerchiefs."[21]

In the fall of 1967, despite all the organizational changes and setbacks in scheduling and personnel turbulence, the division moved inexorably toward its deployment date. Individual training, including physical training each morning, weapons qualifications, and classes on jungle tactics, Vietnam and its culture, and VC capabilities filled September. In October and early November, units accelerated the pace and pushed through squad and platoon exercises. It was at the squad level where the frenetic level of training could be felt the most. One soldier described the training as "being rather intensive now. We were out in the field all last week, then going out again on Monday for another week. Then we jump into Florida & stay for 2 weeks of swamp & jungle training."[22]

Little training time was available for company and battalion exercises, although the 2/506 managed to conduct one airborne exercise when it made a mass tactical jump into Camp McCall, North Carolina. Maintaining a viable airborne assault capability was important to the 101st Airborne Division. It was the centerpiece of its battlefield capabilities, a link to its storied heritage from World War II, and the focal point for unit esprit de corps and individual pride. To reinforce that point, most units conducted practice parachute assaults as part of their pre-deployment training, and specifically scheduled a jump close to their departure dates, until parachute training could be started again in Vietnam.

Sufficient time for airmobile training, and mastering the intricacies of coordinated artillery and close air support, were noticeably missing from training schedules. In an after-action report covering the entire predeployment period, the lack of airmobile training was identified as "a serious deficiency across the entire division."[23] The 101st Airborne Division was a lean, para-chute-trained assault force headed to an airmobile-intensive battlefield with only an understrength aviation battalion as its primary airmobile asset. Units would eventually reach a superior level of proficiency in airmobile operations, but only through hard lessons learned on an unforgiving battlefield rather that in the safe training environment at Fort Campbell.

In addition to personnel turbulence, organizational and equipment changes, a compressed training schedule, and the requirement for all deploy-ing personnel to have two weeks of leave, there was one more major drain on the paratroopers' time. Each soldier, regardless of rank or position, had to undergo an extensive Preparation for Overseas Movement (POM) process. Ev-eryone had to completely clear nine post agencies to include the library, credit union, post housing, and the post finance and accounting office; complete a physical examination, eye exam, and dental checkup; have a current will and power of attorney; and, worst of all in most troopers' opinion, receive immuni-zations for smallpox, typhoid, tetanus, yellow fever, influenza, typhus, cholera, and the plague.[24] The division would also deploy with no loose ends. Automat-ically upon deployment, "any paratrooper absent without leave (AWOL) from the unit would be dropped from the rolls as a deserter."[25]

THE HOME FRONT

At about this same time on the home front, a tide of disillusionment with the war was building in some segments of the population. US troop strength was approaching five hundred thousand, and US casualties had reached 15,058 killed and another 109,527 wounded. The war was costing the US taxpayers some $25 billion a year. More casualties were reported in Vietnam every day, even as US commanders demanded more troops. Under the draft system, as many as forty thousand young men were called into service each month, add-ing fire to the anti-war movement already festering in the draft age segment of the population. To many, the war was misguided, unjust and not going well. One of the most significant anti-war demonstrations, to date, took place in October as one hundred thousand protestors gathered at the Lincoln Memo-rial; around thirty thousand protestors continued in a march to the Pentagon later that night.[26] Student protests were increasing across the nation's campuses as well.

While the country from which the 101st paratroopers departed to go to war was smoldering with anti-war sentiment, they also left behind a vibrant and engaging society that they would only know from letters, and the infrequent newspaper, for the next year. Left behind was a country whose youth was engulfed by the Beatles invasion from England, the 1967 Grammy for Song of the Year went to John Lennon and Paul McCartney for *Michelle*. Yet, the nation remained true to its musical past, as the Record of the Year went to Frank Sinatra's *Strangers in the Night*. Television viewers could only choose shows from three networks, ABC, NBC or CBS, with *The Andy Griffith Show* pulling in the highest ratings. The St Louis Cardinals had just won the World Series, beating the Boston Red Sox four games to three, and the University of Southern California (USC) was the NCAA Football Champions. Thurgood Marshall had just been sworn in as the first black Supreme Court Justice.[27] While in Lakeside Elementary School in Seattle Washington, a twelve-year-old student, named Bill Gates, was just beginning to show interest in the new field of computers.[28] Gas at the pump averaged $.33/gallon, a first-class stamp cost $0.5, a ticket at the movies averaged $1.25 and the minimum wage was $1.40 per hour. [29]

THE DAY BEFORE

Forty-eight hours prior to deployment, all units set for departure were sequestered in their barracks. However, with the town of Clarksville, Tennessee, filled with bars and women, only a few miles away, some young paratroopers could not resist the temptation. "After we were confined to the barracks, a few of us slipped off base," recalled one soldier. "We went to town and overdid it on alcohol. What was the military going to do, send us to Vietnam? A local deputy took us back to base. No trouble, just told us to straighten up."[30]

The day before the battalion lifted off from Campbell Army Airfield, LTC Grange addressed his assembled battalion one last time. Looking at the ranks of paratroopers, standing tall in their dark, olive drab jungle fatigues, he saw eager faces, strong hearts, and youthful dreams of killing VC. The troops showed no fear and appeared unfazed by the violent, life-altering storm waiting for them on the other side of the world. He spoke of the tough training just completed, and the tougher times that lay ahead. He reminded the assembled paratroopers that everyone was in this together and that "no matter what the Viet Cong and North Vietnamese Army would throw at them, they would prevail if they stood together, took care of each other and trust one another with their lives." As he recalled years later, "I told them that I didn't give a damn about ethnic structure. We were a band of brothers going off to war, and we would remain a band of brothers. We would bring as many of us back as we could by taking care of each other." Grange went on to say, "It got a nice response from the troops. Everyone was shaking their heads yes, giving it the old billy goat nod."[31]

Everything was in place. Tiger Bravo was on the cusp of creating its own chapter in the long and acclaimed history of the 101st Airborne Division. In the next twelve months, the company would experience close combat at its worst, as deadly as the 506th Regiment faced in the hedgerows of Normandy, and first cousin to the savage jungle fighting in the Pacific during World War II. The toll on the company would be staggering: more than 50 percent of its original members would be killed, wounded, or lost to disease or injury. Thirty of its soldiers would be killed in action, and collectively it would amass 150 Purple Hearts.[32]

But on that cool December morning, the young and untested company took off from Fort Campbell with its stolen, tiger banner proudly displayed, ready for what lay ahead.

Battle Maps Legend

▦▦▦	Main highway	⬤ (shape)	US fire support base
▬▬	Road	⬤ (shape)	VC/NVA camp
– – –	Trail	♥ (shape)	Dense jungle area
➤➤➤	US combat assault	〰	Swamp area
LZ	US landing zone	⬇	Rice paddies and hedgerows
➔	US ground movement	*10m* ▲	Mountain or hill height in meters
➔	VC/NVA ground movement	〰〰	VC/NVA trench
▬	US ground unit position	⌒	Stream
▬	VC/NVA ground unit position		
✦	Enemy contact		
●	US location		
●	VC/NVA location		

Map graphics above are common to all the battle maps. Individual battle maps may have specific site details not shown above, labeled for the particular action.

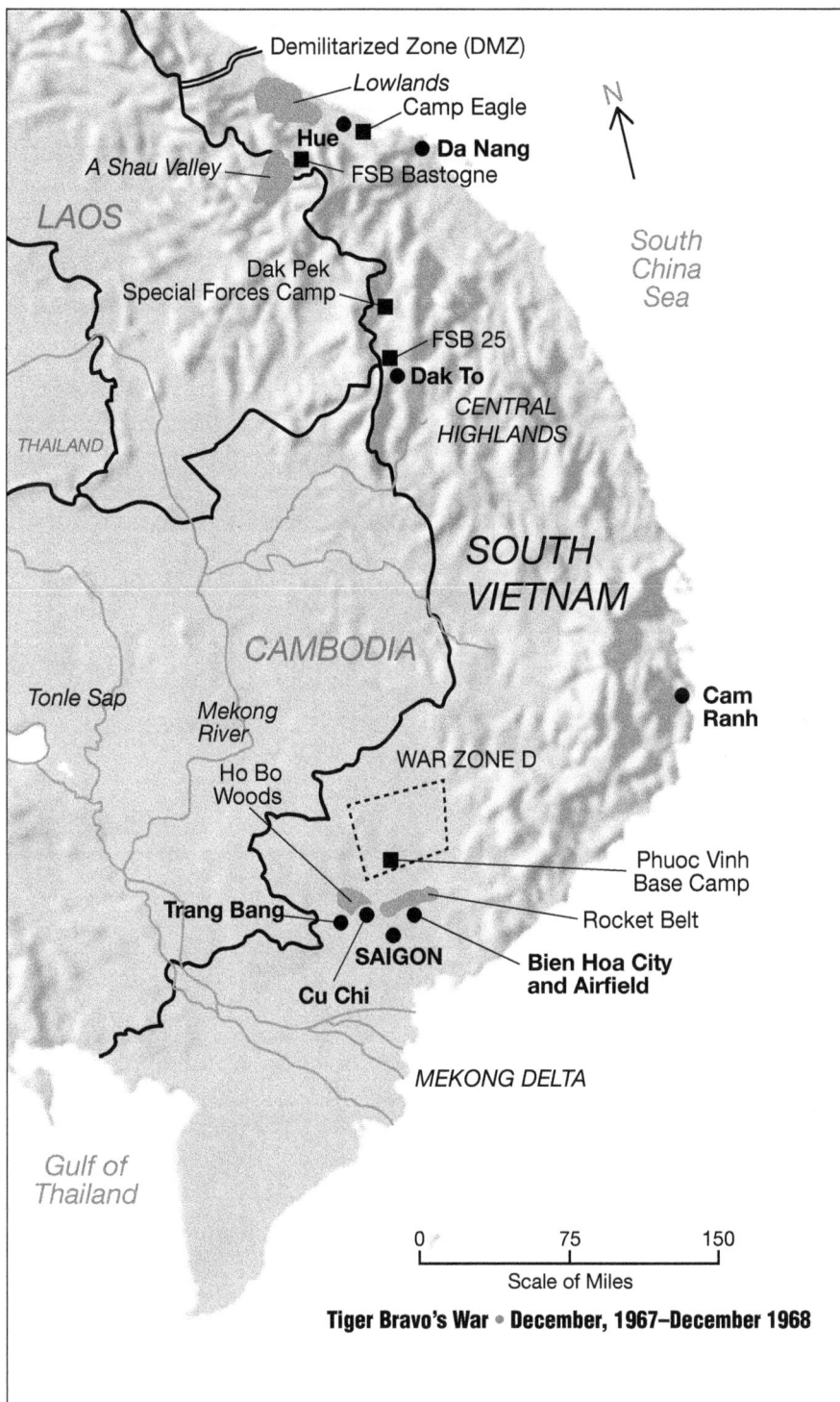

Demilitarized Zone (DMZ)
Lowlands
Camp Eagle
Hue ● Da Nang
A Shau Valley
FSB Bastogne
LAOS

South
China
Sea

Dak Pek
Special Forces Camp
FSB 25
● **Dak To**
CENTRAL
HIGHLANDS

THAILAND

SOUTH
VIETNAM

CAMBODIA

Tonle Sap
Mekong
River
● **Cam
Ranh**

Ho Bo
Woods
WAR ZONE D

Phuoc Vinh
Base Camp
Trang Bang
Rocket Belt
SAIGON
Cu Chi
**Bien Hoa City
and Airfield**

MEKONG DELTA

Gulf of
Thailand

0   75   150
Scale of Miles
Tiger Bravo's War • December, 1967–December 1968

Part I

Operation Eagle Thrust

1

Frontier Enclave

Phuoc Vinh (Binh Duong Province)
December 4, 1967–January 15, 1968

BOMB CRATERS TO THE HORIZON

A vast US Air Force armada of sleek and powerful C-141 Starlifters and lumbering C-133 Cargomaster transport aircraft, split into twin ribbons of gleaming fuselages and white contrails, carried the 101st Airborne Division from its Tennessee base to the Republic of South Vietnam. One curved north through Anchorage Alaska, onto Japan, then Southeast Asia. The other, carrying Tiger Bravo, streaked straight across the Pacific. And so, Tiger Bravo's war began.

For the Tiger Bravo paratroopers, the flight from Fort Campbell, carrying them away from everything familiar and closer with each passing mile to an uncertain fate, seemed to stretch forever. There was little to do onboard, other than sleeping in the rack seats or on the cold metal floor, taking a turn in marathon card games or wondering what it was going to be like in "Nam". First stop was Travis Air Force Base in California, followed by a late afternoon departure across the Pacific and the international date line to Wake Island. Next stop was Clark Air Force Base in the Philippines, then another long leg in the air before the transport banked low over the Song Dong Nai (Dong Nai River) and touched down at Bien Hoa Air Base in Vietnam. For some of the paratroopers onboard, who had never flown in a plane before joining the army, actually landing in a plane was still a novel experience. Every other plane ride had ended with them jumping out of the plane, usually traveling close to 140 miles per hour, with a parachute on their back.

It was a typical sunny, ninety-degree day in the tropics when Tiger Bravo touched down, quite a change from Fort Campbell, and its nighttime lows of below freezing. But, there was nothing typical about the sights and sounds that greeted the new arrivals; the company had landed in the belly of a vast war machine, running at full throttle. A dozen other transports crowded the tarmac, some still discharging lines of paratroopers, others switching crews and refueling for the long trek back across the Pacific. An endless stream of small tractors, pulling trailers stacked with 250 lbs. bombs and rockets, snaked around the arriving troops to an adjacent runway. There, tactical fighters roared off, day and night, for battlefields with odd names like Ho Bo Woods, the Parrots Beak and Rocket Belt, leaving behind the stench of burnt JP-4 jet fuel. Clipboard toting junior lieutenants seemed to be everywhere, recording everything and everyone coming off a plane. Swarms of helicopters darted in and out, added to the tumult. And all around, loomed reminders – barbed-wire, guard towers, sand-bagged bunkers, guard dogs, sentries and roving patrols - that, for all of its fearsome power, this war machine was still in a war zone and vulnerable to attack.

The day after arriving in South Vietnam, Tiger Bravo moved by air thirty-five kilometers due north to a base camp adjacent to the town of Phuoc Vinh, seat of a district capital in Binh Duong Province. One paratrooper, flying by helicopter to Phuoc Vinh, was struck by the effect of the war on the countryside. He wrote that he could see "bomb craters extended out to the horizon."[33] Phuoc Vinh would be the logistical support base of the 3rd Brigade, 101st Airborne Division for the next ten months and an infrequent home for Tiger Bravo on its rare opportunities to stand down and rest.

It was, however, not the first choice for their final destination in Vietnam. The division was originally slated to establish a series of forward operating bases running from points due north of Saigon up to the Cambodian border. This plan was designed to interdict North Vietnamese Army (NVA) infiltration routes and keep them from reinforcing base areas that extended south from Cambodia to an area known as War Zone D, northeast of Saigon. All of this was scrapped when a division staff officer left a copy of the secret plan unattended in his jeep, whereby it was promptly stolen. With the plan very likely compromised, it was decided to deploy the division closer to Saigon. The division headquarters would remain in Bien Hoa, with the 2nd Brigade deployed west of Saigon in the vicinity of Cu Chi and the 3rd Brigade sent to Phuoc Vinh, replacing a brigade from the 1st Infantry Division that had established the base camp two years earlier.

By no means was the Phuoc Vinh base camp in a safe or secure area. It was, in fact, the northernmost point of government-controlled territory along a dirt road called Highway 1A. Convoys, with Military Police security details, could travel north from Bien Hoa to Phuoc Vinh on 1A, but beyond that point it was closed to traffic, and had been since 1954.[34] Its location within War Zone D, one of the most well-established and dangerous of all enemy base areas, made it a tantalizing target for attacks. One such attack occurred in August 1962, when the Military Assistance Command—Vietnam (MACV) compound was attacked by a Viet Cong (VC) regiment and almost overrun. Only a last-minute rescue by South Vietnamese Rangers saved the town, but not before the VC beheaded the Binh Duong Province chief.[35]

The base camp was not the prettiest of places by any stretch of the imagination, with buildings, tents, and trailers erected in no particular order or according to any grand plan. Facility planning and aesthetics took a backseat to mission necessity. If it was needed, build it and get it up and running as fast as possible. Muddy during the rainy season and dusty the rest of the year, Phuoc Vinh was home to more than three thousand American soldiers, airmen, and civilian contractors. One soldier described the base camp as "built on a fine, orange sheet of powdered clay. It was everywhere, and got into everything. During the seasonal rains, the dust turned into a sticky morass of orange mud."[36] But as ugly as it was, it did contain a formidable array of combat power. In addition to an infantry brigade the base camp included an airfield with a thirty-seven-hundred-foot asphalt runway from which US Air Force launched tactical aircraft to support ground troops. It was also home to the 162nd Attack Helicopter Company (known as the "Vultures"),[37] the 2nd Battalion (Airborne), 319th Artillery (105mm howitzers), a battery of 155mm and 175mm artillery, and a variety of service support units. USAF Agent Orange missions were also flown out of Phuoc Vinh. Day and night, the sounds of these war machines conducting their deadly business reverberated across the base. Feeding the insatiable appetite of this beast for ammunition, fuel, spare parts, food, and a thousand other commodities was a daily cavalcade of trucks traversing the dirt highway from the depots of Bien Hoa.

It was a fairly secure enclave on the frontier, beyond which the government of South Vietnam held little to no sway. Walk outside the perimeter and you would be in enemy-controlled War Zone D.[38] Barbed wire, trip flares, guard towers, and antipersonnel mines surrounded the camp, with the perimeter usually guarded by one of the infantry battalions and mortars and artillery readily available to provide fire support. Rather than launching a ground attack against this fairly substantial defense, the VC resorted to the occasional sniper taking a shot at soldiers guarding the perimeter and to nighttime 82mm mortar attacks. While the mortar attacks normally did little damage, they were a constant nuisance, with the shout "Incoming!" sending everyone into a mad scramble for cover.[39] It wasn't long before the troops of Tiger Bravo learned that the mortars were usually fired well over their heads toward the center of the base camp and airfield. Soon there were no more mad scrambles, and sometimes no reaction at all.

By mid-December the sights, sounds, and routines of a war zone had become commonplace: artillery fire all through the night, close air support missions in the distance, periodic enemy probes of the perimeter defenses, reports of VC spies caught in Phuoc Vinh, the smell of burning feces, the sound of helicopters taking off and landing at all hours, and always remembering to never be too far from a bunker or weapon.[40]

The sector for the 2/506 was on the east side of the base camp in an abandoned rubber plantation. The trees were still in orderly rows but had not been producing latex for years, except when used for impromptu bayonet practice by the troops. Tiger Bravo moved into buildings less than a hundred feet inside the perimeter of barbed wire, bunkers, and watchtowers. A and E Companies were not as fortunate, ending up in tents on concrete pads. The B Company buildings, each holding fifteen soldiers, were constructed out of a simple two-by-four frame with screens for walls and a tin roof. Canvas cots lined the walls, with ammunition and combat-ready equipment neatly stacked in the center aisle. For the most part, signs and slogans displayed on platoon hooches at base camp were fairly benign. Not so the one in the mortar platoon, which proudly stated, "You call, we haul. If we can't truck it, mother fuck it." Mike Tarpley remembered it lasted until "someone from Battalion came by and said it disrespected our mothers and we needed to get rid of it."[41]

There was no running water, and no indoor toilets. Latrines consisted of nothing more than a hole cut in a piece of plywood over a fifty-five-gallon drum cut in half. Each day a detail of soldiers would pour diesel fuel into the cutoff drums and burn its contents. The pungent smoke given off by burning feces was a constant irritant, and seemed impervious to even the strongest breeze. "Shit detail" was a powerful incentive to stay on the first sergeant's good side. Showers were available outside, behind the barracks, but were unheated and filled up only every two days.[42] One building for company headquarters, one for the combination supply/arms room, and another for the mess hall completed the company area.[43]

Improvements to the perimeter and company area began immediately. Private First Class Chuck Hanson wrote, "We have been working every day, including Sunday, building bunkers, trench systems between bunkers, placing sand bags around the buildings, etc." The heat was hard to get used to as well. Coming from winter in Tennessee directly into a tropical environment was not easy. Hanson went on to write, "It is very hot & humid & the sun is always out & no wind. We are always wet from sweat, even at night. The effort generated to put on a clean, dry pair of clothes causes us to sweat so we are soaking wet all over again."[44] Hanson, the oldest of four children from northern Minnesota, had dropped out of college in his senior year to enlist. "As the country was in a war at the time, I thought it was my duty to get involved," he wrote. "I specifically volunteered for the Airborne as that would ensure combat and I would be in the company of men who felt as I would."[45]

Initially, the town of Phuoc Vinh was off limits to the 101st Airborne Division. Only 1st Infantry Division soldiers and selected others could visit when off duty. There was not much to the town. It contained a few thousand inhabitants, including a small network of VC spies, with an economy built around the American base. Most locals found work on the base itself as day laborers or maids, or in the bars, massage parlors, and brothels lining the dirt road just outside the main gate. There was nothing unique or memorable about the town, just an assortment of ramshackle buildings made of everything from beaverboard to corrugated tin with colorful, garish signs of misspelled English words.[46]

According to his own account, PFC Bill "Pop" Plemons was behind the decision to finally let the paratroopers into town. One morning Plemons stopped by the battalion aid station and announced to the medics that the town would soon be open to all soldiers because of what he had done. According to Plemons, he had been showering in one of the outside showers in the B Company area when he noticed a full colonel, with a junior officer and clipboard, inspecting the area. He immediately decided to show the colonel that the troops needed an outlet for their pent-up sexual energy, such as the brothels in Phuoc Vinh, by simulating masturbating in full view of the colonel, sound effects and all. "That," Plemons said, "is why we will be going to town."[47]

At thirty-two years old, Pop Plemons was a decade older than the other enlisted men and most of the officers and NCOs in Tiger Bravo, and different from all of them. Leaving his hometown of Crawford, Texas, at age twenty, he served in the US Coast Guard, Peace Corps, and Military Sealift Command before volunteering for the army in March 1967.[48] Irrepressible, irreverent, candid, cantankerous, and a master storyteller, he was known to have waved at the battalion commander rather than salute. It's not that he knowingly broke the rules; he just didn't pay any attention to them. But a more loyal and good friend couldn't be found anywhere, and there was no better soldier in combat than Pop Plemons. He would prove that many times over during Tiger Bravo's first year.

As Steve Lyle, a twenty-one-year-old medic from Amarillo, Texas, told it, "Sure enough, around 1400 hours that same day, Dr. Barrett, the 2/506 Battalion surgeon, came into the aid station and announced that he needed three jeeps and ten to twelve medics to go into town and inspect the prostitutes there and, if necessary, treat them with antibiotics. This was in anticipation that the town would soon be open to the 101st." The first bar/brothel they visited had a large mirror behind the bar, with makeshift shelves full of colorful liquor bottles and velvet upholstered sofas and chairs. Some soldiers from the 1st Infantry Division were drinking warm beer, and a shoeshine boy and a variety of Vietnamese hookers walked the room, soliciting their services. As Lyle recalled, "The shoeshine boy went to one of the medics, PFC Klemish, and asked to shine his boots, and he wouldn't take no for an answer. Irritated, Klemish pushed him aside and said that he was airborne and his boots were already shined." Klemish added, "'If you want to shine boots, go shine that leg's boots,' pointing to a 1st Infantry Division soldier across the room."[49]

The term "leg" had been used by airborne men since World War II. A shortened version of "straight leg," it referred to the common practice of soldiers in nonairborne units to let their trousers hang "straight" down to just above their shoetops, as compared to the tradition in airborne units of paratroopers tucking their trousers into the tops of their jump boots. The term was both a source of pride to the airborne, as a symbol of their elite status, and a flash point between them and other soldiers. It was the primary reason behind many brawls between airborne soldiers and literally everyone else in the US Army. Such was the case in the bar in Phuoc Vinh.

"That not only irritated the kid, but all the legs in the bar as well," Lyle continued. "That's when the kid yelled in broken English, 'He numba one,' pointing to the 1st Infantry soldier, 'you numba ten chicken soldier,' pointing to Klemish." For any paratrooper, under any circumstances, those were fighting words. In an instant, the fight was on. The first to go was the wall between the bar and the bedrooms in the rear, flattened by a 1st Infantry soldier thrown through the air by a paratrooper using skills last practiced in hand-to-hand training at Fort Campbell. Revealed in the act, on the other side of the now flattened wall, was a nude, portly 1st Infantry Division staff sergeant. The large, ornate mirror over the bar went next, smashed by a flying beer bottle. Screaming women and half naked troops scurried all about. "I distinctly remember Barrett yelling, 'Retreat! Retreat to the jeeps!'" said Lyle. "There was a scramble to board the jeeps and I remember the dust cloud we kicked up as we sped down the dirt street back toward the base gate and the madam running after us yelling insults and something about damages as she faded into the distance."

With that escape, the first fight between the two units—the first of many—was over. The town was open to the 3rd Brigade the next day,[50] despite the fact that not one prostitute was inspected or treated.

VIETNAM "SPEAK"

Within the first few weeks, the men of Tiger Bravo quickly became fluent in the specialized lexicon of combat troops. Time and distance were expressed as *mikes* for minutes and *klicks* for kilometers. *Hot* meant enemy contact on a landing zone; *cold* was the opposite. The Viet Cong, a contraction of Viet Nam Cong-san, were *VC, Charlie, Charles, dinks, slopes,* and the all-encompassing *gooks.* Night ambushes were *tigers.* Infantrymen, with the 11B Military Occupational Specialty, were *11 Bravos* or *11 Bush.* Artillerymen were *redlegs. Kit Carsons* were ex-VC who defected and acted as scouts for US units. Operations were conducted in the *boonies, boondocks,* or *bush.* You didn't walk, march, trudge, or move through the boonies, you *humped.* VC could pop out of nearly invisible *spider holes,* within a few feet of alert soldiers, to shoot or throw grenades. Chocolate candy bars in C rations were *John Wayne bars.* A rash, anywhere on the body, was *jungle rot.*

Rome plows were huge D9 Caterpillar bulldozers with blades designed to cut down and clear even the toughest jungle. They were named after Rome, Georgia, where they were built. OD scarfs worn around the neck were *drive-on rags.* Local sandals made with pieces of rubber tire for soles, and worn by most VC, were *Ho Chi Minh sandals. Sitrep negative* meant all was quiet, nothing out of the ordinary to report. Lieutenants were universally called *LT* (Ell Tee). Any small shelter or living quarters, even if just a makeshift poncho shelter, was a *hootch.* They even picked up local Vietnamese pidgin slang: *cockadau* meant kill and *number ten* was bad, the worst. No one wanted to be *tee-tee,* meaning small. Even cash had its own name. Vietnamese piasters were *pees,* and the military payment certificates that were used in US installations instead of dollars were *MPCs.*

The ubiquitous Bell UH-1H troop-carrying helicopter was a *huey* or *slick.* The group of soldiers climbing into the *slick* was a *stick,* usually six soldiers and all their gear. A helicopter used for medical evacuation was a *dust off.* To get a *dust off* you requested *MedEvac*. A *log bird* was another *huey*, only this one was used for resupply. Commanders would circle above the battlefield in helicopters called *C&C ships* that were fitted with a bank of radios. *Gunships* or *hogs* were Bell UH-1C helicopters bristling with different configurations of rocket pods, grenade launchers, miniguns, and M-60 machine guns. Large CH-47 helicopters were *hooks*, referring to the hook on its underbelly where slingloads of supplies were attached. A *loach* meant a light observation helicopter. *Fireflies* weren't insects; they were helicopters with searchlights. *CA* meant a combat assault by helicopter into an *LZ* (landing zone).

LPs were early warning listening posts set up outside the perimeter at night, when the night was so dark that soldiers used only their hearing to differentiate the noise made by enemy movement from normal nighttime jungle sounds. Any company commander, regardless of age, was referred to as *the Old Man*. *Point man* and *slack man* constituted the two-man team out front of the unit on combat operations—the most dangerous job in the war. *Stand-to* meant a time, usually just before dawn, when a unit would be 100 percent alert. To *pop smoke* meant to throw a smoke grenade to mark your position. *Shake 'n' Bakes* were NCOs who were part of an accelerated process to train soldiers to be NCOs, then immediately sent to Vietnam.

Profanity and the F-word were ubiquitous. Replacements were *fucking new guys* or *FNGs*. The least popular C ration was a can of ham and lima beans or *ham and motherfuckers*. Any soldier not humping the boonies with a combat unit was a *REMF*, or *rear-echelon motherfucker*. Lizards that called out in the middle of the night with a human like sound were *fuck you lizards*. But the most pervasive of all terms was a catchall phrase—*FTA*, or *fuck the army*.

Infantry weapons and some of the most lethal war machines on the planet had innocuous, often nonsensical names. An M-60 machine gun was a *hog*. *Thumper* was the name given to the M-79 grenade launcher. An AC-47 aircraft, often called *Spooky*, bristling with 7.62mm Gatling miniguns and carrying its own complement of flares, each generating a light intensity of 2 million candlepower, was *Puff the Magic Dragon* or just plain *Puff*.[51] A high-altitude bombardment of five-hundred-pound bombs and larger, dropped by B-52s from altitudes so high you couldn't see the planes, was an *arc light*. *Foo gas* meant a field expedient mixture of thickened fuel set in the barbed wire surrounding a fire support base (FSB) and detonated in a huge fireball to incinerate an infiltrating enemy, who would then turn into *crispy critters*. *Thud* meant a F-105 fighter jet used in a ground support role. *Willie Pete* wasn't the name of someone's cousin, rather a white phosphorus or incendiary round. *Duster* was the name given to 20mm antiaircraft guns used in a direct-fire, ground-support role. A *beehive* round was an antipersonnel round filled with iron flechettes. *Rock 'n' roll* didn't just mean music the company band played in the mess hall, it also meant firing an M-16 on full automatic. To *drop snake and nape* meant to call in an air strike with 250-pound MK-21 Snakeye bombs and 500-pound M-47 napalm canisters. *H&I* meant harassment and interdiction artillery or mortar fire, usually at night. A *claymore* was not the Scottish variant of a medieval two-handed broadsword;[52] rather, it was a directional, anti-personnel M1A1 claymore mine. Habitually employed on ambushes and in the defense of NDPs, once command detonated, it would shoot steel balls out to a hundred feet in a sixty-degree arc from the device. Troops soon learned that setting a match to a rolled-up ball, the size of a marble, from the C4 plastic explosives lining the back of the claymore could heat a C ration can of coffee in seconds.

LBJ wasn't just the president of the United States, Lyndon Baines Johnson; it was also the in-country penitentiary named the Long Binh Jail. When a soldier hit his *DEROS,* or date of expected return from overseas service, he would climb aboard a *freedom bird* and fly home to the *real world.* Some returning veterans would have to deal with *Jody,* that lowlife back home who *snaked,* or stole, his girlfriend or wife while he was gone. The list went on and on and on. In just a few weeks the soldiers in Bravo could converse in "Vietnam Speak" with such fluency that an outsider in the conversation would be hard pressed to understand half of what was said.

JINGLE BELLS

Once settled in, Tiger Bravo returned to training in earnest, making up for training time lost at Fort Campbell, and became acclimatized to small-unit operations in the tropics. Along with the other companies in the battalion, it rotated through makeshift ranges just outside the perimeter, reviewed squad and platoon formations and tactics, and practiced ambushing at night; only this time, everyone was locked and loaded with live ammunition. Perimeter guard duty was soon added to the mix, as responsibility for the base camp defense and the entire perimeter was shifted from the 1st Infantry Division to the 3rd Brigade on December 16, 1967.[53] For the remainder of December, Tiger Bravo manned its portion of the perimeter, learned the fundamentals of airmobile operations, practiced the tactics and techniques needed to cordon and search a village, perfected how to move securely through the thick terrain of War Zone D, and rehearsed procedures to instantly react if caught in an ambush. Soldiers also sat through boring classes on prevention of heat injuries and tropical diseases, first aid, and currency control. In addition, the 1st Infantry Division and the 25th Infantry Division at Cu Chi provided specialized training on booby traps, mine detection, and tunnel operations.[54]

The pace in base camp slowed for Christmas, and operations came to a halt for a Christmas truce. The holiday spirit led a few infantrymen to rewrite the lyrics to "Jingle Bells" for the occasion, which they printed and distributed throughout the battalion.

JINGLE BELLS, MORTAR SHELLS, VC IN THE GRASS.
  WE'LL GET NO MERRY CHRISTMAS, TILL THIS YEAR IS PASSED.
JINGLE BELLS, MORTAR SHELLS, VC IN THE GRASS.
  TAKE YOUR HOMEMADE CHRISTMAS TREE AND SHOVE IT UP YOUR ASS.
DASHING THROUGH THE MUD, IN A JEEP THAT'S USELESS JUNK.
  OVER THE ROADS WE GO, EVERYONE IS DRUNK.
  WHEELS IN POTHOLES BOUNCE, MAKING ASSES SORE.
  OH, I'D RATHER GO TO HELL THAN FINISH OUT THIS TOUR.
JINGLE BELLS, MORTAR SHELLS, VC IN THE GRASS.
  WE'LL GET NO MERRY CHRISTMAS, TILL THIS YEAR IS PASSED.
JINGLE BELLS, MORTAR SHELLS, VC IN THE GRASS.
  TAKE YOUR HOMEMADE CHRISTMAS TREE AND SHOVE IT UP YOUR ASS.[55]

Needless to say, it was a big hit with the young paratroopers.

All was quiet on Christmas Eve until movement was detected in the barbed wire in front of the 2/506 bunker line. Christmas parties came to a halt as the alert worked its way to the Battalion Tactical Operations Center (TOC), then on to Brigade. It even warranted a visit from the brigade commander, demanding to know why his Christmas party had been interrupted. Soon the sounds of small-arms fire and hand grenade explosions could be heard. Sentinels on the bunker line were taking no chances. It took just minutes for the threat to be identified and contained: a dog had wandered into the barbed wire and spooked a jittery guard.[56]

Three key personnel changes in the company's leadership occurred in December. First, Second Lieutenant Joe Palagyi replaced Second Lieutenant Wigglesworth as the company's artillery forward observer. A native of Cleveland, Ohio, Palagyi came to the company from a stint as fire direction officer in B Battery, 2/319 Artillery. He would prove to be a consummate forward observer (FO), blessed with the ability to remain calm under fire and drop a torrent of hot, exploding metal exactly where needed. He didn't just call for and adjust artillery in support of Tiger Bravo, as would any FO; he "choreographed" the supporting fires, blending in mortars, heavy artillery, naval vessels offshore, flare ships, gunships, tactical aircraft, and any other lethal forms of ordnance that rocketed through the air.

Next, Second Lieutenant Jim Roach joined the company. Roach was a twenty-year-old transfer from battalion headquarters, who appeared more like a skinny sixteen-year-old wearing his older brother's uniform. He had such a baby face that when he first reported to battalion headquarters, a call was placed to the personnel office to make sure it wasn't some elaborate prank. Born and raised in a Catholic family in the suburbs of Philadelphia, his first calling was to be part of the Maryknoll Catholic Foreign Mission Society. But after one semester of Maryknoll Seminary, Jim recalled that "I wasn't ready for college, so I dropped out. Then I realized that I was very likely to be drafted, so I went down to the Army recruiter and signed up." Two weeks into his basic training, all of the high school graduates (only 12 out of a company of 130 recruits) were called into a meeting and asked to volunteer for Officer Candidate School (OCS). "I volunteered, and then asked 'What is OCS?'" Roach explained. "Not exactly an informed career decision."

During the fall of 1968 Jim Roach was serving on the battalion staff where he became known as "Jesus Christ Roach." Everyone was working long hours preparing for the battalion's deployment, and the work pace was hectic, with daily changes. On any given day, his boss, Major Freddie Boyd, the battalion operations officer, would appear in the office, look over Roach's shoulder to see what he was working on, and in his signature way of communicating would roar, "JESUS CHRIST ROACH, WHAT ARE YOU DOING NOW?" Jim was a young infantry lieutenant only a few months out of OCS, and was doing the best that he could, but knew that he still had much to learn. After a minute or two, Freddie Boyd would calm down, and he would spend time helping fill the lieutenant's knowledge gaps.[57]

But Freddie Boyd bellowed "JESUS CHRIST ROACH" so often, that it stuck; James Stevens Roach, a Catholic boy and ex-seminarian from Delaware County, Pennsylvania, had been christened Jesus Christ Roach by a gruff, profane US Army major. But it was "on the line" with a rifle platoon that he made his mark. By the end of the battalion's first year in combat, he would be the longest-serving platoon leader in the battalion—leading three different platoons in combat, twice turning down transfers that would take him off the combat line—and considered one of its best.

Last, I left my job as the battalion's S1 (personnel officer) and replaced Don Goates, the company's executive officer, who was about to be promoted to captain.[58] I was a twenty-three- year-old airborne Ranger, first lieutenant. Like most lieutenants in the battalion, my naïveté about going off to war knew no bounds--all our battles would be victories; my friends would never end up in body bags; the other guy would be hit, not me; and a warrior's welcome home parade, with flags flying and bands playing John Philip Sousa, was waiting for me at the end of my tour. I would be wrong on all counts. I was a member of the Class of 1966 from West Point. By 1968 most of my classmates were either in country or preparing for assignment to the war zone. For a member of our class, the chance of being killed in Vietnam was one out of twenty; the chance of being wounded was one out of six.[59] For an infantryman like myself, the prospects were much grimmer; the chance of being killed or wounded was nearly nine out of ten.

Against my wishes, I had been pulled in the summer of 1967 from A Company, where I was a platoon leader, to become the battalion's S1 just in time for the personnel upheavals of the pre-deployment period. Ever since then, I had pestered the battalion commander to transfer me back to a line company, and it finally was announced that it would happen on December 29, 1967. This long-awaited good news came at the end of what was, up to that point, the worst week of my life. My newborn son and namesake, who was born just six days before I climbed on the plane bound for Vietnam, had been rushed to the hospital for an emergency operation on his neck and was in intensive care. Only two weeks old, his life was in the hands of a surgeon reconstructing a blocked salivary gland, while his father was ten thousand miles away instead of at his bedside. To make matters even more agonizing, somehow, the letter breaking the bad news to me, and explaining everything, didn't arrive until after I had received the second letter that was written, giving an update that the "operation was over, but no one had been able to visit Ricky yet."

Then an angel arrived in the form of a battle-hardened lieutenant colonel with a big heart. Lois Grange, the battalion commander's wife who had a heart as big as her husband's and was a surrogate mother to all the young lieutenants' wives in the battalion, had written to her husband about what was going on and to "check on Rick, his baby is in the hospital." In a letter home that very afternoon, I wrote, "The Colonel saw me sitting on the end of my bunker just holding the letter in my hands. His wife had written and told him, so he tried to cheer me up."[60] The days that followed were a blur of letters and the occasional medical updates sent through the Red Cross. In the end, it all ended happily. The operation was a success, and my son bounced back and recovered completely. And I went on to become the executive officer of the best company in the brigade.

COMBAT READY

January was different for Tiger Bravo right from the start. The company received actual search and destroy missions, no longer going out on quasi-training missions that often turned into the real thing when the VC decided to test the new arrivals. On January 3, 1968, Tiger Bravo participated in the battalion's first combat assault in the Republic of Vietnam[61] (OPORD 1-68). It was designed to be a relatively simple operation. A, B, and C Companies would air assault into helicopter landing zones (LZs) Mike, Pony, and Mule, then sweep back to Phuoc Vinh on foot; driving anything in their path into D Company, which had set up a blocking position just for that purpose. As it turned out the LZs were cold and there was no contact as the ground forces moved along their designated routes, although supporting gunships did fire at ten VC on A Company's first objective.[62] B Company's involvement was uneventful, and not very demanding or time-consuming. Unlike future air assault operations that would be organized rapidly, take place day or night, and often under enemy fire, the battalion's Daily Staff Journal for January 3, 1968,[63] revealed a short, safe, and easily executed first combat mission. The first lift of choppers was airborne from Phuoc Vinh Base Camp at 0822 hours, and the battalion had marched back to Phuoc Vinh by 1739 hours. It was stressful for the rookies in Tiger Bravo, who conjured up VC behind every bush, but it was just an eight-hour walk in the sun for the veterans.

This operation was quickly followed by a second Battalion Operation (OPORD 2–68), which called for the companies, split into Task Force Green and Task Force Brown, to move on two separate axes out of Phuoc Vinh to preselected night defensive positions (NDP).[64] Tiger Bravo's role in the operation was a simple search and destroy mission, with straightforward routes and no significant enemy presence expected. In briefings, the troops had been advised that "if the VC Dong Nai Regiment, which was known to be in this part of War Zone D, decided to engage, then simple and straightforward wouldn't last through the first round being fired." NDPs were temporary, company-size, 360-degree perimeters set up on defensible terrain when the company was on operations away from base camp or a fire support base. In addition to the company perimeter, the company would send out one or two ambush patrols, establish listening posts (LPs) on likely avenues of approach for early warning, ring the perimeter with trip flares and claymore mines, dig foxholes, and register prepositioned artillery targets that could be delivered within seconds if needed.

The operation lasted well into the night. Tiger Bravo had so much trouble passing through such dense terrain that by 2125 hours it was stuck in a bamboo maze. Thirty minutes later Captain Rankin had a flare fired that was spotted by Python 3, the battalion operations officer. Tiger Bravo was only two hundred meters from linkup, but its movement had slowed to a crawl. It wouldn't be until 0220 hours that Bravo Company finally reached its designated NDP location, and by midday it had returned to Phuoc Vinh. One trooper was injured when his M79 grenade launcher fired accidentally and hit his foot, breaking it but not exploding. At about the same time a soldier from Headquarters Company reported a shrapnel wound from grenade fragments. The soldier had stopped during the night to adjust his rucksack when a VC suddenly appeared and tried to grab his weapon and escape. According to the soldier, he threw a hand grenade at the VC, killing him, but in the process he was hit by shrapnel.

On January 9, 1968, the battalion conducted Operation Kickoff (OPORD 3–68). The battalion's mission was to work with combat engineers to open Highway 1A north of Phuoc Vinh to facilitate resupply of future operations to the north. Bravo Company moved out from Phuoc Vinh on foot through Gate 2. Contact with enemy units was not expected, although finding mines and booby traps was a certainty. Soon the reports started. Looker 72 (Engineers) found a booby-trapped hand grenade, Juliet (Recon Platoon) found a minefield, and an airborne forward air controller (FAC) spotted four VC moving with oxcarts. Bravo's move was uneventful, stopping at a suspected tunnel, which turned out to be a well. By 1635 hours all units had closed into the NDP location for the night.[65]

By mid-January the battalion had conducted a series of tactical missions designed to bring the companies to a level of proficiency needed to survive and win on the Vietnam battlefield. It was a measured approach, increasing in complexity with each new mission, as much was completely new to the battalion, or it was a needed skill or capability that was missed at Fort Campbell because of the accelerated schedule. The approach was simple: short, company-size daylight missions in December, and then battalion day and night missions through mid-January. There was much to learn on all levels, from operating with a fourth rifle company (D Company) and a combat support company (E Company), to cross-country movement and land navigation at night through unfamiliar terrain and vegetation. There were also the intricacies of airmobile operations and split-second timing of a combat assault, night ambushes in complete darkness without a sound, and everyone coping with heavy loads, digging foxholes like moles, the heat, the humidity, the strange bugs, voracious leeches, vines and bamboo thickets that seemed impervious to machetes. Everything that would be second nature in weeks was being exercised, mistakes corrected, and then exercised again. It was all carefully planned and controlled by Colonel Grange. He knew the rigors of combat, knew his men, and had a healthy respect for the enemy waiting in War Zone D. He brought all of these to bear in his in-country training plan. Years later, this approach would be described by Mike Malone, considered one of the army's leading experts on small unit leadership, as "Skill + Will + Drill = Kill."[66]

Finally, Tiger Bravo was approved for real missions. The army's standard physiological six-week acclimatization period was over, and B Company was trained and ready. Experience had shown that newly arrived units required a six-week period of adjustment and acclimatization before being committed to the strain of continuous combat operations. It took five whole days to adjust to the abrupt time zone change and to develop a new diurnal cycle. It took another two to three weeks before troops were fully acclimatized to the heat and humidity of the tropics. Also, the army's medical staff had discovered that six weeks were required to develop a "relative biological acclimatization" to the types of infectious organisms encountered in South Vietnam. Commanders recognized that the need for this adjustment period was a physiological and biological reality, and postponed major combat operations accordingly.[67] It was the same across the remainder of the division.

Once commanders at all levels reported their units as combat ready, General Barsanti lost no time in assigning missions to the brigades. Word of the 3rd Brigade's mission quickly spread throughout Tiger Bravo. This was the tough and dangerous mission that everyone had expected since arriving in Vietnam—a mission that would put Tiger Bravo squarely in harm's way. General Barsanti's orders were clear:

DESTROY THE VC DONG NAI REGIMENT AND NEUTRALIZE ITS BASE CAMP

(#359) ALONG WITH OTHER BASE CAMPS AND FACILITIES; DISRUPT VC

INFILTRATION WEST THROUGH AREA OF OPERATIONS MANCHESTER FROM WAR

ZONE D; AND PREVENT THE LAUNCH OF MORTAR, ROCKET AND/OR GROUND

ATTACKS AGAINST THE LONG BINH–BIEN HOA MILITARY COMPLEX THROUGH A

SERIES OF HELIBORNE OPERATIONS AND GROUND ATTACKS."[68]

ALTERNATE UNIVERSE

By army standards, Tiger Bravo was prepared for what lay ahead. The soldiers were acclimated, trained in jungle fighting, able to move quietly both day and night, aware of the enemy forces in the area, and familiar with their tactics. They were full of youthful confidence; in the barracks and along the bunker line there was little talk of death or defeat. There was certainly some apprehension for what lay ahead, but mostly it was anticipation and excitement that carried the conversations. Few could envisage the true nature of combat or foretell the appalling odds they would face to escape unscathed. Chuck Limer, initially assigned to the artillery battalion but later a volunteer to Tiger Bravo, remembered a considerable swing in his thoughts about combat. "I was scared to death and praying to come out alive," he remembered. "Then in the next second, I thought it would be like a John Wayne movie and I would be the hero."[69] Steve Lyle recalled, "I thought [combat] would have some logical sense to it and that I probably would not be wounded, and if I were, it would be minor."[70] Both would soon taste reality firsthand. Each would be seriously wounded in the months ahead.

While there was plenty of talk regarding the Viet Cong, and boasting about what damage Tiger Bravo would inflict on them once the fighting started, there was little thought given to what the VC were thinking or planning. If there was, there would have been far more anxiety and apprehension running through the barracks, as the VC were busily making plans to use its most seasoned, battle-hardened fighters to test this new unit of "chicken soldiers" (a name given to the 101st by locals who were not familiar with the eagle shoulder patch worn by the 101st). The Viet Cong's intelligence apparatus had been collecting information on this American unit it its backyard for weeks, using a loosely knit network of agents, radio intercepts, scouts, and VC sympathizers. Gradually, a picture was formed. Piece by piece, the Viet Cong commanders built an intelligence mosaic of the 3rd Brigade, 101st Airborne Division. Morsels on unit designations came by way of bar girls in Phuoc Vinh from loose talk brought on by too many beers. Troop strengths, names of key leaders, and location of companies inside the base camp were gleaned by maids and laundry workers in the battalion area. Observations on the Americans' fieldcraft, noise, and light discipline, as well as security and other measures of combat readiness flowed in from scouts whenever we conducted training operations. Before long, the Viet Cong's dossier on the 101st Airborne Division at Phuoc Vinh answered all the key questions needed for tactical planning, except the most important ones. When and where would the Americans strike, and could they fight? The answers would not be long in coming.

For the most part, the men of Tiger Bravo went about their business of preparing for "real combat" unaware of the intense scrutiny they were under and with little realization that they were on an inevitable collision course with a powerful and capable adversary. Only the few combat veterans in the company knew that it was just a matter of time before the Viet Cong struck. Rankin, Trent, Matosky, Edge, and Sykes recognized, from their own hard-won battlefield experiences, that the transition from a training environment, even if it took place in the infamous War Zone D, to the world of a rifle company engaged in close combat in Southeast Asia would be the Vietnam equivalent of *Alice in Wonderland*. Little would make sense. Extremes would be the norm; laws of nature upended; stereotypes shattered; core values tested; and hidden, base instincts uncovered.

In this alternate universe, the soldiers from Tiger Bravo would see strong men crumble under fire, while the supposedly weak-hearted goof-offs stand up as men, for reasons they couldn't explain. They would see a young man, who had never been close to anyone in civilian life, give his life to save a buddy. They would see that the thin veil of safety and calm created by bunkers, barbed wire, and base camps concealed their true role, as magnets for mortars, sappers, and sharp-eyed snipers. They would watch the value of a human life plummet to zero as soon as the first shots were fired. They would see their own compassion for the enemy vanish the first time a buddy's poncho-wrapped body was loaded onto a helicopter. It was a world where a strange-looking, flesh-and-blood young man dressed in a black uniform, with his own family waiting for him at home and his whole life in front of him, and who was no older than the Tiger Bravo soldier who had him in his sights, would be little more than a target to be sprayed with M-16 fire.

They would discover that the almighty, all-knowing drill sergeant in basic training was wrong. Where booby traps were concerned, good luck trumped good training. Ten soldiers, eyes darting from side to side and taking slow, measured steps, could walk past the same spot, with the next in line blown off his feet by a booby trap. This was a place where rain didn't fall from the sky; it blasted horizontally in blinding sheets across the ground during the monsoon rains. Here, the same extreme, summer heat that caused basic training in the United States to be canceled because of danger for heat stroke . . . well, that was just another hot afternoon to slog through.

They would realize that this was a different kind of war than the one described in stories from their fathers and uncles, who regaled them with tales of the push across Europe or island-to-island fighting in the Pacific. It would be a war without a front, flanks, or rear, fighting a formless enemy who evaporated after a battle like the morning jungle mists, only to materialize in some unexpected place, usually at a time of his choosing. It was a haphazard, episodic sort of combat. Often, nothing happened. But when something did, it happened instantaneously and without warning. Rifle or machine-gun fire would erupt with heart-stopping suddenness, as when quail or pheasant explode from cover with a loud beating of wings. Or mortar shells would come in from nowhere, sometimes the only preamble being the cough of the rounds leaving the tubes.[71]

Most importantly, they would learn that their foe, waiting to pounce on the first airborne rifle company to venture into the hinterland and outside of the massive umbrella of firepower provided by Phuoc Vinh base camp, was not the black-pajama-clad farmer by day and guerrilla by night, armed with an antiquated weapon, that was portrayed in the newspapers back home. Rather, when the war escalated upon the commitment of US troops in the mid-1960s, the Viet Cong and infiltrating North Vietnamese Army (NVA) units were equipped and armed with the most modern weapons and supplies Communist nations could provide. Some of these weapons, such as the AK-47 assault rifle, were on a par with and sometimes superior to their American counterparts.[72]

Fashioned along the lines of the Chinese Revolutionary Armed Forces, the Viet Cong consisted of three distinct factions: part-time village guerrillas or militia VC; full-time, adequately trained and regionally based local-force VC fighters; and full-time, well-trained, and geographically autonomous main-force VC fighters (sometimes referred to as hard-core VC). Whereas the village guerrillas or militia detachments resembled the Westernized caricature of the poor, dirt farmer who was outgunned and outclassed by the Americans, the main-force Viet Cong most assuredly did not. Main-force units were subordinate to military regions or "fronts," and their organizations mirrored the battalion-regiment-division configuration found in most of the world's conventional armies. Generally speaking, a Viet Cong division possessed an operating strength of 7,350 men. A Viet Cong regiment, such as the Dong Nai Regiment that roamed War Zone D, could field 1,750 fighters. Party membership, literacy, advanced military training, and a very high esprit de corps characterized the makeup of most main-force units.[73] Additionally, at the same time that the 3rd Brigade was settling into Phuoc Vinh, a steady stream of NVA replacements and conventional-style NVA infantry divisions continued to stream down the Ho Chi Minh Trail to stand and fight alongside their Viet Cong brothers.

Tiger Bravo's war was about to begin. Within forty-eight hours of receiving the mission to "destroy the VC Dong Nai Regiment," the company would be in an all-out fight, with heavy casualties on both sides. Soon they would march to the sounds of battle, putting their youthful grins aside. Shoulder to shoulder as brothers, they would face the rising tide. Find the enemy and kill him would be their only task.

2

Forbidden Zone

War Zone D (Binh Duong Province)
January 16-17, 1968

PROWLING THE JUNGLE

Late in the morning of January 16, 1968, Tiger Bravo prowled through a patch of jungle in War Zone D, south of Phuoc Vinh, on a search and destroy mission. Heads swiveling, eyes darting left and right, the soldiers walking point and flank security strained for any sign of Charlie. War Zone D's reputation as a major stronghold of the VC, a veritable green fortress, meant little to the point man and his slack man moving just a few feet behind. They were concerned only with the five to ten feet of visibility they had in the dense undergrowth. They walked carefully, stopping every few minutes to listen, crouching to look for fire lanes cut out of the underbrush below knee level, then moving again. The two men had already begun to sweat as the rising temperature burned off the early morning dew and formed an eerie haze that blanketed the jungle floor. This battlefield choreography would be repeated over and over, sometimes with a swinging machete added to the mix, until exhaustion set in. Then a fresh team would move forward, and the spent team would take its place back in the formation.

Captain Rankin knew that the VC were in the area, perhaps within a few hundred meters, if not closer. Earlier, A and D Companies had encountered light contact there, and less than an hour before, a helicopter had taken enemy fire from a location in the general direction that Bravo was headed.[74]

It was a classic, tactical quandary faced by infantry units on hundreds of battlefields across South Vietnam. Where were the VC? How many? Who would
be the first to engage? Would they engage at all, or meld into the jungle? No
one had the answers—not Captain Rankin, not the battalion commander
or his S2 intelligence officer, nor the forward air controller (FAC) flying
overhead in a small, one-engine spotter plane. No one knew that a sizable
enemy force lay hidden in a concealed bunker complex along Tiger Bravo's
line of march.

The VC had several advantages. They knew Tiger Bravo's original location, pinpointed for them by the resupply helicopters landing at the NDP site
the night before. Plus, they could infer the company's expected direction of
movement, assuming Bravo would return to the general area of the base camp
they had found the day before. They also possessed an uncanny ability to hide
their forces and conceal fortified positions in all types of terrain. War Zone D
was their backyard, and had been for years. They knew every trail, streambed,
and fold in the ground.

It was the opposite for Tiger Bravo. Every step was new territory. But the
company had its advantages as well. First, everyone, down to the riflemen,
could feel an attack coming. The VC were here, somewhere. "I think it was
the first time we really appreciated what was going on," recalled Everett "Doc"
Franks, a medic in the 3rd Platoon. "Until then it had been pretty much like
being on field maneuvers." He also recounted a chilling premonition by his
platoon sergeant: "For once I felt truly uneasy having been forewarned by Mad
Dog [SFC Matosky] prior to the mission that guys were most likely to die this
time out."[75]

For Doc Franks, Bravo Company had become more than just a bunch of
fellow soldiers needing his medical services. Coming from a broken home and
a mother who abandoned him, he found the family he never had as a child
in the burgeoning brotherhood of Tiger Bravo. "In 1955, when I returned
home from school with my older brother," Franks explained, "we discovered
that our Mom had moved out, leaving us but taking my two sisters. All through
high school I was in a foster home. I don't know what happened to my father
. . . never saw him again." According to Franks, the company became "the first
real family that I had, where I was accepted for who I was."[76]

Captain Rankin's briefing to platoon leaders just before moving out had also struck a sobering note when he mentioned to "move with extreme caution, every indicator suggests contact is likely." Also on the positive side was Tiger Bravo's burgeoning skill at moving quietly through the jungle, observing anything out of the ordinary, and sensing the enemy's presence before it was too late. It had nurtured and sharpened the skill since its first amateurish forays into enemy territory more than a month ago. The "scouting" skills of an American soldier in Vietnam is best described in a Department of the Army study authored by the noted military historian Brigadier General (Ret.) S. L. A. Marshall and Colonel (Ret.) David B. Hackworth. "The average US soldier [in Vietnam] . . . was more alert to signs of the enemy than the men of Korea or WWII. The environment had whetted that keenness and quickened his appreciation of any indication that people other than his own were somewhere close by, either in a wilderness or in an apparently deserted string of hamlets. He feels it almost instinctively when the unit is on a cold trail. The heat of ashes that look long dead to the eye, a few grains of moist rice still clinging to the bowl, the freshness of footprints where wind and weather have not had time to blur the pattern in the dust, fresh blood on a cast off bandage, the sound of brush crackling in a way not suggesting other than movement by a man. . . . He gets these things."[77]

Last, Tiger Bravo was not without combat veterans who had been in this situation before, many times, and who had a calming and reassuring effect on the neophytes facing their first taste of combat.

Within thirty minutes one of the two forces would prevail. But for now, the young, yet-to-bleed troopers walked on. Heavily armed, alert, ready to kill— but in their hearts, innocent still.

War Zone D

War Zone D, called the Forbidden Zone by the VC,[78] had been an enemy stronghold and sanctuary since the Viet Minh first occupied it in their war against the French. It was initially made up of five villages in Tan Uyen District, Bien Hoa Province. Starting in 1948 it began to expand, with its center gradually shifting to the northeast.[79] By the time of the arrival of Tiger Bravo it had grown enormously in size, importance, and strength. In 1968, it included portions of Phuoc Long, Binh Duong, Long Khanh, and Bien Hoa Provinces. It was bordered on the east and south by the Song Dong Nai, on the west by highway QL/13, and on the north by QL/14 running through the town of Dong Xoai.[80] It was by every measure one of the most dangerous enemy strongholds in all of South Vietnam.

There had been forays into the zone by both ARVN and American forces over the years, starting with an operation by the Vietnamese Rangers in 1962.[81] But it had never been substantially under government control, and never permanently cleared of major VC/NVA forces. According to Colonel Grange, "War Zone D was a festering wound in the heart of MACV and right in the backyard of everybody. Close to Saigon, the VC could recruit, train, store supplies and rest. I can't believe that the place was allowed to exist as long as it did."[82]

Most of War Zone D was heavily vegetated, with hills in the east and a single mountain, called Nui Chua Chan, rising to 700 meters. Dense broadleaf evergreen forests covered the majority of it except for rice paddies in the southwest corner that were harvested from October through February and a few abandoned rubber plantations. The broadleaf evergreen trees averaged 1.5 to 3.5 feet in diameter and 80 to 130 feet tall, the equivalent of a 12-story building. Trees were generally spaced 30 to 50 feet apart, with their branches forming a continuous triple canopy of vegetation, making aerial observation or location of base camps virtually impossible.[83]

During the rainy season, the area as a whole was unsuitable for cross-country movement by tracked vehicles, except for occasional periods during the dry season on rice paddies or rubber plantations. There were no paved roads, and the few dirt roads that did exist were either partially blocked by trenches dug by the VC across 75 percent of the roadway, or completely blocked. Most bridges were damaged, dismantled, or destroyed. Those few bridges still standing were capable of only bicycle or foot traffic at best. Other than a few existing airfields in government- controlled towns, such as Phuoc Vinh, the area was unsuited for fixed-wing aircraft and offered relatively few areas for parachute or heliborne assaults.[84] The vast majority of the zone was either under Viet Cong control or not firmly held by either side; only larger towns were securely in government hands. Within its borders were VC/NVA combat units, supply caches, service support units, hospitals, and rest areas, often connected by miles of hidden trails.

An indication of how remote and formidable the terrain could be, and a confirmation of the VC confidence that they were relatively safe from discovery, was the identification of a POW camp, with American prisoners, only 35 kilometers east of Phuoc Vinh. In January 1970, Company D (Rangers), 75th Infantry, conducted a raid to rescue the prisoners, but the camp had been vacated just days before the operation took place.[85] This failed attempt was not unusual. US and ARVN forces conducted more than 125 POW rescue missions over the course of the war. While approximately 500 ARVN prisoners were rescued and 110 American bodies were recovered, no living American POWs were freed.[86]

Earlier in the month Lieutenant Colonel Grange and his staff had worked up an audacious foray into War Zone D. It was radically different in concept than anything the army had tried before. The plan exploited the battalion's airborne capability, still razor sharp from training jumps made at Fort Campbell, and the element of surprise. Rather than searching for the enemy and his bases a sector at a time, Grange proposed that the battalion make a combat parachute assault into the center of War Zone D, using an old airstrip in the abandoned town of Rang Rang as a drop zone. Once the battalion had established a foothold, other 3rd BDE units could follow and quickly build a force with substantial combat power. The plan was both tactically simple and brilliant at the same time—an inside-out assault from the enemy's heart. Unfortunately, higher headquarters did not agree. According to Grange, "the plan was well written and doable. It could have been a very good operation, so we sent it up the chain of command. The brigade commander, Colonel Larry Mowery, approved it and sent it to division, who approved it as well. They sent it to Field Force Headquarters, but they didn't want us to do it."[87]

OPERATION MANCHESTER

Tiger Bravo was in War Zone D as part of the 3rd Brigade's Operation Manchester, which commenced on January 12, 1968. The mission given to the brigade was to destroy the VC Dong Nai Regiment and neutralize his base camps,

disrupt VC infiltration west from War Zone D, and prevent attacks against
the Long Binh–Bien Hoa military complex. To accomplish this mission Colonel Mowery had a force comprised of his organic infantry battalions (1/506,
2/506 and 3/187), a 105mm howitzer artillery battalion (2/319), two additional 155mm howitzer artillery batteries, a cavalry troop, a combat engineer
company, a medical support company, and various support and combat support units, to include a detachment of Rome plows. In addition, the 162nd
Assault Helicopter Company was to support with troop lift, combat support,
and resupply missions.[88] To provide continuous support for the operation the
brigade would occupy two fire support bases (FSBs) in the southern portion
of War Zone D. 1/506 would occupy FSB Keene, in a location suited to pre-
venting attacks against the Long Binh–Bien Hoa military complex, and 2/506
would occupy FSB Dave, south-southeast of Phuoc Vinh and that would facilitate operations against the Dong Nai Regiment.[89]

The operation began when the 2/506, minus C Company left behind at
Phuoc Vinh for base camp defense, moved by helicopter to FSB Dave, along
with B Battery 2/319 Artillery and its six 105mm howitzers. By 1110 hours A,
B, and D Companies and the recon platoon had closed on FSB Dave. It was
no small feat, as they had been on a road-clearing mission the day before and
had just returned to Phuoc Vinh at 0800[90] hours that morning. This flexibility
demonstrated their newfound skill at airmobile operations and the fact that
each company was 100 percent mobile, meaning they carried everything they
needed to live and fight with in their rucksacks. By early afternoon, Bravo's
paratroopers were in place, covering its sector of the FSB perimeter. In a letter
home I described FSB Dave as "pushed out of the jungle by bulldozers with
very few trees, no buildings just bunkers, and lots and lots of sand and sun.
No people around except Charles."[91] It was set in a low valley, just south of
the boundary between Binh Duong and Bien Hoa Provinces, and beside a
small intermittent stream (Suoi Dia) with jungle to the north, east, and west.
Just five hundred meters to the southwest was the outer edge of a destroyed
"agroville."[92]

In 1959, the South Vietnamese government started the Agroville Program
to deny the VC access to the local population by forcibly moving the peas-
ants from their ancestral villages into newly constructed farming compounds

protected by government troops. By 1960, there were twenty-three of these communities across South Vietnam, with schools, medical facilities, and even electricity. Each housed up to several thousand people. In a culture like Vietnam's, where people were deeply attached to home and village, such uprooting was traumatic, and many South Vietnamese resented it.[93] The program was abandoned in 1961.[94] By 1968, little remained of the once-bustling planned community near FSB Dave. The perimeter fence was gone; buildings consisted of just a few crumbling walls or were completely razed to the ground; the fields and rice paddies were fallow from years of neglect. Its history meant nothing to commanders on both sides, who saw the destroyed agroville, now covered with dense brush and undergrowth, only as a made-to-order concealed location for an attacker to assemble forces.

The first night, the VC tested the alertness and reactions of this new force in their backyard by throwing hand grenades into the D Company positions, critically wounding one soldier. The next day was spent improving the FSB. Entrenching tools and machetes came out, supplemented by twenty shovels, twenty axes, and twenty pickaxes flown in by battalion. Troopers spent hours filling sandbags in the blistering sun, built or repaired bunkers, cleared brush for fields of fire, and reset claymore mines and trips flares. That night, guards detected shadowy movement but the enemy did not attack.[95]

On January 14, 2/506 kicked off its part in Operation Manchester and sent Alpha and Bravo Companies south of the FSB to look for fresh signs of the enemy. Nothing of note was found—only abandoned tunnels and bunkers—so the companies moved to open areas for pickup by helicopter and flew back to the FSB.[96] The first day of operations in the infamous War Zone D had come up empty.

KILL ZONE

The next day, Tiger Bravo was sent on a similar mission, only this time approximately six kilometers north of the Fire Support Base. It made a combat assault into an LZ across the border into Bien Duong Province, but there was no sign of the VC. That meant little, however, in a war where the enemy always seemed invisible. Following instructions, it started to move south in the

direction of FSB Dave, looking for signs of enemy along the way. It wasn't long
before an unoccupied bunker complex was found. Ordered by battalion to
destroy the complex, soldiers began blowing each bunker with C4 plastic explosives. Taking longer than anticipated with the demolition, the company
moved away from the complex to terrain more suited to a NDP, and set up for
the night.

The next morning, the so-far-uneventful mission quickly turned into anything but. Tiger Bravo was ordered to move back to the bunker complex to
finish the destruction. In the meantime, the VC announced their presence by
firing on a helicopter from a location just a few hundred meters from Tiger
Bravo's position and in the vicinity of where the company was headed. Ordered to investigate, Captain Rankin organized the company into a column
formation of two single files moving parallel about twenty feet apart. The 1st
Platoon was in the lead, followed by the 2nd Platoon, with the company command element made up of the company commander, the first sergeant, and
the artillery forward observer, plus radio operators. Next came the 3rd Platoon and last the 4th Platoon, each moving in two single files. For the first few
hundred meters all was quiet, with no fresh signs of VC.[97] Soon, each man in
Tiger Bravo would discover for himself why War Zone D had earned its reputation as one of the most dangerous of all enemy strongholds.

In an instant, everything changed. Chuck Hanson, walking with the lead
platoon, captured the moment in a letter to his family: "We had gotten in
about 500 yards when we were told that it was about 200 yards more to the
point where the gunships had drawn fire. I started counting paces at that
point. I remember getting to 150 & a little beyond that point when the shit hit
the fan."[98] Tiger Bravo had triggered a carefully planned, totally concealed,
and well-executed VC ambush. Abruptly the orderly, controlled movement of
the company disintegrated into bedlam.

The VC had set up an L-shaped ambush, a basic configuration employed
by both sides in the war, just on the outer edge of a concealed bunker complex. The bunkers were dug deep into the earth, protected by a low roof of
logs and earth and completely concealed by natural ground cover, making
them impossible to detect unless you were right up on them. Only a small
firing slit in the front was visible to the naked eye, and only if a soldier looked

directly at it from a distance well within the lethal range of VC weapons. Perpendicular to the advancing column was the base of the L, manned by VC on heavy machine guns firing down the length of the column. Extending out from the base—parallel to the company's left flank—was the long axis of the L, creating a linear kill zone. In this leg were more VC with light machine guns and automatic weapons, along with snipers in trees armed with AK47 assault rifles. Also attached to the trees were two directional, antipersonnel mines, sited to send metal shards flying into the kill zone.[99]

The kill zone is the epicenter of any ambush. In the perfect ambush no one makes it out alive from the kill zone, or at least without being wounded.

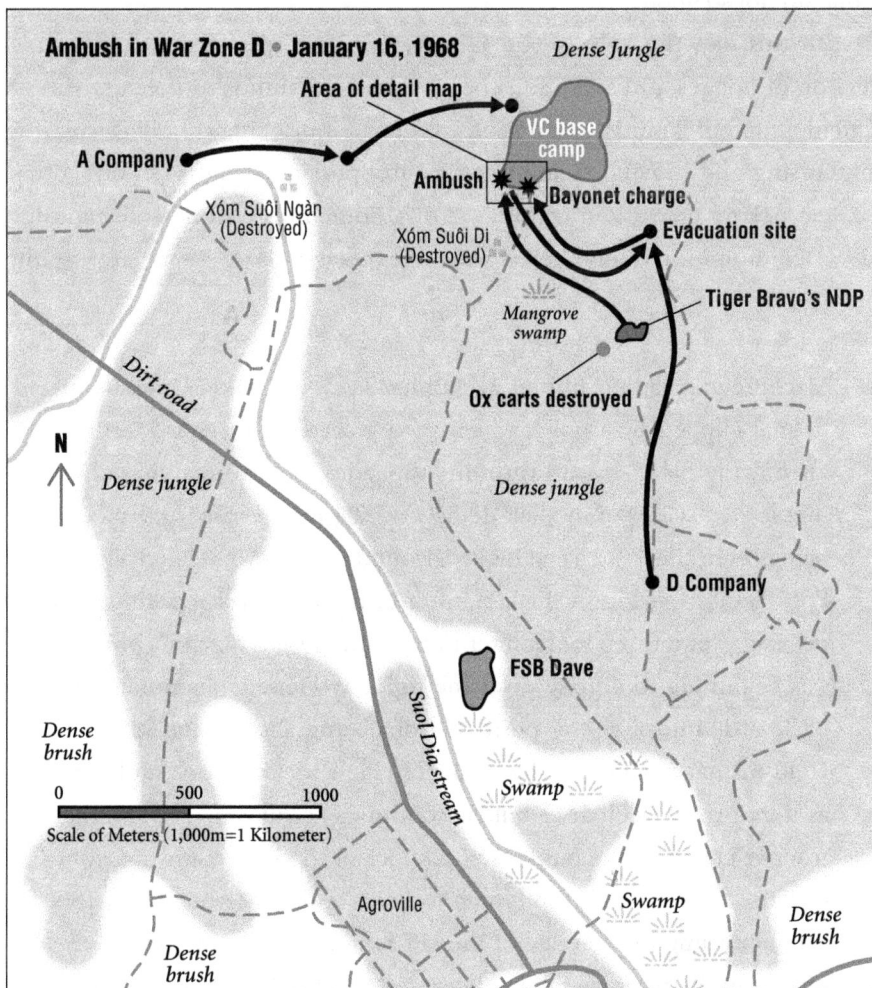

Ambush in War Zone D ● **January 16, 1968**   *Dense Jungle*

Area of detail map

VC base camp

A Company

Ambush

Bayonet charge

Xóm Suôi Ngàn (Destroyed)

Xóm Suôi Di (Destroyed)

Evacuation site

Mangrove swamp

Tiger Bravo's NDP

Ox carts destroyed

Dirt road

N

Dense jungle

Dense jungle

D Company

FSB Dave

Dense brush

Suol Dia stream

0   500   1000

Swamp

Scale of Meters (1,000m=1 Kilometer)

Agroville

Swamp

Dense brush

Dense brush

Every weapon and explosive device is specifically targeted to cover a section of the kill zone, often its sector of fire overlapping with others for complete coverage. Being caught in the kill zone of an ambush has been described as "being stuck in a blender with noise and fear and flying steel. It is fundamentally different than other types of combat because Americans are used to having fire superiority. We are always in control, but being ambushed is a whole new level of chaos."[100] If you are caught in an ambush's kill zone, there is usually no place to hide. Flying bullets and shards of metal hit where you are, where you could go to seek cover, and any escape routes you could take. There is no safe, easy, or sure exit. Your chances of surviving without a scratch range from none, to not so good.

To this day, the ambush on January 16, 1968, remains a random collection of images and sensations burned into the minds of the survivors. A kaleidoscope of sounds—explosions, yelling, shrapnel and bullets whizzing by, cursing and praying —and images—troopers crawling in all directions, bullets kicking up dust and chunks of dirt. Some impressions were captured in letters home or remembered with unexpected clarity years later by the survivors:

> Machine gun bullets hitting like a buzz saw. . . . LT Joe Hillman, his helmet clipped by a bullet which then grazed his forehead, on one knee firing his M-16 and throwing grenades, blood streaming down his face. . . . Cries of medic! Medic! . . . SP4 John Wrisberg, a medic, being hit in the head as he looked around to see who needed aid. . . . Pop Plemons kicking Chuck Hanson in the head to keep him from peering over a three foot tall termite mound just as machine gun fire was beginning to chip away at the top. . . . Homer Pierce knocked backwards into a sitting position from being shot in the stomach. Held in that position by his rucksack, then rocking back and forth as if in a rocking chair as bullets continued to hit him. . . . Another trooper charging the bunkers and knocked to his knees by machine gun fire, then raked from crotch to head by another burst. . . . 1SG Trent crawling to the edge of the kill zone and directing fire on the snipers in the trees, as calm as he was in training back at Ft Campbell

. . . . One sniper blown out of his perch in a shower of leaves and broken branches by a burst from an M-16, then hit with another full magazine on the ground. . . . More cries for a medic above the din of battle. . . . Joe Hillman calling for his men to assemble on him, preparing to assault their way out of the kill zone. . . .[101] Seeing a trooper's body being dragged to safety by two soldiers, one on each side, with his face to the sky and heels dragging in the dirt. . . . Having bushes and tree branches chopped to pieces by machine gun fire just 18" off the ground and inches from your head.

The last two platoons in the column were outside the kill zone, although they too were pinned down by the heavy machine gun fire hitting along the entire length of the company. Dan Bernard, carrying Captain Rankin's Battalion Net radio and walking right behind him, remembered the first minute. "I hit the ground so hard my rucksack came down on my steel pot and it slammed down over my eyes and hit me on the bridge of my nose. When I looked up Rankin was gone, he was already moving up to the front. Stuff was flying everywhere. This is where I got really close to God. I remember saying— God I don't want to die in this shit-hole country."[102]

Almost immediately individuals started crawling towards the front to help and to the left flank to try to relieve some pressure. "I was stunned at the level of violence and the quickness that it unfolded. The explosions and gunfire were beyond description. Immediately, people began screaming for medics," Doc Franks said. "I was with 3rd Platoon just outside the kill zone. I headed to the front of the column and was struck by the number of killed and wounded." He described the scene as "a bloody nightmare. I was ill prepared for. I stopped at PFC Ware who was lying on the ground with his right leg laid wide open by gunshots. Pieces of bone from his leg lay scattered on the ground around him. I realized his leg was beyond saving so I put on a pneumatic air splint to stem the bleeding and gave him a shot of morphine."[103] This type of splint had not been issued to medics in the battalion. Doc brought it with him from his civilian job as an ambulance attendant.

"Moving up the column," Franks continued, "I found my good friend John Wrisberg III laying in the lap of a soldier with fatal gunshots to the head.

His helmet was next to him with two bullet holes in the crown." Wrisberg was Bravo's senior medic—well liked, respected, one of only two graduates of Vietnamese Language School in the battalion and "a gentle soul."

Basil Rivera, a communication specialist turned infantry team leader in the 1st Platoon, was on the platoon's left flank with his team when the ambush started, and saw Wrisberg get hit. When the ambush started," he remembered, "everything went haywire. When the bullets started flying they were coming knee high. You could see it in the trees, it was like a buzz saw. So, I start screaming for everyone to stay low. Stay below that line." Looking to his right he could see Wrisberg staying low like everyone else. "Then we heard cries for medic, medic! Doc [Wrisberg] raised up to look around and I could see his head snap back and his brains come out."[104]

It took only a few minutes for Captain Rankin to reestablish a semblance of control over the platoons in the rear of the column and begin a sequence of actions, practiced over and over in training, to engage the ambushing force. It was normal procedure when only a portion of a unit is caught in a kill zone for those in the kill zone to "immediately return fire and take action to break out," while the platoons not in the kill zone conduct fire and maneuver "to assault the flanks or rear of the ambush."[105] Under no circumstances would a commander reinforce those in the kill zone. This meant that there were only two viable options: attack the enemy's flanks or withdraw. Concurrently, Palagyi requested immediate artillery support from the 105mm howitzers at FSB Dave, and Bernard relayed another request over the battalion net for gunship support.

The ambush lasted just sixteen minutes—from the first report to battalion at 1109 hours to 1125 hours, when Rankin knew enough of the situation to report three killed and twelve wounded. The Battalion Recon Platoon, which had been traveling with Bravo, reported another five casualties.[106] Knowing that he needed to flank the ambushing force with the two platoons not caught in the kill zone, Rankin assembled the two platoon leaders, First Sergeant Trent and myself, to lay out the plan of attack. But at 1132 hours, matters were taken out of his hands when the battalion commander directed Tiger Bravo to withdraw with all its casualties.[107]

Captain Rankin had his back to one of the termite mounds scattered across the jungle floor, with everyone crouched around trying their best not to get shot. The firing had slacked off but not stopped by any means. He issued his orders for the withdrawal, telling me to take a squad and Joe Palagyi and act as rear guard while he lead the company away from the ambush. Then he did something totally unexpected, yet exactly what was needed to calm the jittery nerves of his junior leaders who were tasting combat for the first time. Sensing that what was needed most at this critical point in the battle was a commander who exuded confidence, he took a bottle of beer from his rucksack, popped the top, took a long, noisy drink, then passed it around. His demeanor was as cool as could be, like nothing was going on around him. It was a brilliant piece of combat leadership, signaling that everything was under control.

By this time Basil Rivera and his team had moved left and were engaging the VC. "Next thing I know, there's Captain Rankin moving toward me in that way he had of turning his body sideways as he ran so no one could get a full frontal shot at him." Rankin had Rivera and his team hold the left flank as the company withdrew past them. With the flank protected, paratroopers from the 1st and 2nd Platoons crawled out of the line of fire, dragging their wounded, but carrying only one of the troopers killed in action (KIA). By now the whole company was on the move,[108] the able-bodied carrying the casualties and their weapons and equipment, and firing back at the enemy. Meanwhile, friendly artillery fire dropped between Tiger Bravo and the VC to break contact. Describing the withdrawal, I wrote, "We began to walk the artillery back as we moved back keeping a blanket of steel between Charlie and us. Shrapnel was falling all around us but it was spent so it didn't hurt anyone. We had to move 400 meters through the jungle."[109]

It took an hour for the company to reach the extraction point where the wounded were evacuated. Since there were too many for a single Huey helicopter, an orbiting CH-47 cargo helicopter was pressed into service as well. Chris Backman, who carried First Sergeant Trent's radio, remembered the helicopters coming in and one uninjured soldier trying to claw his way onto a chopper loaded with wounded paratroopers. "He just flipped out," Backman recalled. "He ran up to a chopper, screaming, and crawled in. He wasn't going to get out. It happened that fast, he just snapped."[110] In that moment, he had become a casualty as well. Knowing that he would be of no use to the company on its return to the ambush site and "pretty hard to control," Trent, who had witnessed the meltdown, left him on the chopper[111] with those shattered bodies alongside him.

Once the wounded were out of danger, Tiger Bravo loaded up with ammunition and grenades to return to the ambush site to recover the bodies of the KIAs who had been left behind. No order needed to be given; no questions were asked. There was no discussion of why. Everyone understood that the 101st Airborne Division always brought soldiers back from a battle, dead or alive. Some saw it as a sacred duty to find a buddy. A few went out of revenge, and still others were just following orders or instinctively acting out of a heightened sense of bravado. But all picked up their weapons and headed back. Despite having taken 20 percent casualties, Tiger Bravo was on the move back to the ambush site.

First Platoon had been hit the hardest. Chuck Hanson, 1st Squad, 1st Platoon, and a survivor of the kill zone, wrote, "After we got out & put the dead and wounded on choppers we reformed to go back in & get the rest of our dead. This time I wasn't in front which made me feel a little better. I came out of it very lucky, as I wasn't hit at all, not even by shrapnel from the Chinese grenades that Charlie was throwing all around us. In my squad, there were only 3 of us able to go back in the second time. SGT Tilson, who had a .30 cal bullet stuck in his helmet & Bill Plemons who was hit in the hand & back by shrapnel, but not hurt bad."[112]

FIX BAYONETS

The move back to the contact area, this time reinforced by Delta Company, was uneventful. Captain Rankin's plan was for him to control the company if contact was made, leaving me free to concentrate on finding the bodies. Years later, for those who survived that day, when talking about one fight or another during that year in Vietnam, the return to the ambush site would be called simply "the second time." Writing about the return to the ambush site I wrote, "I was on the left flank when I found three of our dead and we then got attacked again. I ran forward about 10 feet then hit the dirt right next to the fourth body."[113]

At about the same time, we made contact with the VC in bunkers, only this time Bravo was ready. With the bunkers only thirty feet away, Captain Rankin gave the order to fix bayonets, come on line, and charge. "And charge we did," continued my letter. "There are not many people who have seen or even heard of a bayonet charge in Viet Nam, but Tiger Bravo did it. The men were great. This second attack the Company assaulted yelling and screaming Airborne and Currahee, shooting, and throwing grenades. We were determined to make Charlie pay for what he did to us."[114] Chris Backman remembered "lots of yelling and screaming"[115] as well. According to Doc Franks, "It was a textbook bayonet charge. All the guys were growling and screaming. Then they opened up on us with machine guns and we all went to the ground."[116]

My letter went on to describe the company's actions as it broke through the enemy defenses and into the bunker complex itself. "We swept through the whole complex blowing tunnels, bunkers, a hospital, everything. Inside the base camp we discovered a VC hospital. It was apparent that the medical staff had left in a hurry; operating tables were left intact and, to the delight of the troopers, the VC nurses had left behind their black cotton bras and

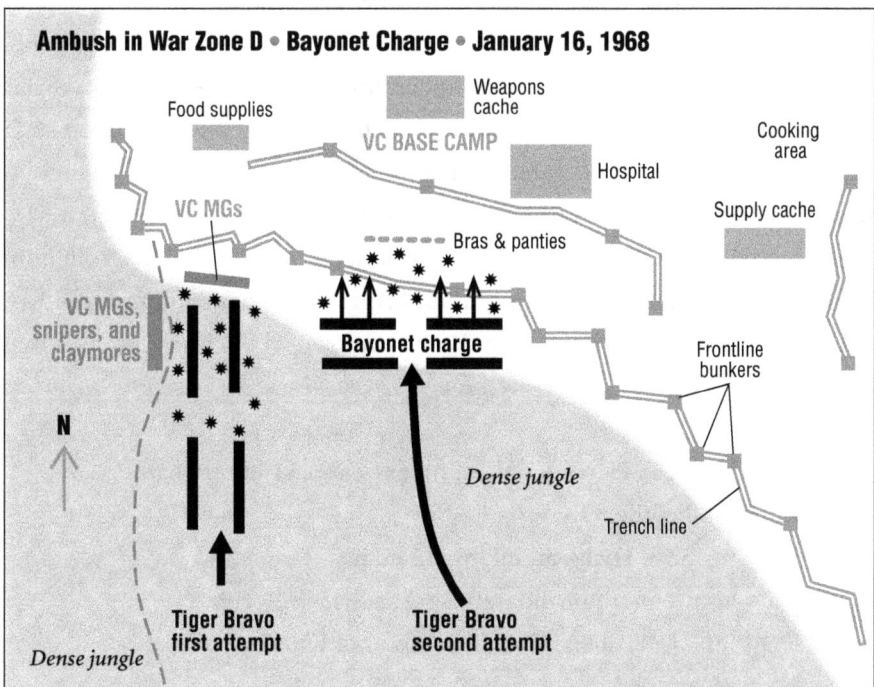

Ambush in War Zone D • Bayonet Charge • January 16, 1968

Food supplies

Weapons cache

VC BASE CAMP

Cooking area

Hospital

VC MGs

Supply cache

Bras & panties

VC MGs, snipers, and claymores

Bayonet charge

Frontline bunkers

N

Dense jungle

Trench line

Tiger Bravo first attempt

Tiger Bravo second attempt

Dense jungle

panties drying on a clothesline. In no time, what was a classic military bayonet charge collapsed into a surreal moment, with paratroopers running around twirling bras and panties in the air on the tips of their bayonets."[117]

It was obvious that the second contact had been with a VC stay-behind force, and not the larger force engaged the first time. But the enemy was still stubborn and had to be pushed out of the base camp, one bunker at a time. Throw a grenade in, then follow it with bursts of M-16 fire. This second time, Tiger Bravo suffered five more wounded in action (WIA), though none seriously. One of the wounded screamed that his finger had been shot off. His platoon sergeant, Sergeant First Class (SFC) Harold Sykes, yelled back, "Shut the hell up! You've got two hands! Use the other damn hand!" Hearing this exchange, Doc Franks, who was just a few feet away, remembered thinking, "This is going to be a hard-assed war." Close by, Pop Plemons did something that was uniquely him, but totally out of place in the middle of a battle with the VC in the jungles of War Zone D. Handing his small Kodak Instamatic camera to a buddy, he jumped to his feet, struck a pose, and yelled, "Quick, take my picture!"[118]

The bunker complex where the VC had been staged was large enough to house a battalion. It turned out to be one of the largest found to date in the area. After a search, troops discovered all the hospital's medical supplies and equipment, an ammunition dump, additional claymore mines, a mortar tube, and a hundred picks and shovels.[119] The complex also contained such an extensive and well-built system of communication trenches and tunnels, it was estimated that it would take up to thirty days to completely destroy it.[120]

Before the day was over Tiger Bravo would take one more WIA from mortar fire into its night defensive position, bringing the day's toll to three KIA and eighteen WIA. The three Tiger Bravo paratroopers to lose their lives that day were:[121]

- PFC Homer E. Pierce Jr., an eighteen-year-old infantryman from Chillicothe, Ohio
- SP4 Eugene S. Hicks, an infantryman, just two weeks away from his twentieth birthday, from Arcadia, California
- SP4 John H. Wrisberg III, a twenty-year-old medic from Mason City, Iowa

That night Tiger Bravo and Delta Company set up in NDPs, and the next day they returned to the bunker complex yet another time to begin the destruction of the VC bunkers and facilities. Unlike the two previous times, this time there was no contact.

TAKING STOCK

On January 17, Tiger Bravo returned to Phuoc Vinh to replace Charlie Company on Base Camp defense. Back in War Zone D the battalion casualty list took another big jump, only this time it was friendly fire and not enemy. In a tragic error, 2/319 dropped an artillery round of 105mm howitzers into Alpha and Delta Companies, resulting in three friendly KIA and twenty WIA. It was a "short round," something every artilleryman dreaded and took exhaustive measures to prevent.[122] Later in the month, when a tally was made of enemy losses during January from operations conducted by the 3rd Brigade, many of the items captured or destroyed in the base camp figured prominently on the list. Among the items was proof that Tiger Bravo, among its many other battlefield accomplishments, had put a serious, perhaps even crippling, dent in the enemy's supply of women's underwear!

91 VC/NVA KILLED AND ANOTHER 30 PRISONERS TAKEN, INCLUDED A WIDE ASSORTMENT OF WEAPONS AND AMMUNITION, AND SUBSTANTIAL QUANTITIES OF FOOD, SUPPLIES AND EQUIPMENT THAT INCLUDED 88.5 TONS OF RICE,[123] 70 GALLONS OF POTENT NUOC MAM SAUCE, 300 PAIRS OF SANDALS, 2 BICYCLES, 27 BICYCLE TIRES, 28 LBS. OF CANDY AND 100 BLACK BRAS AND PANTIES.[124]

3

Mountain of Rice

War Zone D (Binh Duong Province)
January 18-30, 1968

No One

Back at Phuoc Vinh Base Camp, no one observed the lone American soldier slipping between two of Tiger Bravo's platoon buildings and stopping behind a bunker. Nor did anyone notice the small, silver blasting cap from a claymore mine, and its detonator, clutched in his hand. No one knew that just days earlier he had confronted his fear of dying and crumbled, slipping away to hide when the company left for War Zone D. And no one knew he had made up his mind to injure himself and keep himself out of combat, permanently. No one stopped him as he knelt on the blasting cap, took a deep breath, and squeezed the detonator. No one saw his leg explode into chunks of bone and shredded flesh. Most of all, no one, in the circle of hard eyes that collected around him as he screamed and cried in agony, felt pity or leaned in to whisper that "everything will be okay."

In a letter home, I wrote, "We had a real punk give himself a self-inflicted wound so that he could get out of the war. He wasn't even with us when we got hit; he's just heard the war stories and got scared. . . . Needless to say, he nearly blew his leg off. I think he's a little crazy. But, I'll tell you there was little sympathy for him. For one thing, he inflicted this on himself and for another he cried, whimpered, and screamed; when our wounded who had been hit just as bad, didn't."[125]

The soldier had been absent without leave (AWOL) somewhere in the brothels of Phuoc Vinh when Tiger Bravo was ambushed just three days earlier. While his fellow soldiers faced the enemy and stood the test, he chose to hide. By his willful absence from the company he gave up his place in a brotherhood, with an ethos forged in a nameless patch of jungle in War Zone D, that had only one unspoken, unbreakable rite of passage: you fight with us and you are in; everyone else is out. Without knowing it, the young paratroopers of Tiger Bravo were already formulating their response to one of the most profound questions faced by young men of every generation: What would you risk dying for? And for whom?[126] The answer has been called a "community of sufferers," where individuals experience an immensely reassuring connection to others. A community where class differences are erased, income disparities become irrelevant, race is overlooked, and self-interest gets subsumed because there is no survival outside the group.[127]

This would be Bravo's only recorded case of a self-inflicted wound, but unfortunately it was an incident experienced by many combat units in Vietnam. The majority of cases involved soldiers turning their own weapons upon themselves—usually a nonlethal shot to the foot or hand—to escape field duty. But that was not always the case. Records at the National Archives show that 382 soldiers died from self-inflicted wounds during the Vietnam War.[128]

Several days later, Tiger Bravo participated in a more uplifting event. Early in the morning of January 25 the men of Tiger Bravo stood in formation and received the coveted Combat Infantryman's Badge (CIB) from Colonel Mowery, 3rd Brigade commander, and Lieutenant Colonel Grange, battalion commander.[129] Originally to be called the "fighting badge," the CIB was established by the War Department on October 27, 1943, to recognize the unique role of an infantryman who "continuously operated under the worst conditions and performed a combat mission not assigned to any other soldier or unit."[130] The award was open only to infantrymen, from private to colonel, who met specific criteria contained in the US Army General Order: "[A] soldier must be an infantryman performing infantry duties, assigned to an infantry unit while it was engaged in ground combat and actively engage in that combat."[131]

The orders read aloud that morning included the names of 102 Tiger Bravo paratroopers[132] who had paid their dues under fire, and so belonged to that ancient brotherhood to which no amount of money, social pedigrees, or polit-ical connections can gain a man admittance.[133] Each stood proud and straight at attention as a miniature replica of a Kentucky rifle on a blue background was pinned to the left breast pocket flap of his jungle fatigues. The roll call of new CIB holders rang out for the living and the dead, among them:

- Spencer, David W. RA 18845933, who carried a radio all day on January 16 and came out unscathed
- Ware, Bobby R. RA11867786, shot six times but lived
- Hicks, Eugene S. RA18839174, shot and killed in the first minute of contact[134]

This was a celebration of duty above self, bravery over fear, and marked a passage of those in the ranks from fledgling warrior to combat veteran.

CONNELL RICE COMPANY, HOUSTON, TEXAS

Early on the morning of January 26 the company moved silently through the predawn darkness as part of a battalion cordon and search operation of a suspected VC village. At first light, the forces moved in and systemati-cally searched for any signs of VC suspicious activity. Finding nothing out of the ordinary, the units melded back into the forest. Later in the day, Tiger Bravo, along with Delta Company, made a combat assault (CA) into another area that had not seen US units for a while. Their mission was to search and destroy—a standard catch-all operation with no particular target in mind, simply moving systematically through a designated area looking for a fight. Almost immediately, Tiger Bravo's point team, PFC Mike Tarpley and his slack man, PFC Eugene Davis, came across telltale signs that the enemy had been in the area. A piece of bamboo leaning against a tree next to a narrow trail just didn't seem right; it had to have been put there deliberately as an indicator of something. As Tarpley later told a reporter from the *Screaming Eagle*, the official, in-country division newspaper, "After a while your nose is able to tell you Charlie is around. I saw a big rectangular looking thing and told the squad to hit the dirt." He and Davis inched forward on their stomachs and found bags of rice, each weighing approximately eighty pounds, stacked in layers on logs to keep them off the ground. The dark green plastic sheet is what caught Tarpley's eye.[135]

The son of an oil rig worker, Tarpley moved to Snyder, Texas, in the third grade after living "in approximately 40 places chasing drilling rigs in the oil fields." Never one to back down, he left high school two weeks before graduation after a "discussion" with his English teacher. After working his way up from sacker in a local grocery store to produce manager he volunteered for the army, upholding a family tradition of military service; while his father never served because of polio, ten of his uncles were veterans.[136]

This find was not unusual for Tarpley and Davis. They set the standard for point teams in the company. An odd couple at first glance, they were extremely effective in the jungle. They could sense when something was out of place and seemed to know what the other was thinking. Tarpley described the friendship and teamwork this way: "Eugene was a black man from Chicago. I was a white boy from the cotton fields and oil fields of west Texas. We were both from the same dirt poor backgrounds and religious upbringings. During our training at Ft Campbell we talked about how different our childhoods were, but were the same in the outcome. When we got to Vietnam we started noticing how each man conducted himself in the jungle; Eugene and I were very similar. There was a trust between us that was almost spooky."[137]

At about the same time Delta Company made another discovery on Tiger Bravo's flank. This one was initially a cache of thirty-eight five-gallon cans of medical supplies. Soon, however, both companies found more as they expanded the search area: tons upon tons of rice and supplies, with no end in sight. This was a well-planned, fully stocked major supply hub for the VC. As one soldier from Delta Company put it, "It was all systematically laid out." By late afternoon a total of forty tons of food and medical supplies had been uncovered, with seventeen tons lifted out by Chinooks to warehouses in Phuoc Vinh.[138] The discoveries included thirty-three tons of rice, ironically in bags marked "Connell Rice Company, Houston, Texas" or simply emblazoned with a USA logo and the words "Hands Across the Sea."

The canned milk they found was even more intriguing than the rice. Each case, of forty-eight cans, was made of new-looking cardboard that clearly had not been exposed to rain or other weather. A label on each of the cases read, "Canned in San Francisco, California 1/68." More than one paratrooper wondered if the milk had just arrived in Vietnam and why it was already in the hands of the VC. Steve Lyle theorized, "I took it to mean it was no older than New Year's Day 1968. I found that to be unbelievable because it had to be hidden in the middle of nowhere, without a road anywhere near, halfway around the world and discovered by Tiger Bravo, all in less than 4 weeks."[139]

Only two months into its first year in combat and Tiger Bravo had an infusion of bodies to replace those killed and wounded in the costly War Zone D battle. They came mostly as new arrivals from the army's vast replacement system, but a few were transfers from within 2/506. One of the battalion replacements was Steve Lyle, reassigned from the medical platoon to replace Everett "Doc" Franks, who had been promoted to senior medic when Wrisberg was killed. In an ironic twist, Steve Lyle and John Wrisberg were close friends and the only two trained Vietnamese speakers in the battalion. Both had been sent to Vietnamese Language School at Fort Bragg, North Carolina, in the fall of 1967.

Steve came to the army by way of Venezuela, where he had lived for fourteen years "in the Amazon Basin with the snow covered Andes in view." His family returned to the United States in 1958 to escape a violent military rebellion that ousted the ruling dictator, Marcos Perez Jimenez.[140] Fourteen years old at the time, Lyle recalled that "there was gunfire all about during those last days before the regime collapsed. In my mind, Venezuela remained home even at the time we went to Vietnam."[141] Lyle's reasons for volunteering for the army could be traced back to the Gulf of Tonkin incident in 1964. "I was angered by the news that two US destroyers had been fired on while in international waters," he remembered. "By the following summer, LBJ announced that America would be escalating our troop levels in Vietnam. That is when I decided to join the Army."[142]

Having grown up in the tropics, Lyle was one of the few in the company who felt at ease in tropical Vietnam from the very first day. "The heat which poured from the just opened tailgate into the C-141 on that December day in Bien Hoa was wonderful," he recalled. "I walked down the ramp into a world which to me, felt like home. Other than language and food, Nam was very much like where I grew up. The smells, sounds and feel of the tropics as well as the look of rural, impoverished, rain forest villages are pretty much the same around the world. To someone who grew up in America, it may be hard to imagine." But he did notice one significant difference: "The extent of bombing and effect of war on the countryside, as seen from the air, was impressive."[143]

The work of moving the tons of supplies, on trails Tiger Bravo had to cut out of the jungle to a central point for pickup, went on all day. It was backbreaking, stevedore-type labor under a brutally hot sun. Temperatures reached 110 degrees by noon. First Sergeant Trent, along with Staff Sergeant (SSG) Richard Lamb, from Fayetteville, North Carolina, and Chris Backman, from Seattle, Washington, spent the day manning the central collection point and attaching the sling loads to hovering Chinook helicopters.[144] A sling was a heavy rope net that contained the material to be lifted. It was pulled taut by a metal ring, called a donut, which was then attached to a metal hook on the underbelly of the helicopter—thus the nickname of "hook" for a Chinook helicopter. When attaching the donut to the hook, it was important not to touch the hook with one hand while holding the donut with the other. "I learned a good lesson on the first sling load," recalled Backman. "I reached up and grabbed the hook with my bare hand, the donut in my other hand and was knocked on my ass from the static electricity created by the helicopter's rotors."[145]

Word received by the troops on the ground was that this was one of the largest finds to date by the 101st Airborne Division, and one of the largest by any unit in the war. It was surpassed only by a similar find by B Company, 1/506, of forty-six tons of rice and six additional tons of miscellaneous food supplies.[146] All of this occurred over the course of a very long day, without a shot being fired or even a sighting of the enemy. "After drinking Charlie's milk and shipping his rice off, which was really our milk and rice, we trekked over to our NDP site," Lyle recalled. The company dug in, set out claymore mines and trip flares, sent small listening posts out on likely avenues of approach, and settled in for the night.

It would be anything but a quiet night. According to Lyle's account, "I was sound asleep when I was awakened to find a hand over my mouth, my nose pinched shut and a moist mouth on my ear whispering 'don't make a sound.' I was glad to learn it was Soto." What had alarmed Dan Soto and others along the perimeter was "a terrible commotion off in the woods near the listen post." Short, dark-haired, and not able to see very well in the jungle without his glasses, the twenty-year-old Soto had graduated from Mingus Union High School in Cottonwood, Arizona, where he wrestled, ran track, and played football. He had come to the army by way of a short stint in the US Forestry Service in Coconino National Forest near Flagstaff, Arizona.[147]

Immediately the entire 1st Platoon was on alert. Lyle recalled, "I was convinced that it was a herd of at least a hundred elephants heading right for the platoon. Then the whole perimeter lit up with automatic fire, claymores, Foo Gas, and M-79s." Once the firing stopped, all was quiet "as though the elephants had stopped and were plotting their next move." Then the voice of Bill Plemons, one of the listening post soldiers, broke the silence:

"Currahee, it's Plemons. We're coming in."

The herd of elephants had been the two-man listening post team walking back to the perimeter, in the dark, without anyone being forewarned. At daylight, the trees, bushes, and vines in front of the perimeter, to a depth of nearly fifteen feet, were splintered and blasted apart. This prompted Lyle to ask the question, "How could anyone stumble around in the dark and survive what must have been a half million dollars worth of ordnance discharged in one short minute?"[148] The answer: pure luck.

Later that night, the VC, alerted to the company's location by the sudden explosion of firepower earlier in the evening, hit the Tiger Bravo NDP with mortar fire. Luckily, there were no casualties, just a lack of sleep[149] for the rest of the night.

The company learned valuable lessons that night. First, a combination of pitch black darkness, fear, exhaustion, dehydration, and a likely electrolyte imbalance could amplify one's imagination and create a threat out of normal sights and sounds. Second, never move around at night, especially outside a perimeter, without everyone being notified. Last, pure "dumb luck" cannot be explained, only appreciated.

Later in the week, the Brigade Civil Affairs Team distributed the rice and other supplies to local residents clustered around Phuoc Vinh, and to hundreds of hungry and homeless[150] left in the wake of an enemy offensive looming on the horizon. Clearly something big was going on. That was apparent to everyone. However, the strategic importance of these discoveries and their connection to the most famous enemy offensive of the Vietnam War, which would occur only days away on the traditional Vietnamese New Year, went unnoticed by everyone at the tactical level.

The VC were known for prepositioning supplies out in front of its forces before the battle, called its "logistical nose," rather than supplying from the rear by means of a "logistical tail," as the US Army would. In this push-forward versus pull-from-the-rear approach, necessitated by the enemy's lack of transport and secure lines of communication, General Creighton Abrams, General Westmoreland's replacement as commander of the US Army, Vietnam, had identified a major enemy vulnerability that would dictate important changes in Allied tactics. Abrams set about preempting future enemy offensives by seeking out and cutting off their logistical "nose." The large-scale search-and-destroy operations that typified the Westmoreland years gave way to numerous smaller operations designed to find the enemy and seize his supplies.[151]

But Tiger Bravo didn't know that when they found this rice cache. Most of the talk centered on how angry this would make Charlie and what he might do to Tiger Bravo. The next day they headed out on another search and destroy in an area south of Phuoc Vinh. That night one of its nightly platoon-size ambushes was finally successful, a clear indication that Tiger Bravo had learned the fieldcraft necessary to take the fight to the enemy and come out on top. According to a divisional summary for the month of January:

AT 1900 ON JANUARY 27, A PLATOON FROM B/2/506 INF EXECUTED A WELL-PLANNED AMBUSH AGAINST 2 VC SQUADS; LEAVING 6 VC KIA BEHIND; THERE WAS 1 US WIA (MINOR).[152]

SHADES OF BRONZE AND MAHOGANY

At about this time Tiger Bravo displayed the first outward signs of its metamorphosis. An untested, green airborne rifle company was well on its way to being a potent, efficient fighting unit and looking the part, as well. Operations were smoother, and reports were more succinct. Movement across rough terrain became quieter, and security became second nature. Standard operating procedures (SOPs) for airmobile operations, NDPs and a score of other tactical scenarios were routinely executed, day or night. Replacements were being assimilated, taught fieldcraft skills, and educated on the peculiarities of fighting the VC units found in War Zone D. Leaders no longer gazed at a tropical landscape of palm trees and lush green forests, quaint farmhouses, and a primitive checkerboard of rice paddies. Now they saw only enemy territory to be crossed or fought over, potential ambush sites, or locations that offered the cover favored by snipers.

The paratroopers' physical appearance, and the loads they carried, were in flux as well. The youthful bounce in their step became a rolling swagger. Nineteen-year-old privates, who had been with the company since it arrived and who had been on the lowest rung of the airborne fraternity at Fort Campbell, held a certain elevated status in the eyes of replacements because they "had seen some shit." Veterans' skin had darkened into shades of bronze and mahogany. Everyone had lost weight, even already-skinny nineteen-year-olds. "You didn't see anymore baby faces," Chris Backman recalled.[153] Their faces had lost the softness of youth and assumed, via some alchemy of sunburn, sweat, relentless stress, and a familiarity with death, the definition of adults.[154]

Anything not needed for survival or fighting disappeared from LBE (load-bearing equipment) and rucksacks. Even underwear was considered nonessential. Compared to a soldier's load carried in training exercises at Fort Campbell, the number of canteens doubled and a basic load of ammunition tripled. Soon everyone had bandoleers of magazines carried diagonally across their chests. It was a strange sight—stick-thin soldiers, bent under the weight of bulging rucksacks. Mike Scott remembered his rucksack crammed with "C rations, four quarts of water, four frags [grenades], one WP [white phosphorus] grenade, claymore mine, poncho and liner, entrenching tool, LAW [light antitank weapon], a hundred rounds M-60 ammo, head net, mosquito repellent, two or three smoke grenades, and two sticks of C-4 [plastic explosives]."[155] This was in addition to the twenty magazines of M-16 ammunition he carried.[156] In contrast, the typical NVA rucksack, not including rations or

ammunition, would include a spare uniform and undershorts, Ho Chi Minh
sandals, a small plastic sheet used as a rain cape, hammock cover or ground
cloth, hammock and mosquito net, a cloth bag with toothbrush, toothpaste,
soap, and comb, a cup, spoon, rice bowl and chopsticks, cigarettes and Zippo
lighter, weapons cleaning kit, field dressing, and often Mao Tse-tung's Little
Red Book.[157]

Not wanting to attract the keen eyes of snipers, chevrons and insignia of
rank disappeared from jungle fatigues. For that same reason, RTOs would
hide their radios in a rucksack, with the antennas bent over a shoulder, raising
it only when needed. Hunting knives, Marine Corps K-Bar assault knives, and
other assorted blades hung on belts and attached to LBE, where they could
be easily reached. Some carried concealed, unauthorized pistols and small
automatics that were preferred to the bulky, regulation .45-caliber automatic
pistols issued to officers. The company also counted more smokers among its
ranks than before. Every C ration meal came with a mini-pack of cigarettes,
and soldiers soon learned that the smoke from Lucky Strikes kept mosquitos
at bay. Cigar smoke was even better.

The soldiers had learned that when the firing started, close to the ground
was the place to be. Speed was all that mattered; good form was for diving competitions. To fall, flop, or fling made no difference; just get down. They were
no longer surprised at the adrenaline surge, pounding heart, and pouring
sweat that came from being shot at. Nor were they surprised at how routine
could change to chaos in seconds and how a semblance of order could be re-
stored shortly after by shouted orders from a trusted NCO or officer.

Individual beliefs in mortality had changed as well. War Zone D pushed
everyone past the ingenuous "it will never happen to me" conviction, universally held upon arrival in-country. Yet most fell short of the fatalistic "it is only
a matter of time before my luck runs out" state of mind that would plague
the company as casualties mounted in the months ahead. The trappings of
civilization were slowly being eroded, indiscernible to even the most introspec-
tive of us. Callousness began to corrode compassion, and a dead VC was just
something to walk by. One day the transformation would be categorical, but
for now, the soldiers were caught in a moral twilight, between the sensibilities
of American culture and the ruthlessness of war.[158]

A close-knit outfit in Fort Campbell, Tiger Bravo had become closer in Vietnam, but in a different way. The old fellowship once had an adolescent quality to it, akin to the cliquishness of a football team or fraternity. Now it was of a sterner kind; Vietnam had fused new and harder strands to the bonds that had united them in training, woven in by the experience of being under fire and shedding first blood together.[159]

CIRCLE OF HELMETS

On January 28, it was back to Phuoc Vinh for the Vietnamese New Year (Tet) cease-fire,[160] which no one expected to last.

There was even time to hold a memorial service for the battalion's casualties.[161] Early in the morning the battalion stood in formation by a circle of helmets formed by pairs of black, glossy, paratrooper jump boots, one for each of the eleven soldiers killed in action during January. Framed by rifles, locked and loaded, with fixed bayonets stabbed into the hard, baked ground behind each pair of boots. Crowned with a helmet displaying hard-earned rank. One paratrooper memorial as simple and stark as the next.

A trip around the circle was a slice of Americana. The black first sergeant, denied service at a roadside café outside Camp Shelby, Mississippi, on his way to train troops bound for Vietnam. A butter bar second lieutenant, notorious for getting lost in the jungle, who did it one last time. At night, in front of his own ambush. The medic who wanted to be a doctor, lasting only five seconds into his first contact with Charlie.[162] The teenage private first class (PFC), not long removed from high school, who didn't survive the helicopter flight to the hospital. "He had to have died on the way," a small, bowlegged sergeant said. "He was hit bad, and there was no way he could've made it, but I still had him alive when the chopper came. I was pumping the breath into him with my hands and my mouth, and he was alive when we put him on the chopper."[163]

Standing alone in front of the assembled troops, Lieutenant Colonel Grange spoke for the dead. "I think there are two points they would make if they could have come back and been here a moment at this memorial service," Grange said. "First, I think they would remind you that human combat is a hard school whose lessons are learned at a high price. I think they would want you to learn those lessons well and go home safely. Finally, I think they would want us not to forget them and the sacrifice they made in their deep sense of duty as soldiers." Then the battalion chaplain, a middle-aged captain with a crew cut, stepped forward. Clerical cloth adorned his jungle fatigues, but Chaplain Leo J. Matz of the Monterey, California, Catholic Diocese was a paratrooper like the rest of them, and he had seen much more of combat than most. As a seventeen-year-old member of the 101st Airborne Division in World War II he had parachuted into the German lines in Holland and fought at the Battle of the Bulge. "Lord, absolve the souls of these your servants," Chaplain Matz prayed. Then he raised his hand and, with a small plastic vial of holy water, sanctified the boots and helmets of those eleven men who represented many religious denominations. At that the ceremony ended and the companies marched away, each man lost in his own thoughts.[164]

4

Kicking in Doors

Bien Hoa City (Bien Hoa Province)
January 31–February 28, 1968

GRAB YOUR GEAR

Dan Soto remembered being jolted awake on the morning of January 31 by someone screaming, "Get up! Get up! Don't turn on the lights!" In seconds, everyone in the platoon was awake, and there was a mad scramble for the gear and ammunition stacked in the center of each hooch. It was the same clamor in every 2/506 company area; the battalion had been alerted. "Someone broke the lock and started handing out the ammo. Then we went to our secondary positions along the perimeter," recalled Soto. In minutes, Tiger Bravo went from a dead sleep to assembled, armed, and ready to move. In less than half an hour, it was moving through the blacked-out base camp to waiting Hueys on the assault strip. As the division's ready force the 2/506 was on the move.

Just hours before, the Phuoc Vinh Base Camp was eerily quiet. Artillery pieces hunkered in silence, with no fire missions to shatter the calm, as were helicopters parked in neat rows at the airfield. It was equally peaceful in the 2/506 Battalion area, where a false sense of tranquility blanketed the barracks and tents filled with sleeping paratroopers. The only ones awake in Tiger Bravo were on duty: the sleepy charge of quarters (CQ), who maintained contact with battalion while the command group slept, and the sentries manning the perimeter bunkers and watchtowers, struggling to stay awake.

Only in the battalion operations center was there any hint that this night was like no other in the war. Staff duty officers and NCOs huddled around a bank of radios listening to sporadic reports of battalion- and regimental-size assaults on South Vietnamese cities and towns, as well as American base camps and military installations coming under attack, one after another. The countrywide Tet cease-fire had been short-lived. It went into effect at 1800 hours on January 29, with the first violation occurring at 2315 hours, when a platoon from 3/187 received six mortar rounds in its area. By 1040 hours the next morning, the brigade commander announced that the truce was no longer in effect. Throughout the day there were random reports of enemy sightings; one reported the 273rd NVA Regiment just 1.5 miles northeast of Phuoc Vinh.[165] All the indicators hinted that something big was transpiring, but no clear, concrete intelligence picture took shape. As a result, the 2/506 continued with its normal base camp activities.

Only one aspect of the battalion's readiness posture differentiated this night from any other. The 2/506 was the division's ready force. It had to be ready at a moment's notice to deploy anywhere in the 101st Airborne's area of operations. Across the battalion, everyone's gear was ready, rucksacks packed, radios tested, a basic load of ammunition in ammo pouches and bandoleers and weapons secured, but readily available. Ten minutes after receiving an alert, Tiger Bravo could be out of their bunks, off the bunker line, and assembled in the company street, ready to go to war.

That moment had come. Gradually a grim picture emerged: The New Year's truce had been broken by a major, countrywide enemy offensive. Soon to be known around the globe as the "Tet Offensive," it encompassed attacks on a scale never seen before in Vietnam. At its zenith, thirty-six of forty-four provincial capitals; five out of six of South Vietnam's autonomous cities, including the capital, Saigon; and sixty-four district towns, would be under attack.[166]

At 0400 hours, calm and serenity yielded to tumult and clamor when the commanding general gave the order to commit the 2/506, his on-call, ready force. The battalion's mission was to mount a counterattack against a VC regiment that had breached the defenses of the Bien Hoa Air Base. The situation was desperate. Without reinforcements, the linchpin of America's airpower strategy in South Vietnam would be lost. The VC attack needed to be stopped at all costs. In every hooch and tent in the battalion, paratroopers were being rousted out of bed and sent scrambling for their gear and weapons.

Entries in the 2/506 daily staff journal captured the urgency of the moment and the rapidity of the battalion's response:

> * 0400 HOURS ALL COMPANIES TO BE ON 100% ALERT.
> * 0407 HOURS A, B, D COMPANIES ARE TO BE READY TO MOVE.
>   C COMPANY WILL STAY BEHIND AND MAN THE PERIMETER.
> * 0538 HOURS CO A & B AT THE ASSAULT PAD.[167]

In a letter home, I recalled that long walk in the dark to the assault pad by the airfield: "As the troops filed by to be lifted into Bien Hoa by helicopter the radio was full of reports of enemy attacks. Hue hit! Bien Hoa hit! Saigon hit! It seemed like the whole country was on fire."[168] Adding to the drama, the battalion's Catholic chaplain stood in the moonlight and blessed the troops as they passed in the dark. With the young soldiers already apprehensive about the battle that lay ahead, Lieutenant Colonel Grange finally told the chaplain to stand behind a tree and bless everyone from out of sight.[169]

LIKE A DAGGER

While there is no doubt that the 1968 Tet Offensive was a colossal intelligence failure at the strategic level, Lieutenant General Fred Weyand, the commanding general of II Field Force, who had responsibility for the defense of Saigon, was not caught entirely off guard. In December 1967 and January 1968, Weyand's intelligence staff began picking up key indicators that the VC intended to attack population centers with some form of major offensive. Battalion-size enemy forces had attacked the towns of Tan Uyen and Trang Bang, while several main force regiments had been detected moving out of the traditional strongholds toward Saigon. In addition, reports were received that the VC had been reequipping their forces with newer versions of the AK-47 and RPG, and there was a surge in NVA filler personnel being assigned to local VC force battalions.

"Something was coming that was going to be pretty goddamn bad," Weyand said, "and it wasn't going to be up on the Laotian border somewhere, it was going to be right in our backyard." With few exceptions, General Weyand controlled all US and Allied war efforts in an enormous "backyard," encompassing ten provinces that stretched from Saigon to the border with Cambodia, and north to the Central Highlands and the border with Laos.[170] As a precaution, Weyand repositioned most of his forces within striking distance of Saigon, which in his opinion was the ultimate target for any large-scale offensive. Of a total of fifty-three US maneuver battalions under his command, twenty-seven were moved to within assault helicopter range of the vital installations in and around the Capital Military District (CMD). However, during that same period there were no such major shifts in ARVN forces. Plus, South Vietnamese soldiers absent from their units for the Tet truce had lowered ARVN troop strength to about 50 percent of authorized manning levels. Although the truce was canceled at 0945 hours on January 30, the inadequate Vietnamese communication system precluded the effective notification of the bulk of the absent soldiers. Thus, the strength of the forty-six ARVN battalions, spread across III Corps, was still at 50 percent when the VC attacks were launched.[171]

Of the fifty-four main and local force VC/NVA battalions in III Corps, thirty-seven were committed to the initial assault. The enemy's overall strategy was to launch simultaneous attacks on government buildings in Saigon and the key installations of Bien Hoa and Long Binh. The objective was to energize public support with one broad, massive offensive, and present the United States with an untenable situation after the South Vietnamese government collapsed under the weight of the attacks.

Enemy local forces launched the initial attacks, while main force units deployed to block reinforcements and prepare to exploit success. But because of Weyand's preplanning and rapid execution of contingency plans, the VC failed to prevent American reinforcements, either overland or by air, from arriving on the battlefield in time to blunt the initial assaults. In just twenty-four hours, the rapid response of US and ARVN forces had brought more maneuver battalions into the CMD and surrounding areas than the VC had in their initial assault element.[172] Tiger Bravo would stand with other counterattacking units and shoulder its part of the load.

What was of particular concern to the 2/506 and Tiger Bravo was the VC 5th Division, which had the mission of attacking Bien Hoa and Long Binh. For this mission it was comprised of its organic VC 274th and 275th Regiments, supported by an artillery group, the U-1 Local Force Battalion, a sapper battalion, and other local force units used as guides.[173] Paul Vann, an adviser to General Weyand, described the two regiments, whose last known location was approximately seven miles from the US headquarters, as "pointing in like a dagger toward Bien Hoa".[174] Intelligence reports estimated the pre-Tet strength of the 274th Regiment to be thirteen hundred strong, while the 275th Regiment was estimated at fifteen hundred.[175]

The huge, sprawling Bien Hoa–Long Binh logistical and command complex had been carved out of a rubber plantation early in the war. Only kilometers from Saigon, it was situated along the banks of the Dong Nai River and split along its east-west axis by Highway 1A, the main road from Saigon north to Xuan Loc. To the south and east of Highway 1A was Long Binh, the largest US logistics base in Vietnam and home to thousands of civilians and military personnel, General Weyand's II Field Force headquarters, a massive ammunition dump, several US hospitals, a POW camp, and the headquarters of the 199th Light Infantry Brigade (Separate). North and west lay the city of Bien Hoa, the provincial capital. Adjacent to Bien Hoa on the north was the Bien Hoa Air Base, which at the time of the attacks was the busiest airfield in the world, with more takeoffs and landings than Chicago's O'Hare International Airport.[176]

VC RUNNING ALL OVER THE PLACE

At 0300 hours, an intense rocket and mortar barrage hit Bien Hoa Air Base and Long Binh. The veteran 275th Regiment assaulted Long Binh's northern perimeter, while the U-1 Local Force Battalion launched a diversionary attack against the eastern bunker line. Meanwhile, the sapper battalion infiltrated the ammunition dump just north of Long Binh.[177] To the north of Highway 1A, forty-five 122mm rockets, each carrying 14.5 pounds of explosives in its warhead, slammed into the built-up areas of Bien Hoa Air Base, south of the main east–west runway. Three US Air Force personnel were killed and another sixteen wounded in the bombardment. Explosions blew the roof off the control tower and damaged or destroyed numerous buildings and trailers. Parked aircraft were damaged by shrapnel, and flames roared skyward from a burning aviation fuel storage tank and two attack aircraft.[178]

Under cover of the rocket attack two battalions of the 274th Regiment—more than half of their ranks filled with fresh NVA replacements—attacked and penetrated the eastern sector of the Bien Hoa Air Base perimeter. The VC executed this surprise maneuver flawlessly. Undetected, the regiment had completed a nine-hour forced march from assembly areas twenty-nine kilometers due east of Bien Hoa to staging areas in Dong Lach, a small hamlet only two hundred meters east of the perimeter fence. From Dong Lach, the attack force slipped through an old French minefield without incident and poured through four gaps cut in the perimeter fence. They were well trained, experienced, and confident of victory. Many of them carried their green dress uniforms folded neatly in backpacks in anticipation of the victory parade through the streets of Saigon that would surely follow.[179] A prisoner captured late in the battle stated that his commanders told him, "Bien Hoa would drop into their hands like ripe fruit."[180] So confident were the VC of their easy victory that no unit was given a withdrawal route, although the 5th Division did assign rallying points.[181]

It wasn't until 0328 hours that the attackers were discovered inside the perimeter by a US Air Force (USAF) canine team, patrolling the inside of the perimeter fence. When the guard dog alerted on movement in the darkness, his handler suddenly saw hundreds of VC streaming toward the interior of the base. He was immediately wounded by enemy fire, but managed to raise the alarm and release his dog on the VC. The dog became the first friendly fatality of the ground attack when it was shot as it lunged at the closest VC.[182] But the sentry and his dog had served their purpose, and alarms began to blare all over the base.

The fight for control of the airfield would surge back and forth for the next two hours. By the time Tiger Bravo arrived on the scene it was evident that a bloody battle had taken place inside the base and that it was far from being over. Bravo, now reinforced with an 81mm mortar, its crew, and thirty rounds of ammunition, had lifted off the assault pad in Phuoc Vinh before dawn.[183] The choppers flew directly to Bien Hoa at top speed. Approaching Bien Hoa, with the lights of Saigon in the distance, the troops could see tracer fire crisscrossing underneath them.[184] Looking out the sides of the Huey choppers on final approach to a landing zone inside the 101st Airborne Division's headquarters compound, the paratroopers could see two fighter jets on the tarmac that had been destroyed, and a large TWA 707 with wing damage from enemy fire.

Off the right side of the choppers, in a rice paddy that was outside of the perimeter fence and the closest runway, were three Huey gunships circling like buzzards. No more than two hundred feet off the ground, they fired their M-60 machine guns at a hundred-plus black-clad VC caught in the open. Black smoke from the M-60s' red-hot barrels poured out of the chopper's doors.[185] Once on the ground, Dave Spencer remembered seeing "VC running all over the airstrip."[186] Walking from the choppers the men could see bloody flak jackets hanging on posts. As Steve Lyle remembered, "It was obvious that they had been attacked, that this was a big deal, and it was clear to each of us why we were there."[187]

Later in the morning, the momentum of the attack inside the perimeter shifted in favor of the defenders. True, the air base perimeter had been penetrated, but a quick and courageous response by USAF security elements, attack helicopters, and perimeter guards had kept the base from being overrun. Full flight operations were disrupted for a short time, but the US forces never lost control of the base. The VC 274th Regiment had failed in its mission. Many of the attackers had been killed, captured, or escaped back through the gaps in the fence that they had streamed through with high hopes just hours before. However, a sizable portion of that regiment had not been destroyed, and it remained a dangerous threat to the air base as long as it survived. Its general location was known; the regiment had regrouped in the hamlet of Dong Lach and taken refuge in buildings and down narrow streets and alleys in the part of Bien Hoa City that was adjacent to the air base.

As in every war, the only way to defeat an enemy firmly entrenched in a built-up area is to send in the infantry to dig them out, one building and street at a time. Historically, street fighting in modern warfare is some of the bloodiest that infantrymen face. Such was the mission facing Tiger Bravo on January 31, 1968.

BATTLE FOR DONG LACH

Arriving with a force of three hundred paratroopers, Lieutenant Colonel Grange was briefed personally by Major General Barsanti, the division commander. The briefing was surreal; from the patio outside the general's opulent private dining room, Barsanti pointed directly at the battalion's objective, Dong Lach, where retreating enemy soldiers were visible as black dots in the grass just outside the hamlet. Also visible were the Cobra gunships rolling in on them and releasing a torrent of machine gun and rocket fire on the exposed VC. After briefing Tiger Bravo's commander, and First Lieutenant Ron Darnell, commanding A Company, Colonel Grange led the battalion through the east gate. It had been only twenty-seven minutes from their arrival to the start of the counterattack.[188] "Barsanti stood by the gate and saw us off with words of encouragement," noted Grange, whose command group accompanied the lead platoon through the perimeter berm and down the road that cut safely through a defensive minefield and barbed wire barriers. Doc Franks also remembered the division commander's encouragement as he walked through the gate. "He was exhorting us to 'get out there Bravo Company. There's only fifty of them in that building,' pointing at Dong Lach. It was obvious that turned out to be a load of crap because there were a whole lot more of them than fifty."[189]

It was not unusual for Grange to be on the ground with his forward elements. He was "old school," forsaking the relative safety of "managing" the battle from a command and control helicopter and instead "leading" soldiers who were in harm's way on the ground. "When we were in a fight, Grange was there with us," observed Lieutenant Joe Kellogg, the recon platoon leader. "That's why we loved the guy. We knew he wouldn't be orbiting in a helicopter at three thousand feet."[190]

The coming battle centered on the control of Dong Lach and several adjacent, smaller hamlets. The plan was for A Company to assault the north side of the hamlet, driving the VC directly into blocking positions set up by the battalion's recon platoon on the west side. Tiger Bravo would take the south side. Tiger Bravo was also tasked with clearing smaller hamlets and built-up areas as it made its sweep south, destroying as much of the remaining 274th Regiment as possible. The battle started almost immediately outside the perimeter. "We picked up two 12.7mm machine gun teams right away. They had been waiting to fire on planes taking off from the air base," said Grange, who was in the lead with A Company. "They were all camouflaged, in green uniforms, faces painted up green, and were hiding behind mounds of a graveyard near the edge of the wire. Because of the minefield,

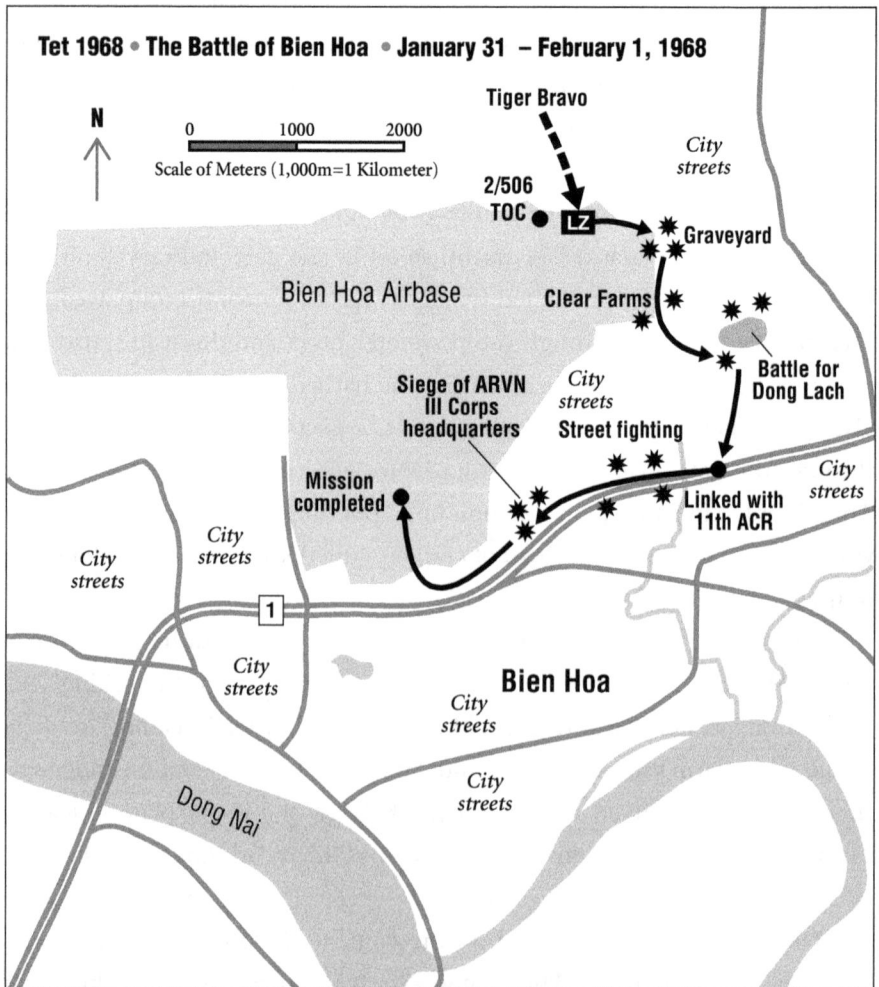

Tet 1968 • The Battle of Bien Hoa • January 31 – February 1, 1968

N

0   1000   2000

Scale of Meters (1,000m=1 Kilometer)

Tiger Bravo

City streets

2/506 TOC

LZ

Graveyard

Bien Hoa Airbase

Clear Farms

Battle for Dong Lach

Siege of ARVN III Corps headquarters

City streets

Street fighting

Mission completed

Linked with 11th ACR

City streets

City streets

City streets

1

City streets

Bien Hoa

City streets

City streets

Dong Nai

no one had been in there grooming it, so the grass was thigh-high and they were hard to see. They were afraid to take us on, so we got the drop on them and eliminated them on the spot."[191]

Pushing south toward the hamlet, the battalion came under heavy fire from the north side of Dong Lach, and from the VC still hidden in the grass outside the perimeter.[192] Tiger Bravo, on the right flank, found itself on a dirt road heading in the direction of its intended blocking position, with an old French graveyard extending a few hundred feet on each side. Suddenly, shooting started from somewhere on the fringes of the graveyard. Everyone scurried off the road to seek shelter behind the tombstones. As the bullets hit like hailstones all around, even General Barsanti was seen running for cover behind an armored personnel carrier (APC).[193]

Then the situation took a cruelly comic twist. With bullets seemingly coming from all directions, some paratroopers were hunkered down on one side of the tombstones as cover from the enemy bullets, while others were seeking shelter on the opposite side. In some cases, there was a paratrooper on either side of the same tombstone, each making the case to the other that their side was the place to be. Steve Lyle, recently reassigned as the medic in Lieutenant Joe Hillman's 1st Platoon, remembered, "In the group I was with, the accusations went back and forth from each side to the other. This struck me as very strange. It made no sense. It wasn't logical and it wasn't ordered; but, it was humorous."[194] Basil Rivera, a 1st Platoon Fire Team leader armed with an M-79 grenade launcher, took cover on one side of the largest tombstone he could find. It turned out to be a lucky choice, as bullets began to splatter against the other side. "I thought to myself, what a hell of a way to die," Rivera wrote. "On someone else's tombstone."[195]

Eager to join the army and, ultimately, be a Special Forces "Green Beret," Rivera's army journey that brought him to Tiger Bravo started with a 4F (not medically suited for military service) classification on his first try to enlist. "I had been knocked off a horse going at full gallop the day before the physical," he recalled. "I still had blood in my urine, so I failed." Just turned eighteen and broke, he returned to Alvin, Texas, where he had lived with his grandmother. "I was disgusted," he continued. "I knew I didn't want to go back to my old job in the grocery store." Then a friend of his father asked if he would like to be on a facilities cleaning crew in Houston, Texas. "Next thing I knew I was working 6:00 PM to 2:00 AM cleaning the NASA building in Houston. I would buff the floors of the astronauts' offices and take my breaks sitting in their leather chairs, just dreaming what it would be like in space."[196]

On his next attempt to enlist he was successful, selecting a communications specialty, thinking it would help on his Special Forces application. "But I was still eighteen and you had to be nineteen to apply for Special Forces. So, the Army assigned me to a Signal Battalion in Fort Polk, Louisiana." Thus, began a series of events that ultimately led to his volunteering for Airborne School; being assigned to the 101st Airborne Division after graduation; finding himself in the Signal Platoon of the 2/506; and finally, being caught up in the reassignment of different military occupational specialties to fill severely understrength infantryman positions. "One day in October 1967 my Platoon Sergeant assembled the Signal Platoon and asked for volunteers to fill infantryman openings in Bravo Company. So, I did. I didn't know what I had gotten myself into."[197]

Soon the firing stopped and the battalion began to move again. A company peeled off to cross an open area for a frontal assault on the north side of Dong Lach, while Tiger Bravo continued straight toward the small hamlets that needed to be cleared before it could establish its blocking position. Just as Tiger Bravo's point team, made up of Mike Tarpley and Eugene Davis—the same team that had found the huge rice cache five days earlier—passed the last tombstone, a VC stepped into the open and threw a grenade straight at Tarpley. "I took a hand grenade in the face," Tarpley recounted. While the grenade was still in the air Tarpley took cover behind a tombstone that protected him from his chin down but left his face exposed. Tarpley recalled, "My steel pot covered my head and eyes so I took the blast in my mouth. It blew me backwards. Then the gook stepped out thinking—yeh, I got him. I had fuzzy vision but I put 20 rounds in him. The grenade took out a bunch of teeth and stuff. The medics wouldn't give me morphine because they thought it was a head wound. They tried to give me APC tablets, but with the roots exposed on my teeth I didn't want to drink water. They sewed me back up and I went right back out with a profile to only eat soft foods."[198]

But Tiger Bravo would not have to go it alone. The cavalry had arrived! At a base north of Bien Hoa, the 1/5 Armored Cavalry Squadron had also been alerted to reinforce Bien Hoa. By 0700 hours A Troop, minus a third of its combat power left behind to guard its base, started a speed march south down Highway 1A. After fighting its way through ambushes, it arrived at Bien Hoa Air Base with only one tank and eight armored cavalry vehicles (ACAVs)[199] able to fight. It was immediately sent to reinforce Tiger Bravo. Tiger Bravo's route took it through a mixed bag—small farms of two or three hooches, and the streets and alleyways of the outskirts of Bien Hoa with small buildings. Each environment presented different tactical challenges that Tiger Bravo had never faced before, nor even trained for. The paratroopers found themselves fighting among working farms and in populated, built-up areas with an armored cavalry force, and under stringent no fire policies for artillery and close air support.

Each new situation called for on-the-spot decisions and unconventional solutions. Initially, the small farms were not much of a problem, and there was no enemy contact. The residents had fled in a panic sometime during the night, leaving behind what few belongings they possessed, a few chickens and pigs, and, to the delight of the always hungry troopers, trays and bowls filled with fresh, succulent fruit. Part of the Vietnamese New Year tradition was to leave a tray filled with five different fruits on the home's ancestral altar to symbolize the family's admiration and gratitude to heaven, earth, and ancestors. Generally, in this part of South Vietnam, the trays were filled with ripe watermelon, custard apples, coconut, papayas and mangoes.[200] While everyone had been warned repeatedly about looting, most paid little attention to the warnings when it came to a few pieces of fruit.

At one point, when the company was clearing farmhouses, the heavily armed and jittery paratroopers, seeing shadows of VC everywhere, opened fire on a sty filled with helpless pigs. Walking nearby, Second Lieutenant Nick Hubbell yelled, "What the hell are you shooting at?!" The trigger-happy private first class, now composed, calmly answered, "Sir, there could be VC hiding in there with those pigs."[201] The troops quickly labeled it the "Great Bien Hoa Pig Massacre."[202]

Clearing the buildings along the narrow streets presented a different problem. Often one squad would be given a street to clear, separated from the other squads in the platoon on adjacent streets. To flush the reluctant VC out of their hiding places, Sergeant Symes, a squad leader in the 1st Platoon, came up with a novel approach. He, and a trooper named Chris Hinman, would jump out into the street and fire over the rooftops. Taking the bait, and thinking they had a couple of easy targets down below in the street, the VC would return fire, immediately giving their position away. A few well-aimed M79 rounds and small-arms fire would quickly silence the threat. Basil Rivera, the M79 gunner designated to put a round through every window where a VC had exposed himself, remembered, "Symes and Chris made it look easy. At times we would look at each other and just grin. We knocked out ten houses that way. They were not given credit for what they did."[203]

With the clearing operation completed, Tiger Bravo assumed a blocking position south of Dong Lach, but contact with the enemy continued unabated. The roar of the battle was constant on both the north and south sides. The VC were fighting for every foot of ground, and few were surrendering. A Troop 3/5 Cavalry remained with Bravo, but by this time two of its remaining tracks were disabled and so many RPG rounds had hit the lone tank that the crew was replaced twice. Three of its men were KIA and another twenty-four wounded. Tiger Bravo had also sustained several wounded in the house-to-house fighting. A Company on the north suffered serious casualties as well— four KIA and another dozen wounded.

Since most of the villagers had escaped the battle and left the area, and friendly casualties were mounting, Lieutenant Colonel Grange decided to call in air strikes to take care of the entrenched enemy. Air operations had resumed at Bien Hoa, and Grange trusted the USAF F-100s to put their high-drag bombs right in the slot between A Company and Tiger Bravo. Initially, the brigade commander wanted artillery to do the job, but when a dud artillery round landed just twenty feet in front of the position he shared with Grange, he acquiesced. In minutes, four F-100s flashed in low and fast in the waning light of the setting sun. "They used high explosive bombs and napalm, and they were right on the money," said Grange. It was only a little town, and it was obliterated when the napalm set fire to the thatch roofs."[204]

Napalm, a jellied gasoline concoction that burns at 900 to 1,100 degrees C and adheres to whatever it touches, was created by Harvard University scientist Louis Fieser in 1942 to help the Allied bombing campaigns in World War

II.[205] It hits like an exploding sunset, with the blast sending fiery red, purple, and orange streaked plumes arcing into the air. By 1945 bombers carrying napalm and high-explosive bombs laid waste to the heart of Germany's industrial base and many of its cities. In the Pacific Theater, it was more of the same. On March 9, 1945, the first bombing raid of a ten-day incendiary campaign against Japanese cities saw 279 American B-29 bombers deliver napalm on industrial targets across Tokyo. A staggering 87,500 civilians perished in the ensuing inferno, making it the deadliest nighttime bombing raid of the war.[206]

The war in Vietnam saw a change in both delivery methods and targets for napalm. During World War II napalm was delivered by multiengine bombers and used primarily in a strategic role against cities and industrial targets. In Vietnam, the focus was on close air support (CAS), delivered at tactical targets by smaller fighter-bombers. A total of 388,000 tons of napalm were used in Vietnam, as compared to 32,351 tons in Korea and only 16,500 tons during World War II.[207]

When the flames subsided, A Company and Tiger Bravo swept the rubble, and came up with 16 prisoners and a body count of 106 VC before pulling back to their night defensive positions (NDP) near the smoldering, smoke-shrouded hamlet. The battle dribbled on most of the night, under artillery illumination rounds, as the VC attempted to exfiltrate out of the area. There were flare-ups of automatic weapons fire every twenty or thirty minutes followed by whispered reports over the radio, "We got another one," or "We got two more."[208] The enemy contacts grew fewer by first light the next morning, and finally stopped on the south side of Dong Lach, but not before Tiger Bravo added one more VC KIA to its total at 0843 hours, and three more at 1036 hours.[209]

SIEGE OF ARVN III CORPS HEADQUARTERS

Shortly before noon orders came for Tiger Bravo to move south to Highway 1A and link up with two platoons of ACAVs from L Troop, 11th Armored Cavalry Regiment (ACR), and a tank platoon from the regiment's tank company to relieve the siege of the ARVN III Corps headquarters.[210] The radio call from Wild Raider (Colonel Grange) to Captain Rankin gave the specifics:

EXECUTE PLAN AS DISCUSSED YESTERDAY. MOVE SOUTH TO THE NEXT VILLAGE
TO HIWAY 1A (GRID 023124). AT THAT TIME MEET UP WITH A TROOP FROM
THE CAV (RADIO FREQUENCY 51.25) CALL SIGN LIGHT HORSE 6, AND THEN
BECOME PART OF THAT ELEMENT. THEN SWEEP WEST BETWEEN RR (RAILROAD)
AND HIWAY 1A. MOVE BACK EAST TO HIWAY 1A TO 01 GRIDLINE, THEN MOVE
NORTH-EAST.[211]

This was the second time in twenty-four hours that the ARVN III Corps headquarters had been attacked and needed a rescue force to prevent its capture. The first time, C Company 2/47 Infantry (Mechanized) raced to its rescue and relieved the siege after heavy house-to-house fighting. Just as quickly, C Company moved on to its next mission to protect the Allied prisoner of war camp east of Bien Hoa.[212] The second enemy attack was conducted under the cover of darkness by the VC 238th Local Force Company but had stalled at the walls of the compound. Once again the headquarters was in danger of falling to the VC, who were entrenched in buildings just across Highway 1A. The VC also controlled buildings on two other sides of the compound, with only the rear wall, which abutted the air base, not under continuous AK-47 and RPG fire.[213]

By this time, I had finally made it to Bien Hoa and had rejoined the company. Since the Tet holiday fell on the US Army's end-of-month payday, the battalion executive officer ordered all the company executive officers, who acted as company pay officers, myself included, to remain behind when the battalion helicoptered into Bien Hoa, and to keep the payrolls with them at all times and under guard. I was furious at being left behind, as was the A Company executive officer. Being young, hard-charging airborne lieutenants, we decided the next day to take an early morning resupply Huey into Bien Hoa and find our way to our companies with our rucksacks filled with payday cash. We thought it was a big deal, but with everything else going on, no one noticed, and we certainly didn't tell anyone that the bulges in our rucksacks weren't extra C rations. It would be five days before events slowed down long enough for me to pay the troops. It was a strange feeling, fighting a battle while carrying stacks of cash worth five times my annual salary. I was more afraid of losing the cash than I was of being hit. All I could think was if I lost it, how would I ever pay it back on a lieutenant's salary?[214]

On the main road leading to the city proper, where the 11th Armored Cavalry Regiment unit was waiting, the company came to a small church with three critically wounded VC suspects lying against its outer wall. Hovering over them was a US colonel, who had just landed his C&C Huey in the middle of the road, looking for someone who could speak Vietnamese. Steve Lyle, a graduate of the US Army Vietnamese Language School, was "volunteered" by his buddies. According to Steve, "The colonel was adamant that I interrogate the VC to find out their unit designation and location. He didn't like my answer at all, when I told him comatose men couldn't answer questions. As if to prove me wrong, he began to yell questions—in English—at the men who couldn't speak English, even if they were conscious. Still no answer, only silence."[215]

Farther along the road, and much closer to Bien Hoa proper, the company was confronted with a stream of refugees heading out of the city. Soon it became a wave of thousands of displaced persons pulling small carts and wagons, and pushing bicycles laden with children. Many were carrying what prized possessions they could on their backs or in bags suspended from either end of a pole balanced on their shoulders.[216] There was no way of telling who was truly a refugee or a VC who had ditched his uniform and arms to make his escape; they all passed unharmed. To add to the chaos of the moment, the company also ran into patients from a local mental hospital run by an order of French-speaking Vietnamese priests. These hapless inmates, of both sexes and all ages, had been released by the VC to wander aimlessly around the city. Some had been shot up by a gunship, which had mistaken them for VC trying to escape. In a letter home I wrote, "The ones we found were hysterical. Most of us couldn't understand Vietnamese to begin with, and now we had crazy Vietnamese talking real fast at us. Our Chieu Hoi scouts said that they weren't making any sense. The inmates were just completely out of it."[217]

At noon Tiger Bravo linked up with the cavalry and begun movement down Highway 1A. Shortly, the paratrooper and cavalry force came to a large Catholic church off to one side of the road. Acting on a report of possible snipers in the steeple, the 1st Platoon was sent to check it out. A high-explosive round from a tank's main gun blew apart the ornate main doors of the church. Then Second

Lieutenant Joe Hillman, Pop Plemons, Steve Lyle, and three others quickly ran
into the sanctuary and cleared the steeple. The end result was no snipers, but
the beautiful mahogany doors, with handcrafted hinges and latches, were in
pieces. While they were standing in the smoking rubble of the church doors a
very distraught Vietnamese woman ran up and started screaming. Lyle translat-
ed as best he could and told Hillman that she was the caretaker of the church,
and that she was mad about all the damage that had been done. At that the
lieutenant reached into his pocket and started counting out Vietnamese piasters
into her hand, not stopping until her anger turned to a wide grin. She then removed her rosary, placed the cross over her knuckles, and had each paratrooper
kneel and kiss the cross. As soon as they were out the door and on their way a
grinning Plemons was heard to say, "Well, that right there is enough hypocrisy
to get us all killed dead before we see another sunset."[218]

Contact was light until Tiger Bravo and the cavalry approached the area
immediately in front of the III Corps compound. The VC apparently had decided to make a stand. It was a poor decision, but it might have been their only
choice as "the only path of retreat for them was blocked by two ARVN battalions placed along roads to the south and west of III Corps."[219] It was a battle
down to the last man for the VC. The 11th Armored Cavalry Regiment's tanks
started by firing canister rounds at point-blank range into the VC-occupied
buildings. Round after round thundered into the block of densely packed
homes and shops. Some buildings collapsed from the barrage of tank rounds,
while others remained semi-intact but caught fire from exploding ammunition
stored in the buildings by the VC during their preparations for the attacks. Immediately the VC returned fire, and some started to flee the burning buildings
but were gunned down in the streets. After five minutes, there was no more
return fire, only silence and flames. The surviving VC had moved farther back
into the labyrinth of alleys and streets of Bien Hoa. After gunships rolled in to
further soften the area, Tiger Bravo moved into the attack. As one participant
put it, "It was the type of thing you see in the movies. We'd fire an M-79 into
a building, then the lead man would kick the door in and go in shooting."[220]

The first Tiger Bravo casualty of the day was Staff Sergeant James Pateo,
platoon sergeant of the 1st Platoon, who ended up pinned down a hundred
feet into an alley, on his back, shot in the abdomen. Since the alley was cov-

ered completely by VC gunners in a building at the end of the alley, it was a death trap—no one could reach him without being shot as well. Steve Lyle, the platoon's medic, found this out the hard way. "Since I hadn't heard a shot," Lyle recalled, "and it was noon in the tropics, and we had been on the go since daybreak, I was convinced that he must have passed out from heat or sunstroke."[221] So Lyle shucked his gear and ran down the alley, thinking he could use a fireman's carry to pull Pateo out. He made it to within ten feet of Pateo before he was shot as well. "Had I known Pateo had been shot, I would not have been so stupid," [222] said Lyle. Hit in the leg, Lyle fell against a wall and watched as the battle raged between VC fighters in the buildings and the paratroopers now bent on saving two of their own.

With two men trapped, something needed to be done, and fast, before more were hit trying to save them. Seeing that the alley was a kill zone and that the VC were just waiting for more Americans to attempt a rescue, I ran down the main road to where the remainder of the company were heavily engaged and brought back a tank to fire down the alley as cover. In just minutes the tank was firing its 90mm main gun at the VC positions. At the same time, Plemons charged into one of the buildings and shot one of the VC gunners, then yelling from an open window, "I got him!"[223] By now the alley was filled with gunfire, tank main gun rounds, ricochets, and chunks of cinderblock flying all around. Then Plemons and several others pulled Pateo to safety. Lyle used that opportunity to extract himself by hobbling across the alley to an open door that led him into a hotel lobby. Lyle, trying to keep his weight off a fractured right tibia, hopped out the front door of the hotel onto a surreally quiet Highway 1A, just as its guests had done before the war.[224]

The battle along that stretch of Highway 1A continued until a total of thirty-six VC were killed.[225] Finally, when the few remaining VC slipped away, the siege of ARVN III Corps headquarters was lifted for the second time in twenty-four hours. In addition to Pateo and Steve Lyle, four others were wounded.[226] One of the wounded was the company commander, Captain Freddy Rankin, hit in the hand while providing covering fire during the extraction of Pateo and Lyle.[227]

Tiger Bravo's sole KIA in the battle was Private First Class Gene L. Keahi, an eighteen-year-old rifleman in the 4th Platoon from Ewa Beach, Hawaii—or as we called him, "Big Pineapple." According to his platoon leader, Jim Roach, "Gene Keahi was moving under fire when he was hit. We were moving in a village just outside of the air base. It was part of a company movement, and suddenly we took heavy fire. It took our platoon another thirty minutes to maneuver up to where he had been hit, and he was dead. Impossible to know if he died instantly, or bled out."[228] At the time of his death, he had been with the company for only thirty days.

Enemy fire was not the only thing that could kill you on the battlefield. Friendly fire from other ground units, helicopters, and artillery could be lethal as well. If the VC were seemingly everywhere and popping up at the most unexpected times, and a wide variety of US combat units were being flung into the battle at a furious pace, then mishaps were bound to occur. The 2/506 Daily Staff Journals for January 31 and February 1 told the tale:[229]

31 JANUARY:

1247 HOURS: A COMPANY AND UNKNOWN FRIENDLY FORCE IN LIGHT CONTACT. NO CASUALTIES.

1525 HOURS: USAF UNIT SHOOTING AT PYTHON 3 (BATTALION OPERATIONS OFFICER). NO CASUALTIES.

1740 HOURS: US ARTILLERY HITS FRIENDLY LINES WITH ONE DUD ROUND. NO CASUALTIES.

1 FEBRUARY:

1701 HOURS: MALFUNCTION ROCKET FROM COBRA HITS JULIET ELEMENT (BATTALION RECON PLATOON). 2 US KIA AND 1 WIA.

Also on February 1, but not recorded in the battalion's journal, was an incident with the 11th Armored Cavalry troop that fought alongside Tiger Bravo in the Bien Hoa house-to-house fighting. A Cobra gunship, making a strafing run on some buildings to soften it up for the infantry, opened fire too soon and a spray of minigun fire hit a line of ACAVs waiting for the attack to resume, instantly killing a vehicle commander.[230]

MISSION ACCOMPLISHED!

By 1815 hours on February 2, the battle had ended and Tiger Bravo was back inside the air base perimeter,[231] riding on what was left of the armored cavalry troop past cheering throngs of US Air Force personnel and civilian workers in a scene reminiscent of the Allied liberation of Paris during World War II. It was a flag-waving, crowd-cheering moment; one airman ran alongside the tank I was riding on and handed me an ice-cold can of beer.[232]

On February 4, Tiger Bravo left Bien Hoa. Mission accomplished![233] An article in the *Army Reporter*, which was the official US Army newspaper in Vietnam, reported that "the heaviest engagement, of all those fought by the 101st Airborne Division during Tet, was in the Bien Hoa area near the 101st Airborne Division Headquarters and the Long Binh Post. The 2nd Battalion, 506th Infantry killed 106 enemy on January 31 and another 58 on February 1."[234]

BACK TO SPARTA

After a short helicopter ride, it was back to a dangerous and spartan existence in enemy territory. Bien Hoa Air Base, with its comforts and luxuries common to most large US military installations in Vietnam, but out of the reach of most frontline troops, was left behind. Left behind was the army's largest post exchange, where Tiger Bravo had been denied entry until the troops had stacked their ubiquitous weaponry outside the entrance. Also left behind was the outdoor swimming pool, which had its locked gate broken open by the company executive officer so that the troops could finally wash off days of grime and soot from fighting among flames. Left behind were the air-conditioned buildings with hot and cold running water and indoor toilets. The same for cold beer, ice cream, milk shakes, and pizza. All left behind for someone else to enjoy.

For the first time the marked difference between the life and opportunities of a rear echelon soldier and that of a soldier "on the line" were plainly evident to every Tiger Bravo trooper. The disparity began to gnaw on the battalion's psyche, which up to that point had been entirely focused on finding and killing "Charlie." Colonel Grange noticed a change in the battalion's élan from that point on. Before that, his soldiers would do "anything to get into a fight. They wanted to find the VC." But after a few days in Bien Hoa, he noted, "they never held back . . . until they got to see how the rest lived."[235]

It wasn't just the disparity in creature comforts that struck some as unfair. According to US Army policy both those stationed in fixed installations and those doing the fighting received the same combat pay of $65 per month, called hazardous duty pay. To many, there was no comparison between the hazards faced by a soldier safely ensconced within the perimeter of a large installation and those faced by infantrymen in harm's way every day and night. Colonel Grange was one of the most vocal critics of the policy and the line soldier's most ardent advocate. "I always felt that the 11 Bravo in the line unit facing all the dangers and hardships should get $100 a month and the guy in Saigon should get $40," he argued. "The conditions for the line units were horrible, not to mention facing the enemy each day. That whole situation was terrible."[236]

The debate over special pay for infantrymen was not a new one. Earnie Pyle, the renowned frontline reporter in World War II, wrote extensively of the need for special pay for an infantryman, who faced dangers and extreme living conditions rarely seen by rear echelon soldiers. "Their two worlds are so far apart that the human mind can barely grasp the magnitude of the difference," he testified at a congressional hearing in 1944. "One lives like a beast and dies in great numbers. The other is merely working away from home. Both are doing necessary jobs, but it seems to me the actual warrior deserves something to set him apart."[237]

The twelve-month tour policy suffered from the same inequity. No matter where a soldier was assigned—rear echelon or front line—everyone was eligible to rotate home after twelve months. "Why let everyone go home the same after one year? Even the simple point system in the Korean War was fairer," Grange contended. "It was a point system based first on how long you were over there, then on your proximity to the enemy. Those there the longest and on the front-line got to go home before those not doing the actual fighting. If you were in a rifle company you got to go home before just about anyone else."[238]

For a rear-echelon soldier, completing a twelve-month tour unharmed was not a certainty, but highly probable. Whereas the chances that a soldier in a combat unit would even survive were far less. Many in Tiger Bravo did not expect to make it through the year unscathed. A Department of Defense study, which examined Vietnam War casualties from 1965 to 1972, concluded that soldiers serving in maneuver battalions, such as the 2/506, were fifteen times more likely to be killed than those serving in non-maneuver battalions, such as rear area support battalions.[239] No one faulted the rear troops for this situation. Soldiers served where the army had the need, we all knew that. Envy, yes. Blame, no. We all understood that soldiers do not make US Army policy. Soldiers followed policy, often were trapped by it, though they sometimes benefited from it. But it was not a soldier's position to decide on policy or even have a voice in its application—fair or unfair. Although the soldiers of Tiger Bravo were not above having a little fun at the expense of anyone in the rear who had daily access to the luxuries not available to line units. One time, on a stand-down at a large fixed installation, where the company was prepping for its next mission, the base camp theater was put off-limits to Tiger Bravo soldiers. This didn't sit well with Joe Adams, who calmly walked up to the theater's front door and pitched in a riot control grenade[240] that normally would be used to clear enemy out of tunnels and bunkers. He escaped undetected after the explosion and pandemonium. The Military Police never did solve that incident, although when problems like that occurred the finger was usually pointed at the 101st Airborne Division, and rightfully so.

There was, however, some common ground between line troops and those in rear areas. Nowhere in Vietnam was completely safe from attack. Fixed installations had their own set of dangers, as they were favorite targets for deadly rocket and mortar attacks. In addition, we all suffered the pangs of a long separation from our families—missing birthdays, anniversaries, holidays, graduations, births, deaths. We all longed for the thousands of special moments that added color to our lives—a child's first steps, holding hands, high school football games, Sunday church services, and walks in the woods with a devoted dog. We all shared in a dream to pick up our lives again after our tour was over—plans for college, a job, raising a family, or just discovering the next chapter in our lives. And grief over the loss of a friend caused the same dreadful feeling of hollowness, whether sitting on the edge of a foxhole in a nameless patch of jungle or on the edge of a cot in an air-conditioned barracks.

NEW BLOOD

By mid-February, the US Army replacement system had started to backfill Bravo's losses with green replacements. Many had been in training in the States when Bravo fought its battles in War Zone D and Bien Hoa. Harry Brown, one of the February replacements, described his assignment to Bravo Company in a letter home: "I am in Bravo Company (called the Tiger Company) which everyone says is the best here. All the officers are West Point men with previous combat experience, and the sergeants are combat veterans too. This company in two months here has lost only three men, even though they are the most aggressive."[241] In the same letter home, Brown went on to give a somewhat inflated description of the weight of a typical soldier's load and miles covered in a day that only a new, impressionable replacement would accept as true: "We have to carry a pack (45 lbs.), 5 grenades, 800 rounds (M-16), flares, explosives and 400 rounds of machine gun ammo per man. That's close to 150 lbs. per man, and they hike about 25 miles a day."

Replacements came from within the 3rd Brigade as well. Tom McClear, a nineteen-year-old member of the Brigade Reconnaissance Platoon—the "Phantom Force"—requested a transfer when the platoon transitioned from a fighting role to headquarters security. "I have changed units. This move was voluntary and was initiated by me. The old platoon I had been in was relegated to the lowly status of a security platoon," Tom wrote to his parents. "I am glad I was assigned to B Company as it was my first choice. It is the best company in the Brigade. They have seen the most action and they have an excellent company commander and First Sergeant. . . . When I reported to the 1st Sgt both he and the CO shook hands and asked where I came from, how I was doing, anything I needed, etc." He went on to elaborate on his first impression, "Both seemed like real gentlemen. I believe they are the first I have run up against in my Army career. At least they are the first to treat me like a human being."[242]

With the uniqueness of urban combat behind it and another month of assorted combat missions completed, Tiger Bravo had "come of age" in an operational sense. The company had started slow with its in-country training, but eventually reached a high tempo of combat operations. In January and

February alone Tiger Bravo conducted twenty-six search and destroy missions and thirty-six night ambushes,[243] each making a small contribution to the company's inexorable odyssey from raw and untested to battle-hardened. Some lessons learned were fundamental: stay low, avoid walking on trails, dig in deep at night. But others involved the essence of combat that could never be taught, only experienced: a realization that battlefields were inherently chaotic, that the "fog of war" was an authentic phenomenon, that leaders would never have a complete understanding of the enemy's location, that improvisation was a necessary virtue, that speed and stealth won as many skirmishes as did firepower, that every moment held risk, and that every man was mortal.[244] But the journey came with a heavy price. By the end of February, the company had suffered thirty-three casualties from combat—and another two injured.

March brought a command structure change to the 101st Airborne Division that led to an even greater tempo of operations for Tiger Bravo. Control of the 3rd Brigade, 101st Airborne Division passed from the 101st to the commanding general of II Field Force, who immediately made the brigade his "fire brigade," one that could move quickly to any hot spot in all of central Vietnam. Previously, the mission had been assigned to the 1st Brigade, 101st Airborne Division, which quickly became known as the "Nomads" because of its constantly moving from one critical area to another. For the next seven months, the 3rd Brigade would do the same, conducting operations from the Mekong Delta to the Central Highlands, and would fight under the control of the 9th Infantry Division, 4th Infantry Division, 25th Infantry Division, 199th Infantry Brigade (Separate), and 11th Armored Cavalry Regiment, earning it the nickname of "the Wandering Warriors." The battalion commander described it this way: "We were nomads. The battalion went everywhere, with everybody. When the going got tough, that's who they called." He went on to describe Bravo's participation: "B Company was a great company . . . a fighting bunch of paratroopers. They saw a lot of action."[245]

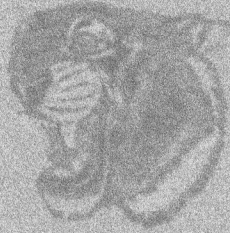

Part II

Wandering Warriors

5

Jungle Turkey Shoot

North of the Song Dong Nai
(Binh Duong Province)
March 1–26, 1968

VC ON THE MOVE

On the early morning of March 1, 1968, Tiger Bravo's night defensive position (NDP) sat in the middle of a bull's-eye. Marching straight at it through pitch darkness was a VC battalion, split into two attack forces, with more than a hundred VC each, from the south and the west. Launched from their safe areas deep in War Zone D, each column trudged silently through light forest and past dense banana groves, guided by local VC militia. On their backs and cradled in their arms they carried the signature armaments of light infantry: small arms, automatic weapons, light and heavy machine guns, grenades, mortars, and enough ammunition to send a torrent of bullets and explosives at the sleeping paratroopers.

Around the Tiger Bravo defensive position, numbering only seventy paratroopers and an 81- mm mortar section, all was quiet; neither close-in listening posts, nor soldiers on guard around the perimeter detected anything unusual. Half the company slept in or alongside their fighting positions. In the center of camp, First Sergeant Trent was trying out his new jungle hammock, set up between two trees. Tiger Bravo's other two platoons manned night ambush positions close to a kilometer away. Neither ambush was close enough to immediately assist in the defense of the NDP, nor large enough to survive independently in a toe-to-toe engagement with one of the VC columns. The platoons' best chance for survival would be a good offense, relying on the surprise and devastating effect of their well-executed ambush to cover their withdrawal to preselected rally points.

Putting a substantial number of combat troops in motion on the battlefield was a serious matter for a VC battalion commander. Before a formidable attack force capable of overrunning a US infantry company could be assembled, a VC battalion commander would have to be certain of his plan's success. He would also require the approval of the area political committee. A mandate from Vo Nguyen Giap, commander of the armed forces of North Vietnam, made the decision to attack, or wait for another day, a simple up-or-down choice. He wrote, "Strike to win, strike only when success is certain; if it is not, then do not strike."[246] Additionally, one of the basic tenets for an NVA/VC attack on a defending US force was that the attacking force would greatly outnumber the defending force, especially at the point of engagement. The ratio of ten VC attackers to one US defender was not an uncommon goal. In practice, it was quite normal for an NVA/VC battalion of five hundred men to attack a US company of a hundred soldiers.[247] The odds this night were all in favor of the VC.

A VC attack—large or small, day or night—followed a standard "one slow step, four fast steps" sequence. Every attack started with a slow, deliberate planning and preparation phase, often including a detailed reconnaissance of the objective by VC scouts, approval by a district political committee, a walk-through of the attack on a sand table mock-up, and rehearsals for everyone involved. Next, the attack force made a quick advance into attack positions close to the objective, normally followed by a quick and violent assault employing every weapon in the attacking force's arsenal. This was followed by a short battlefield clearance step or "mop-up." Often, US soldiers reported seeing the enemy carrying dead and wounded from the battlefield, even suffering additional casualties in the process. The last step was a quick withdrawal of the attacking forces, back to their dispersed locations.[248]

When given ample planning time and the opportunity to select the time and place of attack, the VC were a formidable foe. Tiger Bravo could not have been in a more vulnerable position than it was on the night of March 1.

NDP

Bravo Company had established its defense just north of the Song Dong Nai (Dong Nai River), a muddy, winding waterway often used by the VC to move arms, ammunition, and supplies by sampans—little more than large canoes—in the dark of night. It had been a monotonous four-day stretch of long, tedious marches in the stifling heat, with little to show for it. Day and night, the VC seemed to be lying low, waiting for just the right opportunity. Up to this point in the operation, Tiger Bravo's only contact came on February 28, when two soldiers were wounded by a booby trap and several VC were spotted and fired on with no results.[249] Even a battalion- coordinated Eagle Flight, employing Bravo's 2nd Platoon as the infantry component, came up empty.[250]

Eagle Flights had different meanings during the Vietnam War, but in this context, it was a small, self-contained airmobile unit (command and control, lift ships with infantry and gunship support) that was assembled on a temporary basis. Its employment was characterized by a lack of preplanned landing zones and acceptance of limited fire support. The effectiveness of an Eagle Flight depended upon its ability to react and maneuver rapidly in any combat situation. Often the entire force would be airborne, or on a couple of minutes' standby, to immediately pounce on any target of opportunity.[251] Essentially it was an airmobile combat force flying around just looking for a fight.

Captain Rankin had established Tiger Bravo's NDP just before dark the previous day. Clearly, the VC tracked Tiger Bravo's movements during the day, and watched it stop while it was still daylight, and settle in for the night. Typically, in cases such as this, VC reconnaissance teams of one to three men would shadow American units and send reports back to their commander by runner. Come nightfall they would "go to ground" and keep the American unit under constant surveillance. Once an attack plan was set in motion, these same teams and other security forces would guide the assault and support elements to their attack positions.[252]

On this night the reconnaissance of the NDP by VC scouts, and movement of the attacking forces into position, had been flawless. A US unit stopping in daylight to set up for the night invited an attack or harassing fire from mortars or a sniper. Normal procedure in the 3rd Brigade was for a company to continue search and destroy operations into late afternoon, then, if needed, stop to bring in resupply helicopter(s) with Class I (food and water) and Class V (ammunition). After dark the unit would continue its movement until it reached its nighttime position. Many times this confused any VC who may have been tracking the unit's movement, and negated the opportunity for the enemy to plan a targeted attack.

Additionally, Bravo Company had used this location as an NDP site before. This too would not go unnoticed by enemy scouts. In the fading-light the paratroopers dug in as normal. Some, using old fighting positions from the previous NDP but not completely filled in, scooped out shallow prone positions. As Dan Bernard remembered, "We were digging up our old foxholes. It was the second or third time we had dug up our old holes. The holes were very shallow because we were bored, and knew nothing was going to happen. 1SG Trent even put up a hammock he had picked up from somewhere. He was quite comfortable."[253]

Tiger Bravo was in a trap of its own making, one that had snared every infantry company in Vietnam more than once. Long, physically taxing, boring days and a string of uneventful nights, even in such a dangerous, VC-infested region, north of the Song Dong Nai, fostered complacency. Complacency bred carelessness. In Vietnam, carelessness was often fatal.

Not every tactical advantage was on the side of the attacker. Bravo Company had certain factors in its favor as well. First, there was one glaring omission in the VC commander's understanding of the Americans' disposition of forces. The VC knew where the NDP was, but Bravo's ambushes had moved out in complete darkness, undetected by the VC scouts. They were completely invisible to the enemy. Lieutenant Pat Brooks' 2nd Platoon went to ground in a banana grove eight hundred meters due west of the NDP, and the 4th Platoon, led by Lieutenant Jim Roach, was the same distance due east.[254]

Also, the Tiger Bravo perimeter was well laid out, and used the folds in the terrain and vegetation to maximum advantage. This was no longer a company of fledgling soldiers struggling through a probationary period. Tiger Bravo had been bloodied and hardened since its arrival in country. Establishing an NDP was a skill honed nightly. The perimeter was tight with no gaps that could be exploited, completely covered by intersecting fields of machine gun fire, set inside a lethal ring of claymore mines, and manned by experienced troopers at every fighting position. As per company standard operating procedures (SOP), every fighting position had extra ammunition and grenades laid out and within arm's reach—a precaution that, despite being lax on the shallow fighting positions, the NCOs rigidly enforced.

Last, but vital to any unit's survival, were the defensive concentrations of artillery that Lieutenant Joe Palagyi fired in to cover the NDP and ambush sites. At night, he would preregister targets on all avenues of approach leading into the NDP with fire from the supporting artillery unit. Firing would also be planned on likely enemy attack positions. The firing data stayed on the guns throughout the night so that these preplanned fires, called defensive concentrations or "DEFCONS," could be fired with pinpoint accuracy and within seconds of the call for fire. Depending upon the tactical situation, DEFCONS would often be test-fired with live rounds,

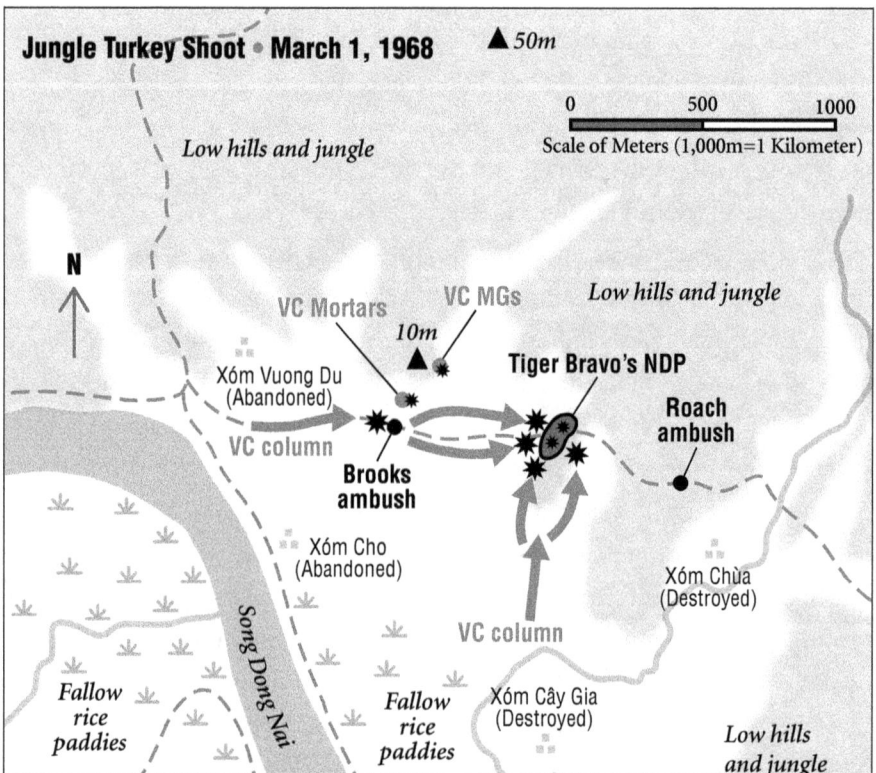

Jungle Turkey Shoot • March 1, 1968   ▲ 50m

0   500   1000
Scale of Meters (1,000m=1 Kilometer)

Low hills and jungle

N

Low hills and jungle

VC Mortars   VC MGs

10m
▲

Xóm Vuong Du
(Abandoned)

Tiger Bravo's NDP

Roach
ambush

VC column

Brooks
ambush

Xóm Cho
(Abandoned)

Xóm Chùa
(Destroyed)

VC column

Song Dong Nai

Fallow
rice
paddies

Fallow
rice
paddies

Xóm Cây Gia
(Destroyed)

Low hills
and jungle

rather than just preplanned, before the firing data were locked in on the howitzers for the night.

The standard procedure was that defensive concentrations would not be placed closer than three hundred meters to friendly positions unless requested by the company commander.[255] This was an instance when line units were hampered by vestiges of standard operating procedures more suited to the training environs of Fort Campbell than the killing fields of Vietnam. Well-meaning policies established for troop safety in training in the United States made little sense in combat and, in the case of artillery fire, actually slowed the delivery of close-in fire support so vital to success on the battlefield. The three-hundred-meter restriction did not match the tactical imperatives brought about by engagements that normally took place less than fifty meters between the opposing forces, and did not take into account an enemy whose basic tactic was to move even closer as protection from American artillery barrages. Even sixty seconds wasted in adjusting artillery fire to where the battle was actually taking place was too long. I habitually gave my consent to bring in these preplanned concentrations much closer. A simple message of "Romeo Sierra Tango Juliet" (my initials, RSTJ, spelled phonetically) sent to the artillery fire direction center and a healthy dose of faith in the abilities of Joe Palagyi, our artillery forward observer, overcame this administrative hurdle.

Only time would tell if these factors worked to the advantage of Tiger Bravo, or would have no effect on the looming attack.

Blowing Up in Flames

At 0330 hours, the *thump* and *boom* of exploding mortar rounds falling on the NDP jolted everyone awake. In the space of several minutes, more than thirty mortar rounds, both 60mm and the larger 82mm, exploded within the relatively confined area of the Bravo perimeter—an oval approximately a hundred meters long and seventy-five meters wide. "Charlie had half our company area pretty well laid out," remembered Tom McClear, who was wounded in the attack. "One mortar tube had zeroed in on the three men position I was in. We received ten or twelve rounds of 60mm mortar fire within ten or twenty feet of our position."[256] In a letter to his sister, he recounted how he was wounded.

"My right arm and part of my upper back got in the way of some shrapnel. Actually, it is not serious. It was like getting hit in the arm by a baseball bat. I am all stitched up now, and will be ready to go again in three weeks."[257]

Luck was with Tiger Bravo this time. When mortar rounds impact, they throw fragments in a pattern that is never truly circular. In some cases, soldiers within a few feet of an exploding round have been spared any injury, while another ten feet away might be blown off his feet. The best estimate of the circular lethal bursting radius of a 60mm mortar round, under ideal conditions, is approximately ten meters. An 82mm mortar round can be lethal up to fifteen meters.[258]

The VC mortars had set up close to Brooks' ambush, to the west of the NDP, and he fixed their location as they fired. He immediately radioed Captain Rankin to report the mortar's location.[259] It was outside the kill zone of his ambush, but easily pinpointed by seeing the muzzle flashes of the mortar rounds leaving the tubes and the movement of the mortar crews. Then, as he shifted his gaze to the front of his position, he was completely surprised by what he saw. It was a reinforced VC company, oblivious of his ambush, walking directly into its kill zone. Catching the VC off guard is not easy under any circumstances. But when you do, the results can be violent, bloody, and oh-so-sweet for the unit springing the ambush.

At the same time, VC machine guns raked the NDP, sending green tracer rounds flying inches over the heads of crouching soldiers. The machine gun fire was coming from two different locations—light machine gun fire from the attacking force close to Bravo's foxholes, and heavy machine gun fire from a small knoll near the enemy mortars. As part of the western ambush force, Harry Brown saw it all unfold. "A .50-caliber machine gun raked our camp."[260] It was chaos. There was no time to return fire—just roll into a fighting position, hunker down, and wait for the attack that everyone assumed would be coming. Watching the rounds impact on the NDP from his location, Harry Brown recalled, "Our whole campsite was blowing up and in flames."[261]

Nothing is simple when bullets start to fly between two opposing, heavily armed forces. Everything moves fast, seemingly all at once. Adrenaline kicks into overdrive. Order dissolve into what looks like chaos to the uninitiated.

The noise is deafening. The "fog of war" keeps anything from making complete sense at the moment. That night was no exception.

In the midst of all this chaos Bravo's training took over. Weapons were fired. Hand grenades thrown. Claymore mines detonated. Officers and NCOs directed fires, encouraged, and kicked butt. Medics moved about. Radio calls went out for supporting artillery fire. A mental picture of the battlefield started to form in leaders' minds. Order started to reemerge as the battle ebbed and flowed. Initially caught off guard, Bravo began to fight back.

Within a few minutes, Bravo's chain of command was in control. It faced three distinct facets of the same battle. All were intertwined and connected, yet each was fought in a different manner. The first was an enemy force walking into the kill zone of the western ambush. Next, enemy supporting fires from mortars and heavy machine guns had to be dealt with by artillery. Simultaneously, a close-in attack on the perimeter, from the VC column approaching from the south, was occurring well within hand grenade range.

Brooks' ambush in the west was sprung by the simultaneous explosions of multiple Claymore mines, followed by a shower of hand grenades. Small arms fire was not used, as it would have given away the location of the ambush force to the much larger enemy force.[262] This took the VC force completely by surprise. In describing the start of the ambush, Brown wrote, "One VC popped up when we blew our claymores. The back blast covered me with mud. Charlie didn't see us. Then the VC shifted fire on us. A 7.62mm Chinese machine gun chopped down banana trees 6" above my head. I was really shaking."[263] Brown survived that fight, and the next morning he was able to see the effects of the claymore on the VC who had popped up directly in front of him: "Today I saw the guy we killed. He got those steel balls between the eyes, tore one eye out, made a 3" hole in his hand, split his left leg from calf to hip, broke both ankles and put 32 holes in his chest. He walked right in front of my claymore. That's one gook who will never kill again. He looked about 20."[264]

After realizing he had ambushed a force four or five times the size of his platoon, Lieutenant Brooks quickly pulled his men out of the banana grove and into a network of rice paddy dikes that offered some protection.[265] The platoon took off running, then stopped and regrouped. Preplanned artillery fire was already hitting the ambush site to cover the withdrawal. The order to withdraw was a judgment call on Brooks' part. Often, ambushing units stay and sweep the kill zone if a small force was ambushed. But in cases such as this one, where the ambushed retains a numerical advantage even after suffering casualties, implementing a preplanned withdrawal is the best course of action. During the withdrawal, Specialist Fourth Class (SP4) Alan L. Gero, from Waterville, Maine, became separated from the platoon and had to crawl more than four hundred meters to safety, dodging fifty VC in the process. As Gero waited for a safe time to crawl up to the Bravo Company perimeter, he saw another ten dead VC being carried off the battlefield. "The Viet Cong were fifteen feet in front of me, and I couldn't make a move,"[266] he said.

During the predawn, three-hour battle, Palagyi called in more than a thousand rounds of artillery from five different firing batteries.[267] According to Palagyi, "The artillery prevented Bravo from having a number of casualties. One of the firing batteries was actually a naval ship. It was firing both interdiction and illumination rounds from two of its big guns—the big 16" guns. Their illumination rounds made it look like noon on the battlefield. That's how we could see them out in the open."[268] His first targets were the VC mortar position, identified by Brooks, and the heavy machine gun on the knoll, both of which were hammering the NDP. Next, he shifted fire onto the still substantial remnants of the reinforced VC company hit by Brooks' ambush. Every time the VC tried to collect their dead from the kill zone, Palagyi called in more artillery strikes, resulting in additional enemy casualties.

The other attacking force coming at the NDP from the south, in groups of no more than twenty each, was too close to Tiger Bravo's fighting positions to safely employ artillery. In a letter home, Chuck Hanson described the attack from the south: "My platoon and one other were in the night defensive position. Charlie shot a whole pile of mortars at us and under the fire his men, between 2 and 3 hundred, we think, surrounded us. They are so skillful that they actually set up machine gun positions 15 feet in front of our positions and we didn't know it until they started firing. We knocked most of them out with hand grenades and M-79 rounds. What probably saved us from being overrun was Lieutenant Brooks setting off his ambush. At that instant Charlie probably thought that we had him surrounded instead of the reverse."[269]

At this point the enemy's well-coordinated, two-prong attack sputtered and turned into a piecemeal attack by groups of VC unsure of exactly where Bravo Company's positions were. Those that had managed to come closest to the perimeter died quickly under a hail of hand grenades. Others started walking around the battlefield in confusion, not knowing where to attack. Both at the ambush site and outside of Bravo's perimeter, they had been hit with claymore mines, hand grenades, and M79 rounds, none of which gave away the paratroopers' exact location, like muzzle flashes and tracers from small arms fire would.

All along the perimeter, soldiers were grabbing grenades and throwing them out into the darkness or at flashes from VC machine guns. One soldier ended up on his back in his foxhole, and couldn't even roll over because of the machine gun fire just above his head. He threw all of his grenades over his head at the enemy while on his back. The next morning there were dead VC within just a few feet from his position, along with one unexploded grenade with the pin still in it.[270]

Reports of the VC walking around, talking, yelling, and even screaming came in from all parts of the perimeter. Captain Rankin said, "It was the first time I've ever been in a night defensive position and had the enemy walk into it."[271] Harry Brown wrote home: "All night they yelled, screamed and cursed us. They ran in open sight and lit cigarettes in front of us. They are crazy! I even saw some women with rifles."[272] Dan Bernard described it this way: "We trashed the bad guys' whole plan. They didn't know what was going on. There was a bunch of Vietnamese talking out there."[273] From Chuck Hanson's letter: "A good percentage of the VC were probably high on dope, as in the light of the flares I could clearly see and hear them a little more than 100 yards from us running around and yelling. It was really quite a turkey shoot."[274]

Such behavior was not unusual for the VC. Similar experiences had been reported by US units before. One study that examined enemy behavior during an attack reported that "a more bizarre, eccentric foe than the one in Vietnam is not to be met. He may blow whistles or sound bugles to initiate the assault; or he may trip the fight with a flare or the beating of a bongo drum. It is in many small ways that the enemy in Vietnam deviates from what we consider normal, sometimes to the stupefaction of our people. Nerves get jangled when in a firefight at close range men hear maniacal laughter from the pack out there in the darkness just a few feet beyond the foxhole. Catcalls, the group yelling of phrases and curses in English, the calling out of the full name of several men in the unit—such psychological tricks are likely to be trotted out at any time."[275]

I remembered hearing the yelling as well. I didn't speak Vietnamese so I didn't know exactly what was being said, but it sounded like a VC leader angrily yelling at his men. I also heard an exclamation in English— "Aww. I've been hit!"—come out of the darkness, just outside our perimeter, only seconds after a flurry of our grenades had exploded. I knew that none of our soldiers was out there, only VC. Yet, somehow, perfect English had come out of the darkness. I never did discover the source.

First Light

At first light Captain Rankin sent patrols to check outside the perimeter, the ambush site, and the VC mortar/machine gun positions. A total of twenty-one VC KIA was accounted for—eleven bodies left behind and ten that had been seen being carried away during the firefight.[276] In addition, Tiger Bravo captured one POW, an RPG-7 rocket launcher, thirteen rockets, and two light machine guns.[277] The bodies were stacked in a grotesque pile near the landing zone used by the resupply chopper. Harry Brown described the scene: "The bodies we piled up on the LZ and had dirt thrown over them. They were missing arms, legs, hands, heads, stomachs, sides and chests were ripped open. One guy's head had an [unexploded] M-79 round stuck in it, ½ his skull was crushed. The sight was terrible. They stunk and flies were all over them. They looked like wax figures in grotesque shapes and positions. They had terrible expressions on their faces."[278] Chris Backman remembered two women among the dead, both heavily armed.[279]

Just after dawn Lieutenant Colonel Grange, accompanied by his driver, Chuck Kudla, who was a former Bravo Company paratrooper, walked the battlefield. Kudla remembered that he saw "a bunch of bodies stacked up; heads half gone and limbs off."[280] Kudla was a twenty-year-old native of Detroit, Michigan, who had joined the army two credits short of graduating from high school. He had insisted on airborne because his father had made a combat jump into Normandy in World War II. He joined Bravo Company in May 1966. In August 1967, he volunteered to be the battalion commander's driver, and go to South Vietnam when the battalion deployed, rather than return home when his enlistment was up, which he could have done since he was on the nondeployable list.

Tiger Bravo suffered only four WIA that night, all from the ranks of the NDP, and only one was serious, with a stomach wound.[281] One of the wounded was a sergeant from the 101st Division Public Information Office doing a story on Tiger Bravo. Little did he realize that he would end up being part of the story. Dan Bernard recalled, "The PIO guy was shot in the arm. The whole time he had been with us we kept telling him he was a pain in the ass, asking us all these questions. After he was shot, the next day he said, 'Damn, this hurts! I don't want to do this job.'"[282]

That same day Lieutenant Colonel Grange moved Tiger Bravo back to Phuoc Vinh for a very well-deserved break. This provided soldiers a rare opportunity to reflect on what they had just gone through, and how success or failure was measured in terms of dead bodies. Unlike previous wars, battlefield success in Vietnam would not be determined by key terrain seized, strongholds destroyed, or villages liberated, but in a tally of human remains. Tom McClear spoke for all of us when he wrote, "After the mortars the VC attacked and we killed 21 and only got 4 wounded, so we did all right. It's rather disgusting to measure success by number of human lives taken, but you find that a different set of gauges accompany any war."[283]

STEAKS, COLD BEER, AND ROCK 'N' ROLL

Their shoulders hunched under rucksacks, jungle fatigues dirty and ripped from crawling through too many vines and thickets, weapons carried with a familiarity that belied their young ages, and tired eyes that spoke of the prior night's battle against superior odds, the soldiers of Tiger Bravo trudged to the company area after landing on the brigade assault pad adjacent to the airfield in Phuoc Vinh Base Camp. Some were too tired to think farther than their next step. The replacements returning after their first operation were too green to know, but many of the original Tiger Bravo cohort cocked their heads and stared through the rubber trees to see if First Sergeant Trent had worked his magic yet again. They were soon rewarded with Bravo's traditional welcome home to the troops: steaks and chicken on a charcoal grill made out of a fifty-gallon drum cut lengthwise, cold beer, and drums and guitars just waiting for the Bravo Company band to start playing.[284] All of this was made possible by the executive officer and first sergeant, and paid for through a system of fines levied on soldiers for petty infractions. The band instruments cost $375 in the Phuoc Vinh marketplace, coming from a combination of the petty infractions slush fund and donations from the company officers and NCOs.[285] After the welcome home meal, it was back to the mess hall for a party, Vietnam style—part jam session, part sing-along, and part just having fun. It was a rare Saturday night in Phuoc Vinh.

In two days, Bravo would be gone again, stalking "Mister Charles." The previous day's battle was fought on payday. On the previous month's payday, the company had fought the Battle of Dong Lach Village in Bien Hoa. Between these two paydays Bravo Company spent just three nights in base camp.

"Here they come, this company hangs together like glue," announced Trent as the mess hall started filling up. Infantrymen, off-duty and torn from letter writing by the sound of guitars, filed in wearing floppy jungle hats, T-shirts, and fatigues. They were all grinning, grabbing bread, cheese, salami, bologna, ham, and a cold beer or soft drink from long tables set up by the mess sergeant, Staff Sergeant Donald White from Fayetteville, North Carolina. Once settled in, they yelled for the band to start.[286]

Starting off, PFC Charles Midgett, from Norfolk, Virginia, grabbed the microphone and swung into a soulful rendition of "My Girl." The next two songs, one country and one soul, were about that far-off world we called "home." Everyone joined in, clapping, whistling, and cheering. Trent, who used to do some singing around Nashville, jumped in with his version of "Green, Green Grass of Home," followed by Midgett again, with the "The Letter." Then, a cook, PFC Richard Dietrich of New York City, started playing "Your Cheatin' Heart" on the harmonica. PFC David Maymon, from Fairfield, Illinois, on drums and PFC Ralph Aponte, from Santa Monica, California, followed his lead on guitar, and the party was on. Sweaty soldiers on their feet gyrated around, everyone singing and yelling.[287] The mess hall throbbed with pulsating waves of energy and sound. The soldiers not dancing were clapping, stomping their feet, and slapping the tabletops in time to the beat of the drums. The pounding music bounced off the camp's rubber trees in every direction, washed over the perimeter bunkers manned by an unlucky few, and collided with the fringes of jungle outside the wire, where VC scouts surely wondered how to describe this strange scene in their next report.

Then Trent grabbed the mike for one of his signature songs and improvised lyrics. Starting with the opening verses of "Detroit City," he sang,

> Last night I felt so bad on the Song Dong Nai.
> Didn't feel too much like singing songs.
> But we dug our holes real deep.
> And nobody went to sleep.
> And we fought old Charlie the whole night long.
> I want to go home.
> I want to go home.
> Oh Lord, how I want to go home.[288]

Everyone was singing along with the same pride and fervor that they used to sing the National Anthem at baseball games back in the "real world." Trent then threw in another verse:

> The colonel came and sat down in his chopper.
> He came and told us, Boys you did real swell.
> Bravo's done the job.
> You had to get up on that knob.
> And we went up and gave old Charlie hell.[289]

The party went on for hours, the troops laughing and singing, forgetting for a few hours that War Zone D was just a hundred meters away, on the other side of the bunker line. For a time, the men could forget where they were and revert to a happier time somewhere back home, kicked back with a beer and wisecracking with their buddies. Everyone needed it, no one passed it up. The nondrinkers in the company soon became very popular as their drinker buddies traded or cajoled them for the cold beer sitting untouched on the table. Reflecting on that time in the mess hall, I wrote in a letter home, "As I sat there watching the company I picked out those who have been wounded, those who have come close, and those who have killed. I couldn't help but compare our troopers with people their own age back in the states. . .. Some come from poor homes, rich homes or broken homes. Some are constantly in trouble, others are straight. They're young; they're alive; they enjoy things. . .. But, they endured things, both physically and mentally, that I hope my son will not have to endure. "[290]

The Tiger Bravo band would stay together for a few more months, with band members coming and going as casualties mounted and troops were reassigned. By the end of the spring, it too had been gutted by casualties.[291] For a time, the nucleus included Chuck Kudla on drums, Ralph Aponte on lead guitar and vocals and Chuck Limer on rhythm guitar and vocals. Limer, the son of a World War II British commando, was born in Scotland and immigrated with his parents to the United States at an early age. A self-described juvenile delinquent by age seventeen, he was given the choice to "either join the army or go to reform school." Once in the army he volunteered for airborne school because he wanted to be like his father. "I loved playing in the band," Limer recalled. "Up on stage I would look down and see everyone laughing, singing, and dancing. Getting drunk and forgetting about the war."[292]

Two days after the mess hall party, Tiger Bravo was back in the field. Filling two lifts of ten Huey choppers, Bravo replaced D Company and commenced search and destroy operations in the southern portion of War Zone D, well within artillery range of FSB Concord. The first night out the enemy probed Bravo's NDP, with movement close enough for Captain Rankin to pull in a listening post (LP) expecting an attack any minute. But it never came. The enemy force melded into the darkness as quickly as it had materialized.[293]

After an empty day searching for the VC, Bravo established another NDP, and once again detected enemy movement just outside its perimeter. At 2110 hours thirty-five VC were spotted moving southwest of Bravo's line of foxholes and, at the same time, an LP detected movement of a second group. Palagyi quickly called in preplanned artillery and illumination on both locations, and the perimeter erupted with M-16 and M79 fire.[294] Describing this action in a letter home, Harry Brown wrote, "We had another fight with that battalion again. Got hit all day and night. I came close to getting my head blown off with an AK47. From now on I won't stand up to eat. The VC when they attack are stupid. They rely on numbers to get us. We got quite a few with the M79s. He usually moves between 7:00 PM to 6:00 AM. Only when you run into their base camp do you see them in the daytime."[295]

The 2/506, as part of "the Wandering Warriors," was now under the operational control (OPCON) of the 199th Infantry Brigade. Tiger Bravo spent the next six days in search and destroy missions during the day and ambushing at night, with meager results. They detained two suspected VC, fired at one other with negative results, and destroyed three sampans.[296] It was all done under a blistering tropical sun, with one recorded temperature of 102 degrees Fahrenheit.

This was followed by two days back at Phuoc Vinh for base camp defense and a break from operations. When back at base, mail call was always a bright spot, and not just because of sweet letters from loved ones. Food packages were everyone's favorite. One package from an innovative mother consisted of crackers, cheese, caviar, a candlestick and candle, cloth napkins and silverware, with a note to select a friend to share in the bounty.[297] The first night at Phuoc Vinh there was an outdoor movie shown at battalion headquarters for all the troops. Halfway through the show Charlie lobbed a mortar round at the movie crowd but missed by a hundred meters. It didn't attract much attention from soldiers who had been hit with mortars much closer, and the show was delayed for only twenty minutes.[298] The next morning it was back to FSB Concord for five days of FSB defense and night ambushes just south of the Song Dong Nai, and several miles outside the heavily defended perimeter of Bien Hoa.[299]

The considerable legend of Pop Plemons grew exponentially one night during this mission. A narrow dirt road, safely traveled only in daylight, ran south from Concord to Bien Hoa through an area of no-man's-land where both sides played out an ambush vs. counter ambush game each night. Plemons had convinced the first sergeant that he needed to go to Bien Hoa to send a much-needed money order home to his family, in dire financial straits. Permission was granted and Plemons hitched a ride on the daily supply run to Bien Hoa, promising to return the same day.

Daylight came and went, and still no Plemons. Then out of the darkness came a figure weaving back and forth and singing a loud, off-key, drunken version of "Old Man River." It was Pop Plemons. Drunk. Unarmed. Out for a pleasant nighttime stroll. His sad story of a family in trouble had been a ruse all along. Pop just wanted to get drunk and have a good time, and that's exactly what he did. After a day and part of the night carousing in the bars of Saigon, where he had somehow managed to end up, he decided to rejoin Tiger Bravo as he had promised the first sergeant he would.[300]

REAR DETACHMENT

Even though the company was fully engaged in combat operations, the administrative requirements of the army continued unabated. On March 24, the 101st Airborne Division's inspector general inspected the Bravo rear detachment.[301] This annual event reviewed a company's adherence to the hundreds of army regulations and policies that covered everything from personnel records to mess hall procedures, barracks cleanliness, arms rooms, and ammunition bunkers. Captain Rankin sent me in from the field two days prior to prepare, understanding full well that we were not ready—and couldn't possibly be ready—for a Stateside inspection in a war zone. Much like students cramming for a final exam, I had the rear detachment work feverishly to clean up the company area and bring the company records up to par. It was an impossible task.

Bravo Company's rear detachment was typical of those in the rest of the brigade. It included anyone in a support role that could only be accomplished in base camp such as the mess team, supply sergeant, the company clerk, and often the company armorer. There were usually another three or four soldiers on hand who were either in transit, going to or from R & R, or recovering from injuries or disease that kept them from field duty. The soldiers spent their days cleaning up or filling sandbags and their nights pulling guard duty along the bunker line. Some companies also kept either the executive officer or the first sergeant in the rear, but Captain Rankin had a different philosophy. He practiced the "fighting XO concept," meaning that both the first sergeant and the executive officer stayed in the field and participated in combat operations unless one or the other had to return to Phuoc Vinh for a specific purpose.

The area in worst shape for the inspection was the company ammunition bunker. It was crammed with all types of ammunition, some inherited from the 1st Infantry Division. The setup was extremely functional, but way behind in its record keeping. It was a major deficiency for ammunition to be on hand and not accounted for in duplicate, sometimes triplicate, reports. With no time to document the unaccounted-for ammunition, I came up with a typical "Vietnam" solution to the problem. The night before the IG was due to arrive we dug a huge hole next to the bunker and buried all the ammunition and explosives not accounted for in the existing paperwork. The next day, we passed the inspection, and I was on the next chopper out to rejoin Bravo in the field.

While I was away, twenty-three replacements arrived in Bravo Company to take the place of casualties suffered over the past sixty days. In a first for Tiger Bravo, all twenty-three were nonairborne.[302] This was the beginning of a steady transition from the fully airborne company that had deployed in December to a mixed company of airborne and nonairborne. It was solid evidence that the airborne school at Fort Benning could not churn out graduates to keep pace with the mounting casualties in Vietnam.

For the most part, the replacements were accepted and judged on how well they performed in the face of the enemy, rather than whether they had gone through three weeks of airborne school. After a few weeks, the only way to differentiate one from the other was to look at the company roster. All nonairborne soldiers had "NAP" by their name—nonairborne personnel. The replacements also arrived with misconceptions as to the enemy they would face. A 3rd Brigade report on the readiness of newly arrived replacements stated that:

IT WAS APPARENT THAT THE MAJORITY OF REPLACEMENTS ARRIVING IN VIETNAM FOR THEIR FIRST TOUR ARE UNDER THE MISCONCEPTION THAT THE VIET CONG IS A RAGGED, POORLY TRAINED AND ILL-EQUIPPED GUERRILLA. TO UNDERESTIMATE THE ABILITY OF AN ENEMY IS A MISTAKE THAT HAS LOST NUMEROUS BATTLES THROUGHOUT THE HISTORY OF WAR. HAVING TO WAIT UNTIL ENGAGEMENT WITH THE VIET CONG TO DISCOVER HE IS A WELL-TRAINED, HIGHLY MOTIVATED SOLDIER CAN BE COSTLY.[303]

CHANGE OF COMMAND

On March 26, Tiger Bravo returned to Phuoc Vinh for three days of base camp defense.[304] Two days later I replaced Captain Rankin as company commander, which followed standard US Army policy requiring commanders to complete six months in a combat, line unit, then rotate out. Rankin was a popular commander who had led the company to its well-deserved reputation as one of the best in the brigade. Selecting a company's executive officer as the new commander was atypical for the times. Normally, as the old commander departed with his hard-earned experience, a fresh, inexperienced captain was brought in to begin the six-month command tour clock all over again, resulting in a dip in the unit's combat effectiveness as the new commander settled in, learned the enemy's strengths and weaknesses, and gained the trust and respect of his men. Initially uncertain about what kind of commander would replace him, the men were relieved to hear that their new leader was one of their own. I had fought alongside them from the beginning, knew the enemy, was proficient in all manner of tactical situations, and understood how to operate in War Zone D.

With an uncommon name such as St John, it was widely assumed that I was related to the voluptuous pinup star Jill St. John, a fact that I never disputed. Shortly after I took command, I took the con to the next level and admitted to the whole company that she was, indeed, my sister. It was all in fun, just something to pass the time. By this time, practical jokes had become an art form in the company, and this was mine. My friends knew the truth, but for a time during the spring of 1968, the "my sister is a movie star" lie had everyone fooled.

Joe Palagyi was a master practical jokester, one of the best in the company. On one search and destroy mission, according to Doc Franks, "He bet me $5 I couldn't eat five, small green Vietnamese peppers he easily held in the palm of one hand. So, I took the bet and popped them into my mouth, thinking it was an easy $5. Immediately, I began to choke and couldn't breathe, while Lieutenant Palagyi was rolling around in the dirt, laughing uncontrollably."[305] Doc had fallen into the trap of eating a pepper with ten times the potency of the hottest jalapeño he had experienced growing up in California. Turns out even the Vietnamese eat them sparingly, with a small portion of just one used to season a large cooking pot of food.

On March 27, with its new, twenty-three-year-old "old man" in command, the company made a helicopter combat assault into a cold LZ near FSB Concord. In addition to myself and First Lieutenant Hank Matlosz as the executive officer, the new chain of command included:[306]

- First Sergeant Herman Trent, Bravo's senior NCO
- Second Lieutenant Joe Palagyi, the artillery forward observer
- Second Lieutenant Ron Vandertuin, 1st Platoon leader
- Second Lieutenant Jim Roach, 2nd Platoon leader
- Second Lieutenant Joe Hillman, 3rd Platoon leader
- Second Lieutenant Pat Brooks, 4th Platoon leader

CAT-AND-MOUSE

Tiger Bravo was heading back to the Rocket Belt. The battalion was tasked with stopping a rash of rocket attacks on the Bien Hoa/Long Binh complex, launched from the same area that Bravo had prowled before and after Tet. Across this wide swath of jungle, light forest, and abandoned rice paddies and fields north of the Song Dong Nai, battalions from an NVA rocket artillery regiment would set up under the cover of darkness and fire salvo after salvo of 122mm rockets at Bien Hoa/Long Binh, destroying aircraft, hitting ammunition dumps, and creating general havoc on the sleeping soldiers and airmen. A cat-and-mouse game between the hunter (Tiger Bravo) and the hunted (rocket-carrying parties of three hundred to four hundred NVA) was about to begin.

6

Rocket Belt

North of the Song Dong Nai
(Binh Duong Province)
March 27–May 23, 1968

LINE OF WEARY BROTHERS

The long line of olive drab ants sloshed quietly along, one muddy boot print at a time. Backs curved like question marks under heavy rucksacks. Their heads and shoulders hunched over, watching the boot prints of the soldier in front slowly fill with brown rice paddy water. Their feet were wrinkled and rotting from days upon days of wading in cesspools. Each soldier was wrapped in his own thoughts, trying to escape the mind-numbing heat and the monotony of looking for an enemy that rarely came out in the daytime.[307]

One hated the boredom when Charlie was gone and he had no one to shoot. One missed his home, fishing for bass, Jennie's thighs, and cruising downtown. Another regretted time not spent with his children, and loving his wife. One was back at the courthouse, remembering the judge's words— "Your choice, son. Sixty days in county jail or join the army. Hurry up, now."—and regretting not taking the sixty days. One feared himself and how he thought less and talked less about friends lost. Yet another saw Charlie all the time, mostly looking back at him. But for me, the long wait for hell's door to open was the worst of it.[308]

Harry Brown, who walked every mile in that line of weary brothers, wrote, "It's just a dull routine now. Walk all day, dig a hole at night, pull guard, and wake up just to do the same thing."[309] The days of the week seemed to blend into a long stream of nondescript twenty-four-hour periods. Occasionally you would hear arguments spring up over what day of the week it was. Doc Franks recalled, "The debate would go on and on. . . . It was just a little thing, but also disturbing not to even know what was the day and date when you might lose your life."[310]

The days were hot and hotter, one as suffocating as the next. What the thermometer said didn't matter. The only valid measurement was what the heat could do to a soldier, and what it could do was simple enough: It could kill him. It could bake his brains or wring the sweat out of him until he dropped from exhaustion. For troops on the line there was nothing they could do about the heat except endure it. Relief came only at night, but nighttime also brought out swarms of malarial mosquitoes and hordes of VC.[311]

Despite the monotony of searching for days across a landscape devoid of any signs of the enemy, the fear that Charlie was just beyond the next wood line or lurking outside the perimeter at night was omnipresent. A study of combat stress in Vietnam concluded that it is hard to equate normal, civilian experiences to the stress and fears in a combat situation. "In combat, danger is imminent and ever present. It is a constant companion every hour of the day, every day of the week. The enormity of this fear is hard to portray and, without such an experience, hard to imagine. Also, dispersed within prolonged periods facing this fear, are long periods of sheer boredom and frustration—yet, always with the knowledge that the enemy has to be faced again."[312]

Then after long periods of boredom the chronic stress felt each day spiked enormously when contact was made. "Firefights always seemed like a bad car accident to me," Jim Roach explained. "Did what you could to prepare—but they were always jarring."[313]

VC HEAVY ARTILLERY

Finding the enemy was just half the problem. If Tiger Bravo did detect a rocket unit in transit, or at a launch site, and made contact, it would be heavily outnumbered. A typical rocket-carrying party numbered 350 enemy soldiers.[314] Casualties, disease, and injuries, coupled with an R & R program that took soldiers off the line for a much-needed seven-day leave, had reduced Bravo's field strength to an average of 125 able-bodied men. Any fight would be seriously one-sided until the US superiority in fire support could be brought to bear, or another company brought in to reinforce. This was not unique to Bravo's situation. All the 2/506 companies faced a similar imbalance. On one occasion, Joe Palagyi remembered, "a platoon ambush was so outnumbered that when it reported over three hundred enemy soldiers moving past them, it was ordered not to engage, but to call in artillery. That decision saved a number of American lives that night."[315]

The 122mm rocket was the newest and most sophisticated rocket artillery piece supplied to North Vietnam by the Soviets, and had recently made its appearance in the South. It was a lethal, highly portable weapon system that packed the destructive power of heavy artillery—equal to the Soviet 152mm howitzer or the US 155mm howitzer. In a typical attack, an NVA rocket artillery battalion would set up nine to eleven kilometers from the target and fire an average of fifty rockets in fifteen minutes. Each rocket was 8.1 feet long and weighed 121 pounds, including the launcher. A well-trained crew could set one up in 2.5 minutes, fire, and be ready to load another two minutes later.[316]

Completely securing the Rocket Belt by blanketing the entire area with American troops was an impossible task. There were far more hidden locations, where VC rocket units could lay up in daylight, and too many disparate approach routes and secluded launch sites than American units could effectively cover. Nor was an immediate reaction strategy always effective. Calling in artillery on the launch sites once the rockets launched, or quickly flying a standby reaction force into the area, was not stopping the attacks. The strategy that stood the best chance of success was to employ infantry companies along likely routes that rocket-carrying parties would use, or at possible launch sites that intelligence had identified.

ROCKET HUNTERS

Tiger Bravo crisscrossed the Rocket Belt for the next three weeks searching for rockets. After a short rest period at Phuoc Vinh, it was right back out for another month of continuous operations.[317] During that time all manner of

tactics were employed by the battalion in a futile effort to stop the rocket attacks. None worked. The units would go for days with no sign of the enemy,
only to see rockets roaring into the sky at night from a distant location. At one
point Bravo switched from the standard, daylight company-size search and destroy operations, with one or two platoon-size ambushes at night, to "roving
tigers"—the name given to the entire company moving silently at night, all
night—designed to intercept rocket-carrying parties on the move. On other
days, in an effort to cover more ground, Bravo conducted multiple combat assaults at the company level, or a combination of platoon-size heliborne combat assaults and cross-country movement. I depicted one of these operations
in a letter home on May 1, 1968.[318]

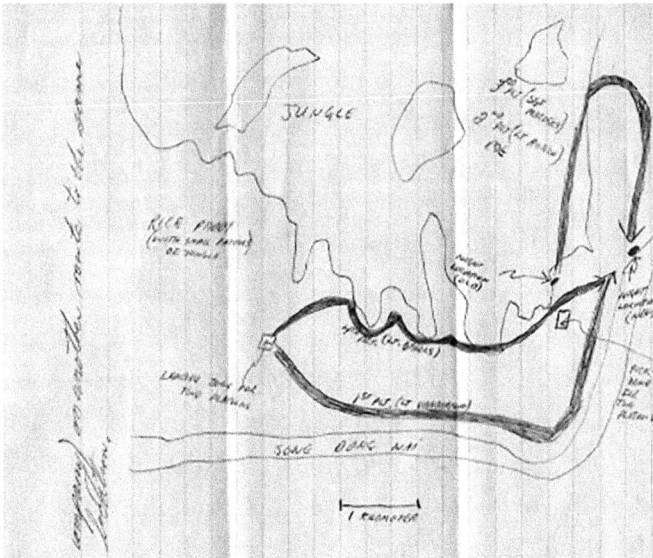

*Sketch map of a
typical company
operation in
Rocket Belt.*

For one week, when intelligence sources reported that a VC rocket unit
would be moving through an isolated, small valley, the company, reinforced
with platoons from A and C Companies and an ARVN platoon, blanketed
the valley with small, mutually supporting night ambushes for five straight
nights.[319] This tactic, called an "area ambush," required the ambushers to
pull back into hide positions in daylight, then back into ambush positions at
night. It required complete secrecy to be successful—no resupply choppers,
no routine movement of soldiers into and out of the field, and no one leav-

ing the cover and concealment of the hide positions for any reason. Again, no luck.

The nights were spent lying still on the ground, soldiers breathing the smell of their own unwashed bodies and the pungent earth of the jungle floor. Time passed painfully slowly on ambushes. After waiting for what should have been an hour, you would check your watch and see that only five minutes had passed. Ambush sites were inhabited by shadows flitting down dark trails, by bushes that assumed human form, and by a viscous darkness that crept in and thickened until everyone was blind.[320]

Since the VC had a radio intercept capability, deception was also part of the battalion's plan. The battalion staff concocted an entire false identity for a fifth rifle company, G Company. Soon it materialized on the battalion radio net, giving false locations for resupply drops and sending reports of the movement of patrols that didn't exist. The subterfuge included false insertions throughout the area of operations to keep the VC guessing—empty lift ships flying in a standard combat assault configuration and landing briefly, as if off-loading troops.

Occasionally, units operating in the Rocket Belt would be jolted awake by the flash/bang of rockets launched in the distance, signaling both a successful VC attack and another missed opportunity by the American rocket hunters. The rockets' fiery plumes could be seen arcing into the sky, then disappearing from sight as they streaked across the Song Dong Nai on their way to Bien Hoa. The 2/506 "Daily Staff Journal" of April 1, 1968, recorded a typical sighting of a single launch by two separate units. Tiger Bravo and the battalion's recon platoon reported the following:[321]

- 0145 HOURS JULIET REPORTS ROCKETS FIRED ON AZIMUTH OF 130 DEGREES, RANGE OF 7,000 TO 8,000 METERS. TEN VOLLEYS WITH 3 TO 5 ROCKETS PER VOLLEY.
- 0146 HOURS B CO REPORTS 30 ROCKETS FIRED AT RANGE OF 3,500 METERS.

One of the few significant contacts occurred early in April, when both B and C Companies made a combat assault into cold LZs, then went on separate search and destroy missions. An excerpt from the 2/506 "Daily Staff Journal" on April 4, 1968, summarized the fight:[322]

> AT 1435 HRS, VIC GRID YT 112265 CO C WHILE ON A CLOVER LEAF PATROL (WEST) SPOTTED AN ESTIMATED VC COMPANY. 1 PLT SWEPT SOUTH TO SUPPORT AND WAS TAKEN UNDER SMALL ARMS, RPG AND AUTOMATIC WEAPONS FIRE. ARTILLERY AND AIRSTRIKES WERE CALLED IN WITH ALSO 1 LFT (LIGHT FIRE TEAM OF GUNSHIPS). THE VC BROKE CONTACT AT 041445. RESULTS OF CONTACT; FRIENDLY CASUALTIES 1 KIA 22 WIA, 1 VERY SERIOUS. VC CASUALTIES EST 25 KIA, EIGHT BY ARTILLERY & AIRSTRIKES. AFTER CONTACT WAS BROKEN CO C MOVING FORWARD FOUND ONE BODY AND ONE M16 RIFLE SERIAL NUMBER N827729. TOTAL VC KIA 33. CO B SWEPT NORTHEAST AND LINKED UP WITH CO C TO FORM AN NDP.

As with all combat journals the "Daily Staff Journal" chronicled the facts but did nothing to explain the urgency of the situation or the intensity of the battle. In this fight, C Company was in trouble. In just a few minutes they had taken substantial casualties. They were fighting an enemy their own size or greater and they needed reinforcements, quickly. Minutes after contact was made and the call for help went out, Bravo Company was on the move. Little navigation was needed; the company headed toward the sounds of gunfire and explosions, and the sight of fighter jets and gunships swooping and swirling in the air over the embattled company. In a letter home Harry Brown wrote, "We shelled those bunkers with 155mm and 8 inch guns. Five airstrikes with 500 and 1000 pound bombs and gunship attacks. Did no good. They dug deeper and fired more."[323]

Tiger Bravo was in on the tail end of the fight, and assisted in evacuating C Company's wounded by hoisting them into a helicopter hovering seventy-five feet above the dense jungle. In terrain that was too thick to allow a helicopter to land, the helicopter would hover above the trees and drop a sling, called a jungle penetrator, to attach to the wounded; then the wounded soldier would be hoisted up through the canopy. Tiger Bravo suffered a total of three wounded from the C Company fight and light contact around the night defensive perimeter later on that night.[324] Mike Brown wrote, "At night (we fought all day) they turned our claymores around wounding some men, and took off."[325]

Often, during heavy contact the priority in resupply by choppers is for ammunition, medical supplies, and anything else vital to winning the fight. Food and water can wait. Often, what the army doesn't provide, the rucksacks of dead enemy soldiers will. "We were in sad shape," Mike Brown wrote, "We didn't get water or food all day! I had only a pecan cake roll (C ration) and coffee. Scott and me got a bag of captured rice and dried shrimp and ate it. Not bad. Scott is Japanese so he knows how to make it taste good."[326]

The 2/506 "Daily Staff Journal" was not confined to just intelligence and combat entries. Administrative and logistic events or situations were recorded as well, as on May 1, 1968, when a report from the 101st Airborne Division's Military Police Battalion was entered:

1540 HOURS 2 MEN FROM B 2/506 WERE PICKED UP IN AN OFF LIMITS AREA IN BIEN HOA: SGT E5 RIVERA, LIDIO A AND SGT E5 MARTZ, LESTER D.

What they were doing has long since receded from anyone's memory, but it does prove one axiom about infantrymen: "Soldiers will always find a way to have a good time." The next entry in the journal reminded all that this was still a deadly combat zone:

1550 HOURS B 2/506 POINT ELEMENT CAME ACROSS TRAIL WITH 5 BUTTERFLY BOMBS IN ONE POSITION AND THREE IN ANOTHER POSITION, ARE BLOWING (DESTROYING IN PLACE WITH EXPLOSIVES) THEM NOW. LOCATED AT GRID YT081208, US MADE, ALL WERE BOOBY TRAPPED[327]

BEYOND BRAVO'S WORLD

To serve in a line unit that spent most of its time in the field focusing solely on finding the enemy and surviving meant one had a myopic point of view about everything else. For the soldiers and leader of Tiger Bravo, their world was anything within small-arms range of the company, to the exclusion of other matters, such as the strategic direction of the war, dissent back home, or even the unit's part in large-scale operations.

However, letters from home and the occasional newspaper were temporary connections to the outside world. The information was usually sparse and outdated by the time it reached Tiger Bravo in the field. One bit of news that did catch everyone's attention, and was discussed at length, was the president's surprise announcement on March 31 to suspend bombing north of the twentieth parallel in North Vietnam to urge the North Vietnamese to start peace negotiations.[328] Chuck Hanson's letter home summed up the reaction of most soldiers to the news: "I don't know what the feeling is in the States regarding the bombing halt. Over here it is mixed with the exception of the Infantry, who understandably are all against it."[329] Certainly none of us was aware that this was precipitated by the president receiving advice on March 26 from a group of distinguished elder statesmen and retired generals that the United States should withdraw from Vietnam.[330] In one of the greatest paradoxes of the war, Tiger Bravo and hundreds of other line units in Vietnam would continue to perform their perilous duty, day after day and week after week, while the leaders in Washington, both military and elected, had concluded that the war was a lost cause.

News of the 101st Airborne Division's battlefield exploits was another matter entirely. The weekly *Screaming Eagle* newspaper, which was published in the division rear area and flown to every company with the mail, regularly reported on an endless stream of battlefield victories. One never read the words "defeat" or "retreat" or "ambush" linked to a 101st unit; it was always the VC or Reds on the losing end. Typical of this type of reporting was the *Screaming Eagle* published on March 29, 1968, which carried titles such as "2-506 Patrol Traps Reds in Ambush," about Tiger Bravo's ambush of a VC battalion on March 1, 1968, "Currahees Kill 415 VC in Phan Thiet Battles,"

"Firefight Nets 36 VC Near Song Be City," and "Black Angels Rip 25 NVA in Song Be Battle."[331]

A DANGEROUS BATTLEFIELD

A soldier in Vietnam faced as dangerous a battlefield as any other generation of Americans and, in some respects, more so. Multiple wounds were more common in Vietnam than in previous conflicts because of the high velocity of lightweight rounds, the rapid fire at close range, and the extensive use of mines and booby traps. Mike Scott's encounter with a VC was typical of a short-range contact: "I'm back on point, sweeping through a few houses near the firebase when a VC, with an AK and magazine pouch, comes around a bamboo bush. I bring up my M-16 and fire," he wrote. "We are so close to each other that I can see the bullets hitting him. He goes down like a rag doll." But nothing is ever that simple in Vietnam. "I pick up the AK, still alert for trouble, when I hear my backup yelling at me," Scott continued. "He's yelling at me to look up. Damn, there is another VC up in a coconut tree."[332]

Mike Scott was one of the new breed of veterans, not part of the original cadre that arrived in December 1967, but still a battle-hardened soldier with several months on the line. He was one of a kind among Tiger Bravo soldiers, and not just because he was a Japanese-American. Unlike the other soldiers, whose fathers and uncles were part of the Greatest Generation and fought in World War II against Germany or Japan, Mike's father was a fighter pilot in the Japanese Air Corps, flying missions against American forces in the Pacific Theater. Although an American citizen with a stepfather who was a retired US Army sergeant major, his father's war record restricted what Mike could do in the US Army, keeping him out of Officer Candidate School and from attending critical training on the classified Starlight Scopes.[333]

Mike enlisted in the army in June 1967 out of Clayton County, Georgia, where his stepfather had moved after retiring. Mike joined up for "no particular reason except he had read about World War II and was curious about war." Once through basic training he volunteered for the airborne after hearing that John Wayne was making the movie *Green Berets* and "thinking it would be fun to jump out of airplanes."[334]

Wounds were also dirtier than in any previous conflict. A tremendous amount of dirt, debris, and secondary bits and pieces of metal were hurled into the typical wound. Massive contamination challenged the army surgeons to choose between radical excision of potentially salvageable tissue and a more conservative approach that might leave a source of infection. The ratio of killed in action to wounded in action in Vietnam, as compared to other wars, was different as well. In World War II, the ratio of KIA to WIA was 1 to 3.1. For Korea, it was 1 to 4.1. In Vietnam, it had risen to 1 to 5.6.[335] Sixty-five percent of wounds in Vietnam were caused by fragments, but 51 percent of the deaths were caused by small arms—compared with only 32 percent in World War II and Korea.[336]

Snipers were a constant threat. They were especially dangerous when a unit was not in a firefight and the sniper opened up indiscriminately at the most visible, vulnerable target. The target could be anyone in the company, and the shot could come from anywhere. Early in April, Tiger Bravo lost Mike Tarpley to a sniper in this exact way. The day began with good news for Tarpley; the morning resupply chopper delivered a Red Cross telegram for him. "The day I was shot, I received a telegram that my son was born," recalled Tarpley. "He was eight days old at that point." Minutes later he was hit in the leg by a single shot from an unseen sniper. "To this day I don't understand what happened. I don't remember a firefight or anything. I just remember POW!"[337]

Sniper fire was unpredictable and many times unavoidable. You could do everything right, take all the necessary precautions, and still fall from a sniper's rifle. On one level, Mike Tarpley understood this, but he always had the nagging feeling that he had let his guard down. To this day, it remains an unanswered question for him. "If my mind had been on Nam and not my son, maybe I would have not been shot."[338] The sniper was never found, nor did he fire another shot. In less than thirty minutes, it was over. The sniper was gone and so was Tarpley, and the company was on the march again.

In addition to an enemy bent on killing Americans, a soldier had to contend with the possibility of injury or death from a multitude of other sources. The boonies teemed with stinging and biting creatures that carried diseases, left lesions that invariably became infected, or from bites that could be fatal. Mosquitoes, flies, dry land and water leeches, biting centipedes, scorpions, fire ants, and thirty different types of poisonous snakes were everywhere. Early one morning at daybreak, a king cobra that had spent the night in a small clump of bamboo inside Bravo's perimeter decided to show himself just as two soldiers were brewing coffee. Surprise! They scrambled for their M-16s and started firing, while the equally frightened snake slipped back into the thicket. We packed up and moved out a lot quicker than normal that morning for some reason.

Harry Brown wrote to his parents about the different faces of danger in the Rocket Belt: "This morning I almost got killed. Someone left a grenade in a hole, threw trash in it and made a fire. BLAM! I was 5' away. That was close. 3 days ago, a cobra almost got me, but I killed him as he was striking. Mean snake."[339] His description of conditions, after an extended period in the jungle, was typical of us all: "It is very hot here and a lot of guys are getting sun stroke. The mosquitos at night are terrible. I have over 100 bites and my lips and eye were swollen shut. Snakes and flies galore. I smell so bad and my clothes and boots are rotting off me. My feet are in bad shape. My hands and face are cut from vines with stickers 1 ½ inches long. The ants are eating everything and I have a bad case of diarrhea."[340]

Dave "Sparrow" Spencer had his own run-in with a deadly snake. On a hot, muggy day, along the banks of the Song Dong Nai, Sparrow was sitting in the shade of a bamboo thicket airing out his feet, after a morning of wading through rice paddies. His pose was no different from the remainder of the resting company—boots off, laying back against his rucksack, with his eyes half closed—when a bamboo pit viper dropped into his lap from the branches overhead. Immediately it slithered down his leg and, being a snake, bit him on the foot.

Pandemonium broke out. Spencer was yelling and jumping around, while everyone nearby scrambled for entrenching tools and rifles to kill the invader. Who exactly did the deed, no one remembers. Spencer's foot began to swell, and the pain, which had started in his foot, moved rapidly up his leg. In twenty minutes a chopper arrived to carry him, and the dead snake, which he insisted on keeping as a trophy, to the nearest medical facility. At the hospital, Sparrow was coherent enough to proudly declare that he had been bitten by a bamboo pit viper. But the doctor didn't believe him. "It couldn't be a pit viper, if it was you would already be dead," he pronounced. Spencer, never one to back down, replied, "Well, sir, it was one and here it is." He reached into his pocket and waved the snake's limp body in front of the startled medical team, sending a trauma nurse, who thought she had seen it all, screaming from the emergency room.

Skin diseases and rashes, commonly referred to as jungle rot by the soldiers, were rampant among soldiers in line units, and were brought on by damp, humid conditions and constantly wading through rice paddies or crossing streams. Feet were especially vulnerable to skin conditions such as immersion foot. The only option available to soldiers "in the boonies" was to remove their boots at every opportunity and let their feet dry as much as possible. Usually this did little good. I once described my feet as "raw meat, that attracted flies every time I took my boots off."[341] Solutions offered from higher headquarters were of little assistance as well; usually they were more compatible with life in base camp than life in the field. An example came out of a 25th Infantry Division report covering a period in June and July when the 3rd Brigade was under its operational control. The report recognized that "personnel who are required to spend extended periods of time exposed to the effects of water and moisture frequently develop skin infections and other skin disorders."[342] The recommendation for field commanders only served to highlight the wide disconnect between higher headquarters' perception of life in the field, and the unpleasant reality of the life of an infantryman in Vietnam. Commanders were advised "to have their men remove damp clothing at every possible opportunity and that all personnel should be issued athletic shorts and shower shoes, that can be worn whenever possible to permit the drying of the skin by sun and air exposure."[343] Imagine the amazement of the VC peering through the bushes and seeing those crazy Americans cavorting around in athletic shorts and shower shoes—they wouldn't know what to do, laugh their asses off or shoot!

Other than wounds or injuries, malaria was another major cause of admissions to hospitals.[344] Counterintuitively, though, the most common causes of hospital admissions were not because of wounds suffered in action, nor even the total of all soldiers wounded and hospitalized due to injury; the primary reasons were venereal diseases.[345] Psychiatric casualties, common in most wars, were present in Vietnam as well, but at lower rates than were experienced in World War II and Korea.[346]

On May 3, Tiger Bravo had its first reported case of malaria[347] —a remarkable record for a company that had been in a mosquito-infested country for
close to six months. At one point, the division's vulnerability to malaria, and
the rising number of soldiers being hospitalized for the disease, prompted
the division commander to publish his rules for combating the disease in the
Screaming Eagle.

- EACH MAN WILL TAKE THE ORANGE (CHLOROQUINE-PRIMAQUINE) ANTI-
  MALARIA TABLET EVERY MONDAY.
- PERSONNEL IN FORWARD AREAS WILL ALSO TAKE A DAPSONE TABLET DAILY.
- SLEEVES DOWN AND COLLARS BUTTONED AT 1830 HOURS DAILY. APPLY
  MOSQUITO REPELLENT, CALLED BUG JUICE BY EVERYONE, ON EXPOSED
  AREAS AND REAPPLY AT LEAST EVERY TWO HOURS.
- ALL MEN WHO CAN SLEEP UNDER A SHELTER OF ANY KIND, PONCHO, TENT,
  BUNKER OR BUILDING, ETC., WILL ALSO SLEEP UNDER A MOSQUITO NET.
  THE NET WILL BE CLEARED FIRST WITH INSECTICIDE. THE NET CAN AND
  SHOULD BE USED IN ALL BUT THE MOST EXTREME SITUATIONS. THOSE WHO
  ABSOLUTELY CANNOT USE A FULL NET WILL SLEEP IN HEAD NETS.[348]

Such rules were easily followed in base camps or large installations, but
not compatible with life in the boonies. These and other administrative de-
crees from division reinforced the perception in line units that higher headquarters was out of touch with life outside the perimeters of base camps and
large installations.

The *Handbook for US Forces in Vietnam* was similarly disconnected from the
harsh realities of life in the "bush." The section on personal hygiene was obviously written by someone who had never been in the field. In part, it admon-
ished soldiers to:[349]

• **Drink more water and carry two canteens to the field.** Most of us carried
double that number. Carrying only two was a sure way to fall out from dehydration or heat exhaustion. We also were issued iodine tablets to purify the water
from any local source, regardless of the contamination or what was floating in

it. Chuck Limer remembered on one occasion, "I had to drink swamp water with iodine pills, and had to clinch my teeth to keep the muck out."[350]

• **Use insect repellant freely because insect-borne diseases common to Vietnam are malaria, dengue fever, encephalitis and plague. The only sure way to prevent contracting any of these diseases is not to get bitten by disease-bearing insects.** Telling someone in the jungle not to "get bitten by disease-bearing insects" is like standing someone in a downpour, without an umbrella, and telling them not to get wet.

• **Avoid leech bites.** Impossible. Leeches were everywhere, on the ground that you slept on and in the water that you spent hours each day slogging through. It wasn't uncommon to see soldiers picking leeches off each other, much like a monkey picking fleas off another monkey.

• **Take measures to prevent fungal infections. Clean and dry feet, armpits, and groin whenever possible, and change clothing and socks frequently.** During the rainy season, you were wet day and night. Clean water was too precious to use to clean up. You either drank it, sometimes with Kool-Aid added to the canteen, or made coffee. When the barrel of a machine gun overheated from constant firing you would more likely to see a soldier urinate on it rather than use precious drinking water. Opportunities for a bath were few, and then only in a dirty, muddy stream or river. Changing clothes frequently was out of the question. We would go days or weeks without clean uniforms. Periodically, a resupply chopper would drop off a bundle of clean jungle fatigues. Soldiers would strip off their stained, ripped, and torn uniforms right on the spot, then the chopper would back haul the smelly, dirty bundle of uniforms.

Exposure to the extremes of weather was a constant problem as well. One night while crossing a rice paddy during a drenching monsoon rainstorm, the company was struck by lightning. Seven soldiers were knocked off their feet, and two had to be evacuated. One of those evacuated was knocked unconscious when the radio on his back was split in half by the strike. It hit so suddenly, and with such a loud noise, that many thought someone had tripped a massive booby trap.[351] Even a chance encounter with a farmer's domesticated livestock could be unsafe. One Bravo soldier searching a village was trampled by an enraged water buffalo and had to be evacuated with a broken leg.[352]

The whirling blades of the ubiquitous helicopter, used in so many ways to enhance a unit's fighting capabilities on the battlefield, could also be dangerous to even the most experienced field soldiers. On April 7, Captain Bill Whitehead, who had previously commanded C Company and was now on the 2/506 staff, was struck in the head by the main rotor blade of an OH-21 helicopter. He died of his injuries less than an hour later.[353]

Drugs, although readily available in base camp, were not a significant problem in Tiger Bravo, especially while in the field. One study, which analyzed the Vietnam era soldier's environment and how it affected his performance in combat, revealed that the lack of a widespread drug culture in Tiger Bravo was not unusual for the times. It read in part that "there is at least circumstantial evidence that very few soldiers who were actively engaged in combat were under the influence of drugs; rather, they showed 'drug sense' and avoided its use when exposed to combat situations."[354] Sadly, this same study uncovered substantial evidence of dangerous levels of drug use in rear areas and base camps, particularly after 1970. It also revealed that fear of battle or of becoming a casualty was not the major reason for drug use, as had been assumed by many; boredom, routine tedium, the desire to "kill time," peer pressure, and coping with an unfamiliar physical and military environment were the major reasons.[355]

Though most of Tiger Bravo didn't partake in drugs, they learned to seize every opportunity to have a good time—rules or no rules. In a hospital halfway around the world in El Paso, Texas, where he had been sent to recuperate from his leg wound and the lingering effects of the hand grenade blast to his face during Tet, Mike Tarpley upheld this Tiger Bravo tradition. The hospital was just a short border crossing from the celebrated, good-time border town of Juarez, Mexico. As Tarpley recalled later,

"One night we snuck out of the hospital to go to Juarez. Me on crutches, one guy with an arm missing, one guy with a leg missing and one in a wheel chair with both legs missing. We went to the "Cave Club." . . . When we got back to the hospital, I was caught sneaking back into my ward by a Major. She could have played linebacker for the Corn Huskers. She took my civvies away, stripped me down to my underwear, took my crutches away and carried me back to bed and said, "Sergeant Tarpley, let's see you sneak out of here now."[356]

We Have a Mission!

May brought a number of promotions and reshuffling of leadership. First Sergeant Trent was officially promoted to First Sergeant E-8. Joe Hillman's time on the line as a platoon leader came to an end. He went to C Company as the executive officer. Pat Brooks also left, to become the assistant S-4 (logistics officer) on the battalion staff. The resulting reshuffle put two officers (1st Platoon, Second Lieutenant Vandertuin and 4th Platoon, Second Lieutenant Roach) and two NCOs (2nd Platoon, Platoon Sergeant Sykes and 3rd Platoon, Platoon Sergeant Matosky) in command of Tiger Bravo's platoons.[357] A few of the original enlisted men left as well. Chuck Hanson succeeded SP4 Gregory Harris, whose enlistment was up, as company clerk. "I was the only one in my squad who hadn't been killed or wounded," Hanson recalled. "So, I was pretty happy to be selected." Hanson had mixed emotions about the change but left with a clear conscience. "Most of us in the infantry have the feeling that if, sometime during our tour, we are able to kill at least one VC, then we have done our job and we can go back to civilian life with the feeling that we have done our share. On that score, I have already qualified several times over."[358] Pop Plemons made it off the line as well. First Sergeant Trent was able to find him a job in the supply section of Battalion S-4.

A terse radio message before dawn on May 23 to the Bravo Company CP ended Tiger Bravo's operations in the Rocket Belt:

DISCONTINUE CURRENT OPERATIONS, MOVE ALL YOUR ELEMENTS TO DESIGNATED PZ'S FOR IMMEDIATE PICKUP.

No more long, drawn-out search and destroy missions in temperatures of more than a hundred degrees. No more battling dehydration, heat exhaustion, and leeches that came out of nowhere and attached to everything. Good-

bye to bugs, booby traps, and boring, mind-numbing days and nights punc-
tuated by brief bursts of fear and adrenaline the few times the VC appeared.

It was the same "call to arms" to all companies across the brigade: "Drop
what you are doing, we have a mission!" Within an hour every company in the
brigade not already at Phuoc Vinh was on the move to pickup zones throughout the Rocket Belt. By the end of the day, the 3rd Brigade was fully assembled
at Phuoc Vinh, ready to be deployed to whatever hot spot in South Vietnam
that needed the "Wandering Warriors." Tiger Bravo would not have long to
wait. In just twenty-four hours it would be plunged into yet another desperate
situation, only this time a long way from War Zone D and the Rocket Belt.

7

Plateaux Montagnards

Mountains in Vicinity of Dak To
(Kontum Province)
May 24–June 14, 1968

OPERATION LUCAS GREEN

On a cool, overcast afternoon in May two columns of Tiger Bravo soldiers, loaded down with rucksacks, weapons, and a full load of ammunition, climbed down the ramp of a USAF C-130 transport plane and onto a remote airstrip in the Central Highlands. Nothing around them looked familiar. At the other end of the airstrip was a cluster of low buildings and a USAF tactical control tower. Just across a dirt road adjacent to the runway were the remains of a fortified camp, with crumbling bunkers and half-strung barbed wire. The cool air and mountains on either side of the airstrip confirmed the general location briefed to the company that morning. "Your destination is somewhere in the Central Highlands," said the briefer. "The exact location is classified." It was the afternoon of May 25, 1968. In the previous forty-eight hours, the 3rd Brigade brought a far-flung, decentralized operation across the southern half of War Zone D to a full stop; assembled at Phuoc Vinh for a day of concentrated maintenance and rearming; and deployed to the Central Highlands.[359]

Tiger Bravo's orders had been sketchy at best when it loaded the aircraft at the Phuoc Vinh airstrip: "This is a secret move to the Central Highlands. There is a situation up north and the brigade has been alerted for immediate deployment. Higher command doesn't want the enemy to know we are on the way so everything is hush-hush. Go to radio listening silence. Take all the screaming eagle patches off your uniforms and load that C-130. A battalion representative will meet you when you arrive and provide more details."

Naturally, there was no battalion representative to meet the aircraft, nor anyone, for that matter, with answers to questions. Spotting an old Vietnamese man riding by on a bicycle, Tiger Bravo's Kit Carson scout waved him down and asked where we were. The answer from the scout, struggling as always with his English, "Here, Dak To." Chuck Limer was on the same plane and remembered his first impression: "The first thing I noticed was Hill 881 (a mountain with an elevation of 881 meters) where the 173rd Airborne was nearly wiped out. There was nothing left but some burned out tree stumps no taller than two or three feet and ashes. It looked like a picture of the moon." Several 4th Infantry Division soldiers in a nearby bunker served as his welcoming committee. "They were glad to see the 101st show up because they were catching hell from the NVA. The NVA were digging trenches into the 4th Infantry's trenches and there was some hand to hand combat. You could smell the NVA corpses . . . a foul smell that hung over the firebase."[360]

Shortly a jeep roared up and out jumped a captain from brigade headquarters. "Welcome to Dak To. Head down the road and your battalion rep will meet you and guide you into the 2/506 area." He then jumped back in the jeep and roared off.

In the province of Kontum the 4th Infantry Division had established a base camp at Dak To, which at the time was a remote collection of buildings that on a good day would struggle to be called a village. But it had an airstrip and was strategically located in a valley close to the intersection of Cambodia, Laos, and South Vietnam. Just seven months earlier Dak To had been the epicenter of a bloody series of engagements with North Vietnam Army (NVA) regular units, known collectively as the Battle of Dak To. During November 1967, the 4th Infantry Division and 173rd Brigade (Airborne) fought off four NVA regiments (24th, 66th, 174th, and 320th), totaling twelve thousand infantrymen, of the reinforced NVA 1st Division. That division was led by General Nguyen Huu An, who had commanded the enemy forces against the US 1st Cavalry Division in the Battle of the Ia Drang Valley in 1965. In the fights that took place up and down the surrounding mountains, American casualties totaled more than six hundred killed, wounded, or missing. Enemy losses were estimated at a thousand killed. Despite heavy US losses the Battle of Dak To was considered an American victory. Three enemy regiments scheduled to participate in the Tet Offensive were so mauled that they had to be withdrawn to refit.[361]

Known to the French as the "Plateaux Montagnards," the Central Highlands was a twenty-thousand-square mile plateau and mountainous region, four times the size of Connecticut, that occupied the center of South Vietnam. It was home to a population of approximately a million, mostly Montagnards[362]— a French word meaning "mountaineer"— living in tribal villages in the hills and mountains. It was a mostly rugged, mountainous area with elevations ranging from forty-five hundred to seven thousand feet in the vicinity of Dalat and from three thousand to eight thousand feet in the area west of Quang Ngai. Steep slopes, sharp crests, and narrow valleys characterized the mountains. Numerous razorback ridges ran in all directions. It was impossible to follow them in any one direction for more than a few hundred yards.

The forested areas in the foothills rose to three thousand feet and were covered with tall trees that formed a dense, closed canopy that blocked direct sunlight from filtering down to ground level, giving a sense of twilight even at high noon. "The jungle has triple canopy," wrote Chris Backman. "Was too dark inside to take any pictures. Many places where you couldn't see the sky, it was so thick."[363] The undergrowth was very thick, comprising an almost impenetrable mass of smaller trees intermingled with thorny shrubs and vines. It rivaled the worst of the undergrowth we had found in War Zone D. Most streams were swift-running and bordered by high, steep rocky banks. It rained almost every day—normally a light drizzle in the morning, followed by afternoon showers that left everyone wet and cold at night. Low cloud cover in the mornings, often down to just a few hundred feet above ground level, made resupply and precise close air support impossible until later in the day. Cross-country movement was tortuous and slow. The rate of march was from one half to two kilometers per hour, with numerous rest stops.[364] In this terrain, where troops had to cut and slash their way through thick jungle and climb up steep slopes, then slip and slide down the other side, "humping in the boonies" took on a whole new meaning.

This classified operation was code-named Lucas Green.

325C NVA Division

Lucas Green secretly reinforced the 4th Infantry Division in the Central Highlands to counter an expected push into that area by the 325C NVA Division,
which had been unaccounted for since pulling out from around Khe Sanh,
farther north. Captured NVA soldiers revealed that the 325C Division's primary objective was to destroy the US Special Forces Camp at Ben Het.[365]
Already stretched thin manning a string of fire support bases covering approaches into South Vietnam from Laos and Cambodia, the 4th Infantry Division possessed neither the numbers nor the heavy weapons to counter this
potent threat. Quickly, the NVA launched a series of multibattalion ground
attacks across the division's area. These were neither probes, nor harassing
attacks, but full-out, overwhelming attacks by hundreds of fresh, well-trained
NVA regulars.

The results of an attack on FSB 29, manned by two companies of the 1st
Battalion, 8th Infantry Regiment, epitomized the size, ferocity, and lethality
of Bravo Company's new adversaries. On the night of May 25, as Bravo was
settling into its first night in the highlands in the relative comfort and security
of the 4th Infantry Division's Dak To Base Camp, hundreds of NVA infantry,
supported by heavy machine guns and mortars, slammed into FSB 29. Fighting raged around the perimeter. The enemy at one point penetrated the south
side of the US defenses and briefly occupied six of the defenders' bunkers.
When the NVA finally slipped away into the night, they left 129 bodies scattered around and in the FSB. The US casualty toll was staggering—14 killed,
56 wounded, and 1 missing in action.[366]

Even the US air superiority was challenged: on May 22 and again on May
27 the NVA shot down US helicopters.[367] The enemy launched massing artillery
fires against an FSB southwest of Ben Het, employing concentrated fires from
82mm recoilless guns, 140mm rockets, 100mm guns, and for the first time in the
Central Highlands, their 105mm howitzers.[368] On another occasion in late May,
FSB Brillo Pad received 1,100 rounds from recoilless rifles, mortars, and rockets
in four days, averaging 14 to 15 rounds impacting every hour.[369]

The appearance of the 325C Division was not a typical reinforcement of
existing enemy units in the area, nor an exchange of one NVA unit for another.

Rather, it was a massive infusion of a potent NVA division that, in terms of its sheer size and heavy weapons, shifted the balance of combat power in the area from the US forces to the enemy. A 4th Infantry Division report from that period determined that "after the infiltration of enemy forces into Kontum Province was completed, friendly forces were opposed by the largest concentration of enemy units ever assembled in the Central Highlands."[370]

A rarely seen, conventional armor capability of the 325C was especially troubling to 4th Infantry Division commanders, and raised considerable alarm when the brazen NVA built a hard-packed earthen road, measuring four to five meters across, to move its tanks from across the border into the 4th Infantry Division's sector. According to one intelligence report, "The road follows terrain contours and has no drainage ditches. Due to heavy bombing the road is impassable from YB784227 to YB 803238. But, if repairs are made, this route could provide a high-speed armor approach to Ben Het."[371]

Bravo Company would now face a conventional, large NVA force with its own organic infantry, artillery, and armor—much different from the invisible, quick-striking VC it had earlier battled. Intelligence reports estimated that the 325C NVA Division entered the area with two regiments, totaling 6,000 soldiers, with artillery support provided by 100mm and 105mm howitzers and an array of tanks in support.[372] The division also possessed antiaircraft artillery; flamethrowers; large-caliber mortars; and, according to one intelligence report, the capability to deliver a CS riot control gas[373] similar to that of the United States.

This would not be easy.

TASK FORCE MATHEWS

By May 26 the 3rd Brigade had completely closed on Dak To and become a part of Task Force Mathews, along with the 1st Brigade, 4th Infantry Division. Assembled to counter this new enemy threat, the task force's mission was to protect the Ben Het Special Forces Camp and destroy the 325C NVA Division.[374] The 1/506 Battalion immediately deployed to the high ground south of Dak To, blocking possible attacks from that direction. Two companies of the 1/187 Battalion went under the operational control of the 1st Battalion,

8th Infantry Regiment, hit hard in recent engagements, and one company secured the 3rd Brigade Base Camp, named Camp Whitehead after Captain Bill Whitehead, the popular 2/506 company commander killed recently by a helicopter blade. The 2/506 assumed responsibility for the perimeter around the airfield and the 4th Infantry Division headquarters.[375]

Still under radio silence, the companies resorted to using old-fashioned company runners to relay orders and status reports, and ground lines running between bunkers. Even once on the ground, units could not use radios, nor let anyone see insignia or markings with a screaming eagle on it. Tongue in cheek, one of Bravo's men made up a sign for the company command post that proudly identified the company as the "442nd Light Bicycle Company (Airborne) LT Spoke Commanding." Along with everyone else in the battalion, Tiger Bravo settled into bunkers near the airstrip and awaited orders, which were not long in coming. All units were alerted and told to prepare for a major combat assault on May 28.

The plan called for the battalion, with Tiger Bravo as the lead company, to make a heliborne combat assault onto a 2,900-foot mountaintop approximately 5 kilometers from the Cambodian border. Since the mountain was covered by triple canopy jungle and dense vegetation, which precluded a standard CA with artillery and gunship fires on the LZ, the plan was to have either B-52 bombers clear the proposed landing zone with 500-pound bombs just before the lead ships touched down or to use a single Daisy Cutter bomb to create an LZ.

The BLU-82B, or "Daisy Cutter," was the largest conventional bomb in existence—17 feet long and 5 feet in diameter. It contained 12,600 pounds of explosives, six times the size of the bomb that destroyed the Federal Building in Oklahoma City in April 1995. It was launched out of the back of a C-130 plane flying at least 6,000 feet above the ground to avoid the bomb's massive shock wave. Once clear of the plane, the Daisy Cutter released its own parachute for the descent. Attached to the end of the bomb was a 3-foot-long detonator. Because the bomb detonates before it actually strikes the ground, there is very little cratering. Rather, it destroys anything within a 600-yard radius of impact.[376] The hope was that the devastating effect of either of these options would create a makeshift landing zone by blowing apart the jungle, and enough of a window for the lead company to land before the enemy could react.

It was audacious, and could possibly work. But from a tactical standpoint either option would present major problems to Tiger Bravo, the first company on the ground. Soldiers would either be climbing into or out of 25-foot-deep craters from the B-52 strike, or crossing hundreds of yards of splintered trees, much like a lumberyard hit by a tornado, while fighting an unknown- size enemy. Once established on the ground, the battalion, along with other 3rd Brigade units, would constitute a blocking position against which the 4th Infantry Division would attempt to drive any NVA units in the area.[377] Luckily for Tiger Bravo, an emergency meeting of company commanders was called at midnight, when word was received that the attack had been canceled. Intelligence reports had begun to filter in that the 325C Division was on the move again, only this time away from the Ben Het area.

NGOK PENG DIENG (PENG DIENG MOUNTAIN)

But now an additional threat appeared. Reliable intelligence sources reported the movement of another NVA division headquarters into the 4th Infantry Division's area of operations in the vicinity of Dak Pek Special Forces Camp. It was the NVA 2nd Division, which had overrun two US Special Forces camps two weeks earlier, just north of the 4th Infantry Division's area of operations.[378] There was no clear picture of the location of the NVA 2nd Division's subordinate regiments, but all indications were that the division could attack the Dak Pek camp in the next few days. The Allies had already lost two camps; losing another, especially when forewarned, would be a disastrous tactical defeat. If these new intelligence reports were correct, it meant that the Americans now faced two NVA divisions, similar in size and firepower, and capable of mounting battalion and regimental assaults on American units and camps, at times and places of their choosing. Even with attached ARVN units added to the rolls, the Americans and South Vietnamese were at a decided numerical disadvantage. The balance of combat power in the mountains had tipped even farther in the enemy's favor.

Leaving the defense of Ben Het, and the waning threat from the 325C NVA Division to the 1/506 and 4th Infantry Division, the Task Force commander assigned the mission to protect the Dak Pek camp to the 3rd Brigade,

101st Airborne Division.[379] Dak Pek was a Montagnard village approximately 45 kilometers north-northwest of Dak To, and only 14 kilometers east of the Laotian border. Its loose collection of huts, housing the fighters and families of the Civilian Irregular Defense Group (CIDG) mobile strike force led by US Special Forces, sat on low hills in a valley surrounding a fifteen-hundred-foot dirt airstrip.[380] The Special Forces camp, established by the 5th Special Forces Group in 1962 as one of a string of outposts along the border to monitor and interdict enemy infiltration into South Vietnam,[381] was a small complex of bunkers and barbed wire on one of the hills just west of the airstrip. Towering, jungle-covered mountains with crests above four thousand feet loomed over the camp from all sides. The strategic advantage of its location near the border was offset by the tactical disadvantage of being in a valley with unoccupied high ground in all directions.

This new threat called for a different strategy. The Americans could not risk going toe-to-toe in a battle with such an overwhelming force, especially since the terrain and vegetation diffused the artillery and tactical air support that normally gave the Americans a decided advantage. The decentralized approach for operations in War Zone D, with companies conducting semi-independent search and destroy missions to find the enemy, would not work here. The enemy concentrations were too large and well-armed. Close-in sweeps to clear the enemy from around fire support bases were feasible, but ground attacks by US forces against major concentrations of enemy battalions invited disaster. A company-size unit, on its own, ran the real possibility of being isolated by hundreds of enemy foot soldiers and overrun before other friendly units could arrive.

Commanders had to find another way to destroy the NVA divisions without spreading infantry units throughout the area, as in War Zone D or the Rocket Belt. The solution came from the adoption of a classic guerrilla axiom: hit the enemy at the time and place of its greatest weakness with your greatest strength. In this case, the enemy was most vulnerable when its forces were massing in preparation for an attack. It was the enemy's soft underbelly, a window of opportunity where the application of the Allied forces' massive firepower could do the most damage. This strategy led to a threefold plan. First, US units would occupy and defend key terrain around Dak Pek with infantry, thus denying its use to the enemy. Second, commanders would use all available 4th Infantry Division, II Field Force, and US Air Force reconnaissance and radio surveillance assets to find the enemy when it was moving in large numbers, and especially where it assembled for the attack. Finally, the commanders would utilize massed air strikes and artillery fire, far beyond normal supporting fires on the battlefield, to pulverize the enemy formations and attack positions.[382] As with the aborted plan to use B-52s in a tactical role against the 325C Division, this plan too included the use of B-52 strikes, this time on large concentrations of enemy forces.

The next day, Task Force Mathews put phase one of the new plan into motion. By 1030 hours the 2/506 Tactical Operations Center (TOC) was operational just off the Dak Pek airstrip, and the companies began flowing in, Chinook load after Chinook load.[383] The battalion had to move quickly to seize the key high ground, between the Laotian border and the vulnerable SF camp, before it was occupied by the NVA 2nd Division's infantry, mortars, and artillery. A Company launched first, making a combat assault into a cold LZ onto Ngok Kof (Kof Mountain). Bravo Company was next. Carrying a double load of ammunition, the company made an unopposed combat assault onto the crest of Ngok Peng Dieng (Peng Dieng Mountain), more than forty-one-hundred feet in elevation, southwest of Dak Pek.[384] In a letter home Harry Brown described landing on the mountaintop, "What an LZ! About big enough for one chopper and sitting on a 10-foot-wide ridge line."[385] By midafternoon the battalion had won the race: the key terrain between Dak Pek and the enemy's location, somewhere to the west, near the Laotian border, was in American hands. The first step in the strategy had been taken. Now it was time to dig in and hold. But before the battalion could reinforce the two companies perched on mountaintops, the afternoon rains halted all further air operations. Alpha and Bravo would have to dig in and hold their respective peaks until the weather cleared.

Tiger Bravo was perched on the highest point of Peng Dieng Mountain. It was a small, bare peak, with a radius of less than fifty meters. It was surrounded to the west, north, and east by steep, nearly impassable, jungle-covered slopes that would slow the advance of any attacker to a crawl. On the south side a narrow ridgeline extended for approximately three hundred meters before it too dropped off into the same jungle-covered slopes as on the other sides.[386] It was all new, defending a mountaintop against a numerically superior enemy; it was the first NDP where it wouldn't be necessary to throw grenades at an attacker, just roll them downhill and let gravity take over, and it was the only time the troops had dug in above a layer of clouds that blanketed the valley below. Standing in the way of an NVA division accomplishing its mission left everyone uneasy. As Harry Brown explained, "The reason we are up here is to stop two NVA divisions, plus some 50 tanks. We were purposely set in their path to make contact. . . .. Imagine two companies sitting on a hill waiting to get hit by 10,000 NVA."[387]

The first night was quiet and cold, much colder than the warm, humid nights of War Zone D. Out came an odd assortment of sweaters, rain parkas, and Currahee sweatshirts that were last used during physical training runs at Fort Campbell. The morning brought a break in the weather, allowing the battalion to reinforce Alpha with C Company, and to send a 4.2" mortar section and D Company to wedge into an already crowded mountaintop with Tiger Bravo on Peng Dieng.[388] During the day, there were no detectable signs of enemy activity from the company's vantage point, but division intelligence and surveillance assets had more success. Without warning, a mountain ridgeline five kilometers west of Bravo's position erupted in a spectacular display of US airpower. Five arc light missions dropped their loads of bombs, one after the other without a break, engulfing the far ridgeline with explosions on top of explosions. The show went on for ten minutes or more until the bombers, still invisible to the naked eye, turned and headed back to base.[389]

The planes had originated from Andersen Air Force Base on Guam, a twenty-six-hundred-mile one-way flight to Saigon. The round-trip missions lasted twelve to fourteen hours. The B-52s, called Buffs, flew in loose, three-ship formations to a refueling point north of the Philippine island of Luzon and refueled again over the South China Sea. On return, they flew directly east across the Philippines to Guam. Later in the war, a second arc light base was established, at the Thai Navy airfield at U Tapao on the Gulf of Siam. With this addition, B-52s took off from Andersen, flew their missions, recovered and rearmed at U Tapao, flew another mission, and then returned to Andersen.[390]

B-52 air strikes in support of Tiger Bravo and the other companies created the most unusual combat power dichotomy of the war, an oddity not often seen on a modern battlefield. In this remote section of the Central Highlands, a B-52, the most destructive and technologically advanced weapons system in the US arsenal, dropping up to sixty thousand pounds of high-explosive bombs from an altitude of thirty-thousand feet, was in support of an infantryman, the army's most elemental and limited weapons system, employing a rifle, grenades, and bayonet at ranges from a hundred meters down to arm's length. The former was a huge, streamlined, and noisy machine, carrying a well-rested crew in a relatively clean and climate-controlled shell that was impervious to environmental extremes. The latter was chronically tired and sleep-deprived from constant day and night operations. They were plagued by bugs, snakes, leeches, and the cuts and slashes of an unforgiving jungle, and spent months on end living, eating, sleeping, and fighting in an environment where some unseen enemy was doing everything in its power to kill them.

By all accounts the three-phase strategy achieved its objective. Intelligence reports indicated that the NVA were forced to mass their forces, which became lucrative targets for an unprecedented number of B-52 strikes. By the time the enemy defeat was completed, 178 B-52 strikes had been unleashed against them. The withdrawal of enemy forces began in early June in the Ben Het area and ended in late June in the Dak Pek and Polei Kieng areas. The 325C Division, which had been hit the hardest, withdrew west into the triborder region. After failing in the Dak Pek area, the 21st Regiment and the NVA 2nd Division headquarters withdrew northwest, into Laos. The exodus would continue until late June, with the NVA 1st Division withdrawing southwest, into Cambodia.[391]

FIRE SUPPORT BASE 25

With the success of the arc light bombardments came new orders for Tiger Bravo. For the third time since arriving in the Central Highlands, the company was on its way to a trouble spot. Early on the morning of May 31 Tiger Bravo lifted off from Peng Dieng Mountain and was deposited at the Dak Pek

airstrip, then ferried to Dak To. At Dak To, the company was officially de-
tached from the 2/506 and attached to the 4th Infantry Division. One platoon
moved out to guard a bridge, while the remainder, numbering only eighty soldiers, climbed on choppers for Fire Support Base (FSB) 25. The company's
mission was to reinforce the battalion at FSB 25, which had been in sporadic
contact with the enemy for days and which included two different mortar attacks that morning. Nondescript and muddy, it perched on two small hilltops
that crowned Ngok Peh (Peh Mountain), at an elevation of 9,438 feet. Occupied by two understrength infantry companies in a patchwork of bunkers surrounding an artillery battery, its only distinction from others in the area was
the grisly name given to the largest hilltop on the west side of the base: Suicide
Knob! It was by airborne standards a "mud hole," with bunkers in need of re-
pair and weeks of trash thrown into the perimeter wire.[392] By 1840 hours that
night Tiger Bravo had settled into its new home and was manning its half of
the perimeter. Other than lights being observed in the jungle below, the night
passed without incident.

At dawn on June 1, 4th Infantry Division soldiers on FSB 25 climbed
down the mountain to sweep the jungle at its base, leaving Tiger Bravo to fill
in the vacated positions on the perimeter. It started as a typically mundane
day for Tiger Bravo, with some soldiers pulling guard duty, others cleaning
weapons and filling the ubiquitous sandbags to shore up collapsing bunker
walls and trench lines. Others just sat quietly, lost in their own world, reading letters or writing home. It was a peaceful interlude where, despite the
surroundings, it felt as if the war were somewhere else. The company in the
jungle below reported no enemy activity. The base artillery battery's guns
were silent, and no helicopters were landing or taking off. Even the noisy
generators had been shut down for their daily maintenance. In many ways,
at moments like this, it could be any of a thousand isolated camps and fire
support bases scattered across a thousand remote locations and manned by
thousands of bored soldiers.

As with any respite from the war, the moment was fleeting. An explosion
brought everyone back to reality. Within the perimeter were several sump pits,
filled with trash and garbage that had been set on fire by the 4th Infantry Division unit before it left. Unbeknownst to Bravo a grenade was in one of the pits

and it exploded, or "cooked off." Specialist 4 Ralph Aponte, sitting on a bunker next to the pit, took the full force of the blast, with hot metal fragments ripping into his head, stomach, and back.[393] Chuck Limer remembered rushing to the scene and seeing one of his fellow bandmates, torn and bloody from the grenade blast. "Around 1 PM I heard an explosion close to an area where the 4th Infantry was burning off some trash. I ran to the area and found that someone had put a hand grenade in the trash. . . . Ralph Aponte was sitting on a bunker just above the trash pile. . . . He was medevac'd to Plei Ku. . . . I saw him later at a hospital in Cam Ranh Bay. His whole right side was paralyzed. He asked me to write a letter home to his mother, which I did."[394]

Notwithstanding that the company was many miles from any administrative headquarters, we were not outside the reach of the army's penchant for bureaucratic investigations and assignment of blame. After an investigation, Lieutenant Colonel Earl Keesling, the newly assigned 2/506 battalion commander, determined that "since the sumps were in another unit's area and the debris in them had been placed by the departing unit, there was not any negligence on the part of B Company 2nd Bn (Airborne) 506th Inf."[395]

FIND THE SONS OF BITCHES

That night Bravo Company received its mission for the following day. Officially, the mission was to sweep the small valley off the north side of the FSB, looking for any signs of the enemy. But in the words of Lieutenant Colonel Mike Malone, the 4th Infantry Division battalion commander in charge of FSB 25, it was to "find the sons of bitches who hit my battalion." This would be a daylight operation, covering four or five kilometers of jungle. The plan was for the company to move north, through an area where two US light observation helicopters had been hit by ground fire, to the high ground on the other side of the valley, then dogleg to the southwest to the site where the battalion commander's command and control chopper had been shot down just five days earlier.[396] From there the company would move southeast to the base of Ngok Peh, and climb back up to the FSB.

At daybreak, Tiger Bravo, numbering only seventy-seven soldiers in total, left the FSB perimeter wire and climbed down the steep slope to a small val-

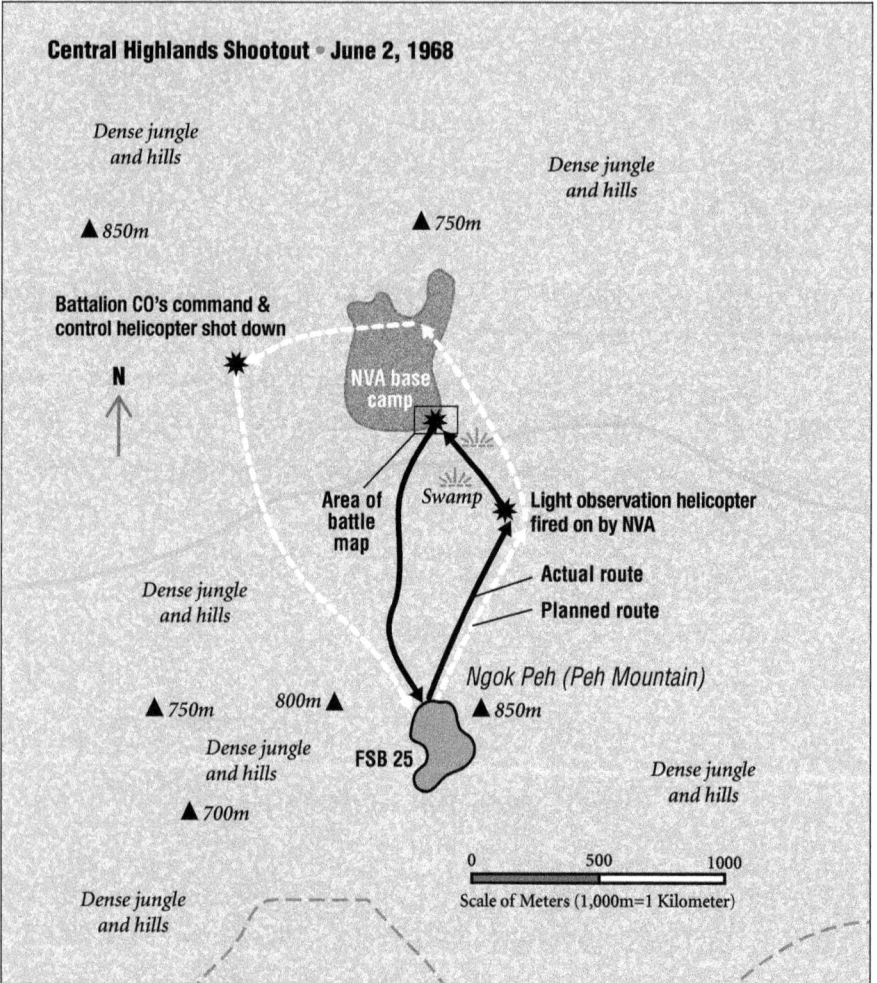

Central Highlands Shootout • June 2, 1968

ley, approximately 1.5 kilometers across and bisected from west to east by an intermittent stream and swamp. "Going through the wire I heard a 4ID officer and soldier talking about us," recalled Doc Franks. "According to one of them they had never sent out less than two companies together, and yet we were sending only one. "Damned Airborne," one of them said. "They have no idea what they're up against." Franks remembered not being impressed. "Instead of making me afraid, that comment made me proud to be with a great company, officers and NCOs. We were indeed, a cut above the rest."[397] That was a typical reaction for Tiger Bravo. Our innate bravado was always a match for whatever situation we faced.

Heading down the slope we passed several NVA bodies, obviously from a recent attack on the fire support base. The company then set out across the valley slowly, cautiously, and expecting contact at any time. The area within a ten-kilometer radius of the FSB was crawling with NVA. Contacts ran the gamut from battalion-size ground attacks, to helicopters being shot down, to mine clearing parties being ambushed, to booby traps and the ubiquitous mortar and rocket attacks. No one expected this to be easy.

At about 1400 hours, Bravo waded the swamp that was midway across the valley and headed up a small hill on the north side, each man checking his buddy for leeches. Immediately, the point team stumbled onto a small, empty campsite with half-eaten bowls of rice left on the ground and food still cooking in fire pits. "We caught them by surprise," Larry Burton remembered. "They were cooking meals on a fire. Their clothes were stretched out on ropes drying."[398] A string of bunkers unfolded behind it. The bunkers fanned out going up the hill, then curved left on the hill's reverse slope before disappearing into dense vegetation at a small jungle clearing.[399] They were not easy to detect, having been dug into the ground with a very low, camouflaged silhouette, and showing only a small firing aperture that was hard to discern. Beyond twenty-five feet they appeared as just another small hummock on the jungle floor.

The jungle clearing had sparse vegetation at ground level, but a canopy of thick foliage overhead made it invisible from the air. It was clear that Tiger Bravo was on the edge of a large NVA base camp, with a very high probability of more bunkers past the clearing. It was much larger than the company could check out in just a few hours, and we were under orders to be back at the fire support base by dark. Bravo composed 50 percent of the defending force, and the battalion commander wasn't about to go through a night with half of its perimeter empty. For the next two hours Bravo checked each bunker with the same results: no enemy soldiers, but all signs indicated that it was an active base camp. So where were the occupants?

According to standard enemy practice, a fortified base camp is roughly circular in form, with an outer rim of bunkers and foxholes that enclose a network of living quarters, usually frame structures aboveground, command bunkers, kitchens, and sleeping platforms. The shapes varied according to the rise and

fall of the ground and the use of natural features to restrict attack on the camp
to one or two avenues. Some of the bases, and in particular those used only for
training or as a way station, had minimum defensive works. In all cases, however,
the enemy went to considerable lengths to defend from a ground attack. He
took every opportunity to make nature work in his favor: trees, shrubs, and earth
itself reshaped to conceal bunkers and trench lines. When he chose to stand
and fight, the choice was seldom, if ever, made because he was trapped beyond
any chance of withdrawal. If he fought, he expected to inflict more than enough
casualties on the Americans to warrant making a stand.[400]

Just as Tiger Bravo began to pull back and mark the location for a future
clearing operation, a weapons cache was discovered in a cluster of bunkers on
the far edge of the clearing, just where the dense jungle resumed. The 2nd
Platoon moved to investigate, while the 3rd and 4th Platoons held in place.
"We started down a part of a hill that had a little dog left with bunkers on both
sides filled with weapons, ammo and food," Chuck Limer recalled later. "As
we came down to a clearing, 1SG Trent noticed a half circle scratched in the
ground with a X in the center." The platoon found two heavy machine guns,
two 75mm recoilless rifles, and assorted small arms in an orderly row, as if set
up for an inspection. Close by was an open shelter with a bank of radios on a
small table—obviously a command post. The X on the ground was either the
outline of a pending attack on FSB 25 or, more than likely and far more sin-
ister, it was the NVA commander's hasty sketch, made only minutes before, of
how he was going to deal with Bravo Company in the middle of his base camp.
First Sergeant Trent bought into the bone-chilling, "surround the Americans"
theory. Pointing at the X he said to Doc Franks, "See that X. That's us! The
half circle going around us is the NVA."[401]

While some troops started to destroy the weapons and radios, the 2nd
Platoon pushed security out twenty meters or so. Meanwhile, Limer, who was
standing a little to the rear with Lieutenant Joe Palagyi and Sergeant Trent,
decided to check out an isolated bunker approximately sixty meters from the
cache point. As he peered into the bunker, an NVA soldier in a tree just above
him dropped a Chicom hand grenade. The explosion hurled Limer ten feet
backward, knocked his M-16 out of his hands, and peppered him with shrapnel. Then another NVA soldier shot him at point-blank range, hitting him in

the left leg and arm. Bloody, seriously wounded, unarmed, and alone, Limer recalled that "I crawled as fast as I could behind the bunker in a hail of bullets. That's when I looked up and saw Doc Franks running toward me."[402] Doc Franks saw Limer getting hit. "I saw this guy way the hell out there by himself," Franks recounted, "I watched him walk up to a bunker, then saw him get shot. He was stumbling backward and refused to go down. So, I took off running. There were so many bullets flying around I felt like my head was in a bee hive."

"Doc Franks was the first to arrive at my side," Limer recalled. "He was yelling at me to stay down and firing his M-16 as fast as he could at the NVA. He must have fired all twenty clips and the NVA just kept coming. They were coming at us from everywhere."[403] Franks remembered that it was Limer's M-16 that he picked up initially to fire back at the NVA, who were now only about twenty-five feet away. "By the time I got to him," Franks said, "he was down on the ground and rolling around. The NVA then opened up again from the bunker. I knew I needed to take care of the NVA first, before I helped Chuck. I could see Chuck's rifle four or five feet from the bunker so I went up and grabbed it and fired one shot into the bunker and it jammed! So, I half crawled back and picked up my rifle and fired into the bunker until there was no more return fire."[404] For the next few minutes, Franks stayed with Limer and fired at the NVA in the bunkers, and at those heading directly at them from out of the jungle. "All around us I could hear the sounds of a full-scale battle, M-16s, AK-47s, hand grenades, and RPGs. The NVA were running right by our bunker trying to outflank us, Franks was shooting at them at point-blank range. I told Doc to get back and save himself, I was shot up bad and didn't think I would make it. He said he wasn't going to leave me."[405]

Franks had fired all his M-16 ammunition by the time help arrived, and one clip from a .45- caliber automatic he carried in a shoulder holster. "Then First Sergeant Trent showed up. He was firing and yelling orders," Limer recalled. "I saw Trent shooting up into the trees. One NVA fell right next to us, then First Sergeant Trent threw a smoke grenade between us and the NVA and both First Sergeant Trent and Doc Franks grabbed me by my webbing and started running back to the company through a hail of bullets. One bullet hit Doc Franks in the boot heel. I thought we were not going to make it but we did."[406] Limer was carried to the relative safety of the company perimeter for Doc Franks to treat his wounds. But, as Limer remembered, they were still in the midst of a battle. "With the battle still going on and bullets hitting the trees all around, he managed to get most of the bleeding stopped. After that he put me on a stretcher and helped carry me to a place that was cleared for a helicopter to land and evacuate the wounded."[407] According to one account, before he was evacuated, Limer had one important request of his savior medic: "Doc, if my dick is gone just let me die."

Franks's actions not only saved Chuck Limer's life but also highlighted the dual role of combat medics in the Vietnam War. They were both medical aid men and armed combatants. For most of the twentieth century, medical personnel did not carry weapons and wore a distinguishing red cross to denote their protection as noncombatants under the Geneva Convention (the basis for the rules of international law dealing with the protection of victims of armed conflict, noncombatants, and prisoners of war). This practice continued into World War II. However, modern combat medics are armed combatants, and most do not wear distinguishing markings.[408]

As Doc Franks recounted years later, he carried everything needed to fight and save lives, "In addition to my rifle, ammunition, and grenades, I carried battlefield compresses (various sizes), morphine, pneumatic air splints, a jungle litter, Ace bandages, serum albumin (blood volume expander), surgical kit, *Merck Medical Manual,* narcotic-based pain relievers (Darvon and codeine), malaria tablets (handed out daily), general meds, antiseptics, and various bandages."[409]

Medics were always in the line of fire, and, as a group, suffered an inordinate number of casualties. During the Vietnam War, the 101st Airborne Division would have 153 of its medics killed in action.[410] In this contact, Tiger Bravo lost one of its medics, Doc King, who was shot in the leg and had to be evacuated along with the wounded he had just treated. Medics were favorite targets of the VC and the NVA, who followed the medics' movements around the battlefield to find more American targets, knowing that wherever medics went there were sure to be more Americans to shoot. As one 4th Infantry Division after-action report noted:

THE NVA WOULD SELECT TARGETS BY ORIENTING ON THE SHOUTS FOR A MEDIC; PATIENTLY WAITING FOR THE MEDIC AND OTHERS TO ARRIVE, THEN TAKING THAT AREA UNDER INTENSE FIRE. TIME AFTER TIME THERE WAS A DIRECT CORRELATION BETWEEN SHOUTS FOR MEDICS AND INCREASED ENEMY FIRE.[411]

OLD-FASHIONED GUNFIGHT

Not stopping, the NVA flowed around Franks and Limer and went into a direct assault on the 2nd Platoon. This was not the slashing hit-and-run tactics that Tiger Bravo faced against the VC in III Corps. In II Corps, when the NVA attacked they came in waves and pressed as close as possible to the American forces to negate our overwhelming fire superiority. As Franks remembered, "They just wanted to kill Americans."[412]

Franks's solo gunfight slowed the attackers as they poured into the clearing. The NVA bounced off Franks, then hit the 2nd Platoon head-on. The battle

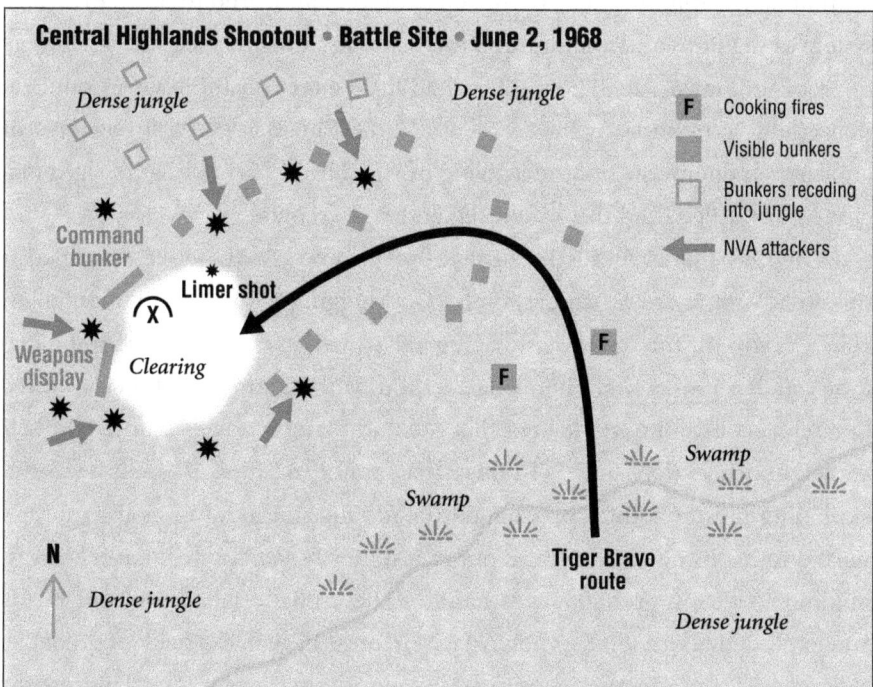

Central Highlands Shootout • Battle Site • June 2, 1968

was at extremely close range, with opposing hand grenades crisscrossing in the space between the two forces. Larry Burton, who was part of the 2nd Platoon point team, remembered, "We killed two right off. Some started firing from bunkers. We threw grenades in and they threw them back out so we would pull the pin, let it cook off for a second, then throw it in. . . . There was so much firing that bark was flying off trees."[413] In just seconds the 2nd Platoon was fully engaged in an old-fashioned gunfight at close range. Then others began popping up from fighting positions dug in on the sides of bunkers and from nearly invisible spider holes on the jungle floor. The fighting was at such close range that shrapnel from 2nd Platoon M79 rounds hit one Bravo soldier.

By this time, Lieutenant Palagyi had artillery from FSB 25 ready to fire, but the NVA were too close to Bravo's ranks to safely drop artillery on them. So, the 2nd Platoon moved back to the 4th Platoon, freeing Palagyi to call in the artillery. Meanwhile, First Sergeant Trent and the 3rd Platoon carried Limer and the other casualties back another hundred meters out of the bunker complex to a one-ship LZ hacked out of the jungle for evacuation. This left the company headquarters, 2nd Platoon, and 4th Platoon, numbering no more than fifty-plus soldiers, on the edge of what appeared to be an extensive system of weapons and ammunition cache points, and engaged with an enemy force of unknown size. The artillery had little effect on the NVA in bunkers that could withstand anything except a direct hit. Bravo was up against an entrenched enemy, with more arriving by the minute, firing AK-47s, light machine guns, RPGs, and throwing Chicom hand grenades.

Then, as I later recalled, "all hell broke loose."[414] My letter described it this way: "Well, there we were; two platoons slugging it out! 2nd Platoon killed 10 NVA right off the bat without taking any casualties. I was right with the Platoon Leader, PSG Sykes. Then Charlie hit us with 82mm and 60mm mortars. In a few seconds, the whole area that the 2nd Platoon was in erupted. Myself and eight others were hit."[415] To this day I am not completely sure if it was the exploding mortar rounds or a hand grenade that hit me. Just as the mortars started to hit I saw an NVA raise out of a spider hole about fifteen feet away, holding a Chicom grenade by its handle. The grenade sailed over our heads and exploded about ten feet behind us. He must have had a stack of grenades because just seconds later he popped up and threw another one. This one hit a branch over my head and dropped just behind me, exploding seconds later. Dan Bernard, who was right beside me, but miraculously came out without a scratch, remembered seeing me "go ass over bandbox," and thought, "Holy crap, he's dead!"[416] I remember flying through the air and losing my helmet, glasses, and M-16 in the blast. After being helped to my feet and feeling no pain, I said to no one in particular, "I'm okay. I'm not hit." A voice from behind me said, "No, sir, you are."

Eight wounded, plus the platoon needed to carry them and provide security, had reduced Tiger Bravo's fighting strength to a dangerous low. Plus, the NVA had begun to move around Bravo's flanks, obviously trying to surround us and cut off the 2nd and 4th Platoons from the 3rd Platoon, which was still a hundred meters back, at the evacuation LZ. Even though I had been wounded in the back with shrapnel, I was still able to function and remembered hearing the enemy fire shift around our flanks. It was time to pull back before the situation became any more critical than it already was, consolidate with the 3rd Platoon, and drop as much artillery as possible on the NVA. I remembered my boots filling up with blood during the withdrawal, but not much else. According to Bernard, "The CO was in and out of consciousness, every few minutes he would straighten up and crank out orders, then his legs would turn to rubber and we would have to hold him up."[417] Tiger Bravo barely made it out of the enemy base camp; it was only minutes away from being surrounded when the order was given to withdraw. In describing the situation Doc Franks recalled, "Had the CO not maintained control, we had a good chance of being wiped out. He kept control and order during the complete chaos of combat. We appreciated that he always knew what to do. We depended upon it."[418]

The artillery barrage worked. Contact with the NVA ground force was broken, but not with their mortars. Doc Franks remembered standing on a hill looking back at the just-vacated base camp area and remarking to Joe Palagyi that our artillery seemed to be creeping back up the hill in Bravo's direction. Palagyi replied, "Not ours. It's theirs."[419] Larry Burton remembered being on the receiving end of it. "When it first started dropping," recalled Burton, "I thought it was our own artillery, because it was too heavy to be mortars."[420] Inexplicably, the NVA mortar and artillery barrages stopped creeping up the hill, then ceased completely. Then, woozy from loss of blood and under protest, I was the last to be evacuated. As usual, First Sergeant Trent had accurately assessed the situation when he told me, "Sir, we can't carry your ass back across that valley. You're being evac'd."

Within minutes of my evacuation, Tiger Bravo started back across the swamp. Three and a half hours later it was still on the move—exhausted, low on ammunition, out of water, and numbering less than sixty able-bodied soldiers. With Mad Dog Matosky and his unerring sense of direction in the lead, the company lurched through the black jungle right up to the base of the mountain where FSB 25 sat, unseen in the darkness. Larry Burton recalled helping others along the way, "Some guys had minor wounds and we had to help them get back."[421] Before attempting the climb in the dark and stumbling into the fire of sleepy, trigger-happy 4th Infantry Division soldiers, Trent radioed the 3rd Battalion, 8th Infantry Regiment headquarters and requested "that every living soul was awake and told that friendlies were coming in." Finally, just before midnight, Tiger Bravo struggled through the barbed wire perimeter and collapsed on bunkers all around its defensive sector. "I was so thirsty I must have drank a gallon of water," recounted Franks. "Then I just collapsed on the roof of a bunker."[422]

Back at the base camp, the NVA had already cleared its dead from the battle site, to be buried in unmarked graves deeper into the jungle. The wounded were being treated in the camp's makeshift aid station, probably in one or two levels underground. In the morning, the seriously wounded would begin the long trek across the border into Laos to other medical facilities. It would take several days and not all would survive.

By now, Tiger Bravo's seriously wounded had passed through a medical clearing station at Dak To and were already being treated at the large, regional 71st Evacuation Hospital in Pleiku, South Vietnam. By the time I arrived in Pleiku, most of the wounded were already in recovery. Limer, the first to arrive, had survived an operation and was being prepped for air evacuation to Japan. The process was the same for all of us; we were carried off the Chinook helicopter on a stretcher straight into the 71st Evacuation Hospital operating room. Multiple operating teams worked side by side on a row of casualties, friend and foe. I was on my stomach, still conscious as a surgeon picked shrapnel from my back and dropped the pieces into a metal pan—clink, clink,

clink—until only one was left, which was too close to my spine. Just across from me, and at eye level, was a naked NVA soldier with six or seven bullet holes in him, being worked on by another medical team. He received the same level of care as I did, only his procedure went on under the watchful eye of a shot-gun-toting MP standing at the head of the operating table.

BLUE PAJAMA BOTTOMS

The next day, with its mission accomplished, Tiger Bravo, minus the company commander and eleven wounded, moved by helicopter back to the vicinity of Dak To, reuniting with the rest of the 2/506. With the threat around Dak Pek gone, the battalion had returned to the hills surrounding Dak To. The battalion plan called for A, B, and C Companies, plus the recon platoon, to establish a base on hilltops in the vicinity of FSB 13 and conduct search and destroy operations on and around their respective hilltops. Tiger Bravo's hill was more of a jungle-covered mountain north of FSB 13, topping out at 1,043 meters high. Beginning on June 5, Bravo patrolled on and around Hill 1043, searching for the NVA. On June 7, Bravo discovered an empty complex of eighty bunkers and fighting positions, but no signs of NVA. It was the same for the remainder of the battalion, with C Company discovering another complex the next day.[423]

By June 9, the tense situation that had precipitated the Wandering War-riors' move to Dak To had eased enough to send the 3rd Brigade back to Phuoc Vinh. Tiger Bravo's battle on June 2, 1968, had been the only substan-tial contact during the brigade's entire foray into the Central Highlands. After a search of the base camp discovered by Bravo, the 4th Infantry Division found evidence of an intense battle and thirty-five dead NVA soldiers.[424] On June 10 a steady stream of USAF C-130s carried the brigade back to III Corps and its base camp.[425] Doc Franks remembered taking off while the airfield was being shelled: "The C-130 that Spencer, Matosky and I were on was hit by ground fire leaving Dak To. We couldn't return to Dak To because, according to the crew chief, the airstrip was under fire. So, we flew all the way to Tan San Nhut Air Base near Saigon with a damaged wing. I guess the Dinks were saying good-bye to us."[426] For the first time in months the brigade was assembled in one location without any operational missions to discharge. At Phuoc Vinh command of the company passed to the executive officer, First Lieutenant Hank Matlosz, while I was hospitalized.

Meanwhile, on a beautiful beachfront setting on the peninsula that juts out into the South China Sea to form Cam Ranh Bay, six of Tiger Bravo's wounded from the June 2 battle were recovering at the US Army's 6th Convalescent Center (CC), which was the army's only rehabilitative care facility in Vietnam. Any soldier requiring thirty days or less of recuperation after being treated for his wounds would end up at the 6th CC. This 1,300-bed facility, averaging 750 to 850 patients at any time, provided full, in-depth rehabilitative care for casualties so they could return to combat quickly and not be evacuated to Japan or Stateside. This lifted the burden of rehabilitative care off the eighteen 44th Medical Brigade hospitals spread across South Vietnam. Anyone with medical complications or needing more medical procedures went to a nearby US Air Force surgical hospital.[427] My second operation, to close up my wounds, took place there on June 7, 1968. Medical records showed that after six months on the line, my weight had dropped drastically, to 135 pounds from my original 165 pounds. This was typical for an infantryman in Vietnam.

Physical conditioning experts at the convalescent center put the patients through a graduated daily routine that included organized physical training, a mile run, weight lifting, sports, swimming, and surfing.[428] All the patients were ambulatory and in uniform; only their jungle fatigue trousers were taken away and replaced with blue pajama bottoms. Since the patients were confined to the convalescent center, anything outside the gates was off-limits to them. The blue bottoms were an easy way for the Military Police to cruise the bars and brothels of Cam Ranh Bay and quickly identify the occasional drunken patient. Inside the compound, it was difficult to look like a rough, seasoned combat veteran when you strode up to the bar in one of the clubs in your baby blue pajama bottoms.

As part of its mission, the 6th CC also awarded Purple Hearts to patients before they returned to their units. On June 13, 1968, a ceremony was held in the commander's office where twenty-three soldiers received the Purple Heart. The reading of the names was a barometer of which units had seen the most recent action. In this ceremony, there were soldiers from the 4th Infantry Division (ten), 101st Airborne Division (nine), 1st Cavalry Division (one), 173rd Airborne Brigade (one), 196th Light Infantry Brigade (one), and 198th Light Infantry Brigade (one). Representing Tiger Bravo in this ceremony were First Lieutenant Richard L. St John, Staff Sergeant Phillip C. Parker, Sergeant Charles W. Wright, Specialist 4 William B. Farwell, and Private First Class Roy C. Webster.[429]

Back at Phuoc Vinh the battalion rested, rearmed, refitted, and rotated through small-arms ranges. Then on June 14 an alert came in from II Field Force for the brigade to once again shoulder its Wandering Warrior mantle. This time the 2/506 was ordered to Cu Chi, the home base camp for the 25th Infantry Division. Once again Tiger Bravo found itself on a C-130 heading to another hot spot. By 1530 hours the entire battalion, minus D Company, which had been left behind to guard the Song Be Bridge along Highway 1A, and one platoon per company left behind for base camp defense, had moved to Cu Chi.[430]

This brief stopover provided no time for Tiger Bravo to sample the base camp's amenities. Cu Chi was a typical rear-area, fixed installation. It was surrounded by a large cleared area and a bunker line that consisted of observation towers, firing positions with overhead cover, an earthen berm, barbed wire entanglements, spotlights, and minefields. The support units inside the Cu Chi base were assigned sectors of the defensive perimeter with very specific, rehearsed plans for reinforcement and counterattack should the enemy be bold enough to launch a ground attack. Although living conditions were austere by Stateside standards, bases such as Cu Chi allowed a permanence unheard of in similar camps in World War II or Korea. Cu Chi had virtually all of the facilities found at permanent installations outside the war zone, including a large post exchange (PX), clubs for officers, noncommissioned officers, and enlisted men, a USO Club, barber shops, a Military Affiliate Radio Station (MARS) so troops could call home, a Red Cross field unit, sports fields, miniature golf courses, swimming pools, and chapels.[431] It also contained a number of semi-official Vietnamese enterprises, such as tailor shops and the notorious steam bath, aptly named the "steam and cream" by the troops.

Another of Pop Plemons' escapades took place at Cu Chi. Pop had been relegated to light duty in the battalion rear area because of a severe case of "jungle rot" on one leg, so bad that Pop hobbled around on crutches for a week. That didn't stop him from spending each night at one of the clubs scattered around this mammoth base camp. Every night Pop would hitch a ride to the club on one of the supply trucks making a run, but no one knew how he made it back to the battalion area after closing time. Surely, he couldn't walk the distance.

The mystery was solved one night when Joe Adams, who was in Cu Chi because of problems with his ears, went to the club with Pop. Joe remembered that at closing time an MP patrol showed up and the NCO in charge walked up and said, "Well, Pop, are you ready to go? Come on, we'll give you a ride back to your company area." Joe Adams was stunned. Here was an MP sergeant offering a ride to a 101st Airborne soldier, when the 101st had a reputation for giving the base MPs a hard time, especially when alcohol was involved.

"That's really nice that y'all make that effort since he's on crutches and all," Adams recalled saying to the MP sergeant.

"Nice, hell!" replied the sergeant. "We just got tired looking for our jeep every night. When he got ready to go, Pop would go out there and hot-wire our jeep and drive it off back to his company area."[432]

The battalion had been selected to lead a surprise night attack on a fortified VC village in the Ho Bo Woods—a small but deadly enemy stronghold due east from Cu Chi and north of Saigon. The next ten hours for Tiger Bravo were filled with planning for the combat assault, troop briefings so that everyone down to the last rifleman knew his part, and checking and rechecking equipment, weapons and radios . . . but little sleep. In a war replete with perilous, complex missions for rifle companies, to execute a night combat assault into VC-controlled territory ranked among the most dangerous.

8

Surprise Attack in Ho Bo Woods

Ho Bo Woods
(Hau Nghia Province)
June 15–July 21, 1968

NIGHT COMBAT ASSAULT

In the pitch dark of a hot June night, nine choppers lifted off from the assault pad at Cu Chi and flew west, disappearing into inky blackness. On board were fifty-four Tiger Bravo soldiers; each one was heavily armed, carried extra ammunition and grenades, and had been handpicked to be part of the initial assault force. They were aboard the first of three flights that would carry the company into a surprise combat assault. After fifteen minutes the helicopters split into three groups and started their final approach into three separate landing zones. Eighteen paratroopers would disembark at each LZ. Counting on surprise to offset modest numbers, the assault forces would seize the landing zones and hold them until the choppers made a forty-five-minute round-trip to Cu Chi with another load of Tiger Bravo paratroopers.

In those last few minutes before touchdown, each soldier sat alone with his thoughts and prepared mentally for what lay ahead. It would either be a quiet landing on a cold LZ in a pitch dark corner of Vietnam, or the chaos and noise of a hot LZ, with confusion magnified tenfold because of the darkness.

The veterans on board sat in silence, wondering if the pilots could find the LZ in the darkness, worrying that eighteen fighters could be lost in minutes if a sizable force of VC happened to be on or near the LZ. Thinking that forty-five minutes is a long, long time to be left alone in a VC stronghold. Wondering if the green replacement on board, commonly called a "cherry," would hold up under fire. Likewise, the nervous cherry, on his first combat assault, clutched his rifle and stared quietly into the black night, trying to recall all of the instructions from his platoon leader at the operations briefing. Remembering only someone behind him whispering, "If this works it will really piss Charlie off." Marveling at how the veterans on the chopper could be so calm. One even had his eyes closed.

Tiger Bravo was part of a brigade operation to cordon and search the village of Thanh An, in the heart of a Viet Cong stronghold called the Ho Bo Woods. By day the village appeared like many others in the area—a sleepy collection of farmers' huts surrounded by rice paddies and banana groves. But intelligence reports had identified it as a VC-controlled, semifortified village that was used to stage ground attacks launched out of the Ho Bo Woods against US and ARVN installations. Sources could not confirm whether a VC unit was inside the village preparing for an attack, but the 101st Airborne companies had to expect the worst.

The Ho Bo Woods was a 360-square-kilometer VC refuge approximately thirty-five kilometers west of Bien Hoa. It was a patchwork of rice paddies, large areas of brush and trees, and Viet Cong–controlled villages where the VC had spent years building bunkers and camouflaged trench lines.[433] Tunnels connected the bunkers and earthworks, enabling the defenders to pop up, shoot, disappear, then fire again from another angle—a jack-in-the-box maneuver that doubled the effect of their numbers. Should they decide not to defend the village, they knew every possible escape route and tunnel leading to the surrounding countryside. Invariably they found a way out, day or night, as if they had a sixth sense for soundless escape routes.[434] To add to the danger, every approach into the village would normally be heavily mined and booby-trapped.

In an attempt to achieve complete tactical surprise, the choppers landed without the normal artillery preparation. There were no artillery rounds exploding around the landing zones to provide cover during the most critical first few minutes of any combat assault—the unloading of soldiers and the movement off the wide-open LZ into the surrounding woods or jungle. According to Major John Sharp, the 3rd Brigade operations officer, "This was the first time this sort of thing had ever been done in Vietnam." The only warning that something big was happening, as the helicopters roared out of the night, was the sudden appearance of one-million-candlepower flares dropped by the air force over the village and the surrounding landing zones.[435]

Hot LZ

Initially, the gamble that surprise could be achieved without an artillery preparation paid off. On the far side of the village the 3/187 landed unopposed, as did A Company from the 2/506.[436] Then the VC woke up, just in time for the first lift of Bravo Company to come roaring in out of the darkness. Red tracers arcing down to earth from the M-60 machine guns fired by the door gunners crossed in midair with green and white tracers arcing skyward from VC machine gunners seeking out the troop helicopters. "I don't know how you could sneak up on anyone in helicopters all flying in formation," Dan Bernard, on one of the lead choppers in the Tiger Bravo formation, recalled. "As we made our approach I saw our door gunners firing, and a whole bunch of green and white tracers coming up at us. We came down right in the middle of a fight. So much for the element of surprise."[437] The first lift of Tiger Bravo had landed under fire at 0328 hours. Just as the helicopters cleared the area, VC popped up in the tree line and sent a hail of small-arms and automatic-weapons fire at the soldiers still trying to clear the LZ.[438] The first casualty was from a sniper, then two more, one serious.

Chuck Kudla was on the first lift as well. "As I left the aircraft someone started shooting so I went to ground. I must have hit a rock with my left knee, which caused a break in my kneecap bone, the patella."[439] He recounted what happened next. "After the shooting stopped I got up in great pain, and helped secure the area. The sun came up and we patrolled. I used my weapon as a crutch. By midafternoon I was taken by chopper to Cu Chi. Was put in a full leg cast and spent seven days in the hospital." This was the end of the war for Kudla. As the driver for Lieutenant Colonel Grange he had been offered a safe rear area job when Grange left command. But that was contrary to his sense of duty; instead, he volunteered to rejoin his old unit on the line. Wounded once before, this injury was more serious and would send him all the way back to America. He would never return to Bravo.[440]

By the time the second wave of Tiger Bravo landed, every VC in the area was awake and shooting at any American helicopter trying to land. Three helicopters were hit by ground fire before they could land and offload the troops. On one helicopter the soldiers were shot while they were still on board. Without landing, the helicopter peeled off from the formation and headed directly to the hospital in Cu Chi. Unfortunately, one of the wounded died in the air. The initial report was that it was a Bravo Company soldier who died, but later he was identified as a member of the battalion's recon platoon.[441]

Combat assaults never went according to plan, especially when the enemy contested the landing. Hot LZs were sheer bedlam, with shouts and explosions and some troops firing and moving according to plan and others not at all. The soldiers had only seconds to jump out of the chopper before the pilot lifted off. Any hesitation by a soldier, or a nervous pilot not touching down or lifting off before his load of soldiers exited, meant a soldier was jumping into midair as the helicopter was lifting off. On this night, it happened to Dave Spencer, who was carrying First Lieutenant Matlosz's company command radio. According to Dan Bernard, who was in the same helicopter, "Spencer was the last one to jump out of the helicopter. He landed like a piece of crap right in the middle of a firefight. The ship was ten or twelve feet up and he landed on his back. He was so shaken that he jumped up and started running straight toward the enemy. We had to scream at him to get him back."[442]

Hesitation was only human, but a soldier refusing to leave the chopper during a combat assault was unthinkable. In the hundred-plus combat assaults that Bravo made during its first year, there was not a single instance of a soldier refusing to exit a chopper. Mike Scott, who made dozens of combat assaults during his time with Tiger Bravo, recalled what it was like to approach a landing zone just seconds before landing: "If I was in the first slick, there wasn't time to think about it. If you heard the first slick get fired on, you had time to think about it. Thinking did not do any good. It's not like you could turn around or get off somewhere else."[443]

The last lift of Bravo flew into another maelstrom, only this time the firing was much more intense than the first two lifts experienced. Just as the helicopters touched down, for that precious few seconds when helicopters and soldiers were the most vulnerable, both sides of the LZ erupted in fire.[444] Bullets and rocket-propelled grenades came from everywhere. Bravo soldiers immediately returned fire, but the sounds of their meager response were lost in the uproar that surged across the LZ. Shouts, explosions, gunfire, and helicopter engines at full throttle were soon joined by the staccato bark of gunship machine guns working the tree lines and the deep crump of artillery landing nearby.

No experienced commander should expect to have a clear picture of the battlefield, nor have the ground tactical plan survive the first few minutes of contact, especially during a night combat assault into a hot LZ. All hot LZs start out as pure chaos; nothing seems to make sense. The secret to success is to understand that this phenomenon will occur and to make quick decisions, not waiting for the "fog of war" to lift, because it rarely does. A passage from the classic military text *Infantry in Battle* highlighted that "In war, obscurity and confusion are normal. Commanders will seldom have a clear insight into the enemy situation. The abnormal is normal, and uncertainty is certain to occur."[445]

As quickly as it had started, it was over. The overwhelming response of those on the ground and from the air had silenced the enemy defenders. Joe Adams, a PFC acting squad leader, remembered "sitting on a paddy dike looking out toward the village and watching the Phantom jets strafing and bombing out in front of us. It was like daylight. Then along comes one of the enemy walking straight out at me, so I got down and took up a good firing position using my rucksack for support. Then along comes another Phantom and that VC stops and shoots at the jet. Well, that Phantom rolled and came back and that VC shot at him again. Just when I was ready to pop the VC that Phantom dropped one right between his legs. There was nothing left of him including the rice paddy dike. I mean nothing." His one-word description of the night CA was the universal "it was a "clusterfuck"—the soldier's way of describing something completely unorganized and out of control.[446]

Adams had been with the company less than two weeks, having just completed the 101st Airborne Division's SERTS (Screaming Eagle Replacement Training School) training in Bien Hoa. "I was in SERTS when we received word on June 5 that Bobby Kennedy had been assassinated. Right after that I flew to Phuoc Vinh, joining Tiger Bravo when it returned from Dak To," he recalled. The SERTS in Bien Hoa was the first stop for any newly assigned lieutenant or junior enlisted soldier. It was five days of how to stay safe, healthy, and alive in Vietnam, taught by 101st Airborne Division veterans plucked from the ranks of the line units. Training day and night, the replacements became familiar with every weapon in the infantryman's arsenal—both US and the enemy's. How to survive booby-trapped infested villages and trails, navigating live fire reaction courses, learning battlefield first aid/field sanitation techniques, and the occasional mortar attack rounded out the curriculum.[447] For once the drill sergeant's mantra of "what you are about to learn will one day save your life," pounded into every recruit in basic training, was true. Also true was the equally ubiquitous drill sergeant prophecy of "look to the soldier on your left and right; one of you will not make it."

By first light Bravo had assembled its forces that had been spread out in the three LZs, and was in position outside the village. By now the village was completely surrounded by six American companies, along with a contingent of South Vietnamese National Police. A radio message from battalion started the search. There was no organized resistance; the VC or NVA unit that had initially occupied the village was long gone. Tiger Bravo did capture one wounded NVA soldier who admitted to being a member of the NVA 101st Regiment. He had been brought to the village to be further transported to a VC hospital deeper within the Ho Bo Woods. The company also came across fighting positions, a cache of penicillin, and assorted VC rucksacks. Villagers admitted that there were thirty VC in the village before the search started but that they had all escaped before daylight. However, not all had left. A and C Companies made sporadic contact throughout the day with the few VC unable to escape the cordon. By the end of the day the battalion had suffered seven wounded and one killed. The cordon and search operation accounted for two NVA KIA, four POWs, three suspected VC, and an assortment of medical equipment, rice, and other enemy supplies.[448]

The next day the battalion started independent, company-size search and destroy missions around Thanh An. A and C Companies made contact in their areas, but Tiger Bravo's was relatively quiet. One VC was spotted and shot at by a Bravo Company soldier, but escaped. Also, just after dark one of Bravo's LPs, just outside its NDP, had movement to its front, but nothing came of it. The two ambushes set that night came up empty as well.[449] That night, shortly after midnight, the word went out to all companies to move at first light to one of four PZs spread out in the battalion's area of operations so that they could be lifted back to Cu Chi. The next morning, it took five hours and nineteen separate lifts for the battalion to consolidate in Cu Chi. Conditions were spartan: buildings with concrete floors to sleep on, a hot breakfast, C rations for lunch, then a hot supper. That night everyone was confined to the battalion area, waiting on orders. The only clue that the operation would have something to do with building bunkers came when twenty sandbags were issued to every soldier, with orders to carry them in their rucksack on tomorrow's mission.[450] No one was happy about this development.

The 2/506 had yet another shift in mission. This time it would go farther west, near the small town of Trang Bang in Hau Nghia Province. Trang Bang had been the scene of epic and bloody battles over the past two years. The terrain would change as well. There were miles and miles of flooded rice paddies and swamps split by a meandering river, a few streams, the occasional cluster of farmhouses and patches of jungle—not to mention a very active VC and NVA threat. From the air, it all looked the same, sunlight glinting off brown water for mile after mile, but on the ground, it wasn't. Walking in brown soup up to your knees meant rice paddies; suddenly dropping into muddy water over your head meant a stream, canal, or river.

AIRBORNE NO LONGER

While all of this was going on, rumors about the 101st Airborne Division losing its airborne status, and having its name changed, became reality. By US Army, Pacific (USARPAC) General Orders 325 and 326, dated June 28, 1968, the fabled 101st Airborne Division was redesignated the 101st Air Cavalry Division. Effective July 1, 1968, its organization changed as well, from an airborne to an airmobile configuration.[451] Although immensely unpopular with every paratrooper in the division, and a huge blow to the airborne mystique, it changed little in a tactical sense for a company in the boonies. Sometime in the future it would mean more helicopters and greater battlefield mobility, but for the infantry in the summer of 1968, nothing changed. It was still combat at its most elementary level, foot soldier against foot soldier.

Regardless of the change in name, the "airborne versus leg" friction continued unabated. Acceptance of nonairborne officers was especially difficult, none more so than the new brigade commander. The outgoing commander, Colonel Lawrence Mowery, an old-time, decorated paratrooper, was replaced by Colonel Joe Conmy, who had previously commanded the 3rd Infantry Regiment— "the Old Guard"—at Fort Myer, Virginia. Word quickly spread that he had never held a command in combat and, infinitely worse, that he was not airborne-qualified. Yet here he was, taking command of three thousand soldiers, the majority of whom were airborne and proud of it. Even before his arrival, changes in the brigade's swagger were evident. There had been signs posted all over Phuoc Vinh that proclaimed, "Repent legs; God is Airborne." One day they were up; the next, gone. Removing the signs was easier than dealing with the underlying dissatisfaction in the brigade over its new commander. In time, this would dissipate as nonairborne replacements arrived in increasing numbers to replace casualties suffered by the original core of paratroopers. For a time, though, a favorite pastime of the paratroopers, while back at Phuoc Vinh Base Camp, would be to walk by Colonel Conmy, salute, and yell, as loud as humanly possible, the standard Airborne greeting of "AIRBORNE SIR!" A small gesture, but one that spoke volumes about how we felt.[452]

Two months later a long-awaited message reached division headquarters proclaiming the redesignation of the 101st yet again, from the hated 101st Air Cavalry Division to the 101st Airborne Division (Airmobile). It was not a solution that appeased the hard-liners, but it was enough of a compromise to allow everyone to continue to wear the same screaming eagle patch that had been worn throughout the division's storied history. In his announcement, MG Zais, the division commander, who had been airborne since World War II, wrote, "I know I share the great pride and happiness of each trooper . . . and recognize the responsibility to keep the high reputation and standards represented by the redesignation as Airborne."[453]

FIRE SUPPORT BASE (FSB) LELA

At 0630 hours on June 18 the battalion received a mission to make a combat assault approximately five kilometers southwest of Trang Bang, in the vicinity of an abandoned fire support base (FSB Lela). C Company would be the first wave into the LZ, and Tiger Bravo would follow. Once the area was secure, battalion headquarters and A Company would move by vehicle on dirt roads leading to the FSB. Once on the ground, the battalion would reconstruct the FSB and use it as a base of operations for company-size search and destroy missions to the south and west, as far as the Song Trang Bang[454] (Trang Bang River).

The LZ turned out to be cold. Quickly, C and B Companies fanned out to clear the FSB and five hundred meters around it in every direction. At 1330 hours the convoy arrived, and reconstruction began. While daylight lasted, there was a flurry of activity. A Company left on a search and destroy mission. C and B Companies started to rebuild their portions of the two-company perimeter, at the same time sending security patrols wading through the muddy, swampy countryside to guard against an attack at a very vulnerable time. In the center of the FSB, the battalion's Tactical Operations Center (TOC) was set up with mortar positions established for close-in fire support. Seeing the extent of repairs that the battalion had to make, the battalion's S4 (logistics officer) called in an emergency requisition for ten rolls of concertina wire, five hundred iron pickets, and five boxes of sandbags.[455]

Concertina wire was critical to the defense of the FSB. It needed to be installed first. Sometimes called razor wire, because it was much sharper than the barbed wire used to keep cattle from wandering away, it was employed in every American fortified position in Vietnam. The wire came in coils that, when expanded out to full length, could protect fifty feet of perimeter. Once staked in place by metal pickets, it was the first line of defense against enemy ground attackers. To make it more formidable, it would be laced with trip flares, booby traps, and claymore mines to explode in the face of enemy infantry trying to breach the perimeter. Time permitting, most units would string triple concertina—two parallel strands of concertina with a third strung on top, all tied together.

A concertina wire perimeter was not impenetrable by any means. Attackers would use bangalore torpedoes—long pipes filled with explosives—to create lanes for the waiting assault force to swarm through. Then there were the enemy sappers. These were elite troops, capable of defeating any concertina barriers put in their path (or so it seemed). The term "sapper" has been used for more than two hundred years in Western armies to denote a military engineer who built fortifications, worked with mines and explosives, or used his skills to undermine or breach enemy defenses. There were three types of enemy sapper units in Vietnam: urban (intelligence- gathering, terrorism, and assassination), naval (attacking shipping, bridges and US/ARVN bases near water), and field (breaching American defenses as the vanguard of a ground attack or carrying out independent sapper raids to destroy American headquarters, equipment, or materials[456]).

It was the field variety that confronted Tiger Bravo. These were highly trained, skilled, and dedicated shock troops, undergoing up to six months of specialized training in the art of infiltration and explosives. Under cover of a mortar attack, and dressed only in loincloth or nothing at all, they would pick their way through American defenses to hurl explosive satchel charges into bunkers or foxholes. Usually there was no warning. Sappers could bypass or neutralize any trip flares or booby traps in their path. The secret to defending against sappers was counterintuitive. When the mortars started falling, one needed to overcome the instinctual reaction to seek cover or curl up in the bottom of a foxhole. Instead, it was a time to illuminate the perimeter with flares, or use the scarce starlight scope, to detect the sappers and engage using everything available—small arms, claymore mines, fougasse, or beehive rounds from the infantry's 90mm recoilless rifles or the FSB artillery.

It wasn't long before the local VC tested their new neighbors. One of the C Company patrols, heading to its ambush position for the night, made contact with ten VC, killing two. Long after dark, three VC came across a paddy dike, headed directly at the C Company side of the FSB. They too were killed outright. Then the VC mortars started. Twenty-five rounds exploded in the C Company sector of the FSB, but all landed just outside the barbed wire, causing no friendly casualties.[457]

The next night, on June 19, the enemy mortar crews had relocated, readjusted their aim, and targeted the opposite side of the FSB. It was Tiger Bravo's turn to be mortared in the dark. Thirty-one mortar rounds came raining down on Bravo Company in five minutes, sending everyone diving for the bunkers. This attack was like the other one; every round missed, falling outside Bravo's perimeter![458] Two nights in a row, the normally accurate VC gunners had completely missed two hundred Americans concentrated in an FSB no more than a hundred meters across. It was atypical from what Tiger Bravo had faced before.

On June 20, I rejoined Tiger Bravo as the commander, with a sore back and the realization that I was mortal. It could happen again, and the next time I might not be so lucky. For the next five days, the battalion launched search and destroy operations from FSB Lela, with little results other than C Company sinking a sampan in the river that was carrying two VC, with a LAW (a single-shot, shoulder-fired light antitank weapon). On one of those days Tiger Bravo was paired with a company of ARVN Rangers.[459] The days passed like every other day had in the Rocket Belt except in this area there was water everywhere—a flat, muddy, shallow lake interspersed with paddy dikes and bamboo thickets. At one point, when the men needed a break from the exhausting effort in hundred-degree temperatures, I stopped the company and sent out patrols in every direction, with one simple order: find dry land where the men could take their boots off. There was none. We spent the night in knee-deep water with only tree roots and a paddy dike to rest on.

On June 25 operations under the 25th Infantry Division came to a halt and the 2/506 Battalion was returned to 3rd Brigade control. The battalion moved to Phuoc Vinh for a much-needed period of base camp defense, which included security for a strategic bridge over the Song Be River, a few miles south.[460] For the first time since it began combat operations in January, the battalion was all together and had time to assimilate replacements. The break let us catch up on needed training and give some respite to worn-out rifle companies that had been in harm's way for six straight months. Leaders studied lessons contained in *The Vietnam Primer*, a study of company operations in Vietnam by the historian S. L. A. Marshall. There was also time for the never-ending bunker repairs, guard duty, local patrolling and ambushes, and work details around camp. But, there was also time for drinking beer in the battalion clubs or in the bars of Phuoc Vinh

SONG BE BRIDGE

Periodically, companies in the battalion would have a respite from the constant strain of search and destroy missions in the Rocket Belt. For a few days every month or so, companies would be assigned to guard the bridge over the Song Be. From July 2 to 6, it was Tiger Bravo's turn for a break.[461]

The bridge was a vital link in the constant flow of supplies from Bien Hoa/ Long Binh, north on Highway 1A to Phuoc Vinh. Across its two-hundred-meter span flowed daily convoys loaded with everything the forces at Phuoc Vinh needed to fight the war. Surprisingly, although always subject to attack, especially considering its high value to the logistical effort, it was in a relatively quiet sector, and rarely saw any sign of the enemy. Duty at the bridge was easy. Soldiers guarded the bridge from bunkers on both sides of the river but did little in the way of population control of the Vietnamese civilian traffic flowing up and down 1A. Each night there were close-in, ambush patrols sent out, more for early warning than anything else. Days were monotonous, boring, and uneventful. It was a time to rejuvenate oneself, catch up on the news from home, and unwind from the stress of combat. Only one trooper, Daniel Kennedy, was evacuated, and that was for a high fever, something that occurred quite regularly for all of us.[462] A fever of unknown origin (FOUO) could strike with little warning and would quickly spike to a dangerous level.

The camp was a small, company-size firebase, with mortars in the center instead of artillery. The perimeter had the usual concertina wire and bunkers for defense, plus a substantial number of claymore mines and trip flares protecting in all directions. The troops made poncho awnings on the bunkers or combined several to make squad-size hooches to beat the sun. Hot chow was brought every day from the company mess hall in Phuoc Vinh, which was just a few miles north. All in all, there was little out of the ordinary about the small camp. Every day troops swam in the river, played horseshoes or cards, wrote letters home, and drank warm Vietnamese beer. Sometimes it was Ba Muoi Ba 33 Beer, brewed as a golden lager in Da Nang. Most of the time the only variety available on the bridge was a local brew, known as Tiger Beer—or Tiger Piss—by anyone who drank it. Each day, Vietnamese would show up at the bridge with a crate of Coca-Cola or beer lashed to the back of a motor scooter.

For a few piasters, you could get a one-liter bottle of Tiger Piss, with rolled-up newspaper stuck in the neck as a cork and sediment floating in the beer. It was bitter, probably had formaldehyde added in the brewing process, but it was beer! The only things missing were women, and that was remedied for some soldiers by the prostitutes who set up shop in a bunker close to Highway 1A. For $5 MPC, which would go to the first sergeant's beer and steak fund, Dave Spencer would provide a condom courtesy of First Sergeant Trent.

It paid to have a well-connected company first sergeant who could occasionally arrange for a few luxuries to be smuggled onto the daily logistics helicopter that was a company's lifeline in the field. Trent was a master at working his connections at battalion headquarters. On one particularly hot day, the late-afternoon helicopter arrived at Tiger Bravo's NDP with the usual load of water, chow, ammunition, and mail. In just seconds the waiting soldiers had unloaded the supplies, and were crouched off to the side and away from the whirring blades cutting the air a few feet above their heads, when out of the open door flew two extra mailbags. As the chopper lifted off, Trent yelled to no one in particular, "Bring those two mailbags to the CP." Two soldiers promptly complied and dropped two bulging, dripping-wet mailbags at Trent's feet. One split open and spilled cans of cold Ballantine and Pabst Blue Ribbon beer and ice all over the ground. Somehow, Trent had accomplished three seemingly impossible feats, all while in the jungle, in the middle of VC territory—find enough beer to fill two mailbags, chill it with ice, which rarely made it out of established base camps, and smuggle it onto the exact chopper heading to Tiger Bravo. But he did it, and that's how cold beer was served in in the middle of nowhere.

Combat seemed far away from the bridge. Guards were posted at all times around the perimeter, and patrols went out each night, but little happened. The only excitement was self-inflicted, and had nothing to do with enemy contact. One night on an ambush led by Second Lieutenant Jim Roach, he decided to crawl through the jungle to check on his men. Not knowing it was his platoon leader, and conditioned by months of combat and dozens of night ambushes to react to any movement, Eugene "Thump" Davis, who rarely missed, turned and fired an M79 beehive round at the crouching figure behind him. Fortunately for Roach, the beehive round was fired at such close range that the blast did not have time to completely expand into a wide, lethal pattern. Roach came away with only the front of his jungle fatigues shredded, along with his dignity. "Thump Davis was punished by the company commander, who had announced that anyone who missed a target within five meters would answer to him. For his punishment, he had to buy beer for the whole company," remembered Roach. "And then he apologized to me for missing."[464]

BACK IN THE FIGHT

Too soon, the downtime away from the supervision from higher headquarters, and the constant presence of the VC, were over. But the respite continued. Tiger Bravo spent the next seven days on base camp defense at Phuoc Vinh,[465] sleeping on cots in platoon billets and enjoying hot meals in the mess hall. On July 14, I transferred command of Bravo Company to another infantry captain, as I prepared to go on seven days of R & R in Hawaii. Counting my XO time, I had been with Tiger Bravo for seven months. It was time for me to take a break.

There were two internal clocks inside us all. The first was the number of days until our tour was up. That magical day when we could leave it all behind, hopefully never to return. The second was the countdown to a week of rest and recreation, or R & R for short, in a far-off corner of the Pacific Rim. Whether it was nonstop drinking and chasing women, sightseeing, eating in upscale restaurants, a reunion with a spouse, or simply enjoying a hot shower in a clean hotel, escape from the war zone went by the name of Romeo and Romeo. Everyone in the war zone received a week's leave and free air travel from Tan Son Nhut Air Base near Saigon to Manila (Philippines), Sydney (Australia), Tokyo (Japan), Kuala Lumpur (Malaysia) or one of a dozen other exotic cities on the Pacific Rim. But the number one destination was Honolulu, Hawaii. By 1968, 25 percent of all R & R junkets were to Honolulu (Hawaii), where soldiers reunited with wives and families, stepping away from the war for a few precious days.

The very next day after I left, Tiger Bravo was back in the fight. While the rest of the battalion remained at Phuoc Vinh, the company was lifted by helicopter to Cu Chi, where it was placed under the operational control of the 1/506. That unit had assembled in Cu Chi in preparation for deployment into Area of Operation (AO) STREAM, northwest of Cu Chi. Intelligence officers reported that the southern portion of this AO was a known sanctuary for Viet Cong forces, and some type of contact, ranging from sniper fire to platoon size, was a daily occurrence.[466] For once, the intelligence reports were correct. Just outside of Cu Chi, Tiger Bravo made a CA into a cold LZ, but had one soldier wounded by a booby trap. The following day it was more of the same. Another booby trap, another wounded in action. July 17 was quiet; the company moved all day on a search and destroy mission but made no enemy contact. The next day and night brought almost constant contact for Tiger Bravo. During daylight, the company made contact with an enemy shooting small arms. That night Bravo's NDP was attacked by an unknown force of VC employing RPGs and small arms, resulting in two wounded. The next night Tiger Bravo shot one of its own. A soldier wandered outside the NDP perimeter, and when he failed to acknowledge a challenge from the troops on guard, he was shot.[467]

On July 19 Bravo found a newly constructed, battalion-size base camp of nearly two hundred bunkers along the bank of a river. No contact was made; the enemy had vacated the complex just hours before. On July 20, the company made another CA. This time it was the pickup zone (PZ) that was hot. Two helicopters were hit by ground fire, but no one from Bravo Company was wounded.[468]

A long, hot, bloody summer had begun. Little did anyone know that Tiger Bravo's worst day in combat was just forty-eight hours away. It would be deadlier than battling an entrenched enemy in War Zone D, street fighting in Bien Hoa during Tet, or marching against NVA regulars in the mountains of the Central Highlands. Everyone, from the novice company commander on down, would be pushed to their limits and tested as never before.

Tiger Bravo officers (kneeling Nick Hubble, 1st row on left Jim Roach & 3rd from left Joe Hillman, 2nd row 2nd from left Rick St John), Phuoc Vinh Base Camp, December 1967.

Mike Tarpley and "General", Phuoc Vinh Base Camp, with company mess hall in the background, January 1968.

2/506 Memorial Service, Phuoc Vinh Base Camp, January 1968.

Tiger Bravo soldiers, War Zone D, date unknown.

Siege of ARVN III Corps Headquarters, Bien Hoa, Tet, 1 February 1968.

Rick St John,
Rocket Belt,
February 1968.

*4th Platoon
cleaning weapons
(on left with
glasses, Mike
Scott) Song Be
Bridge, March
1968.*

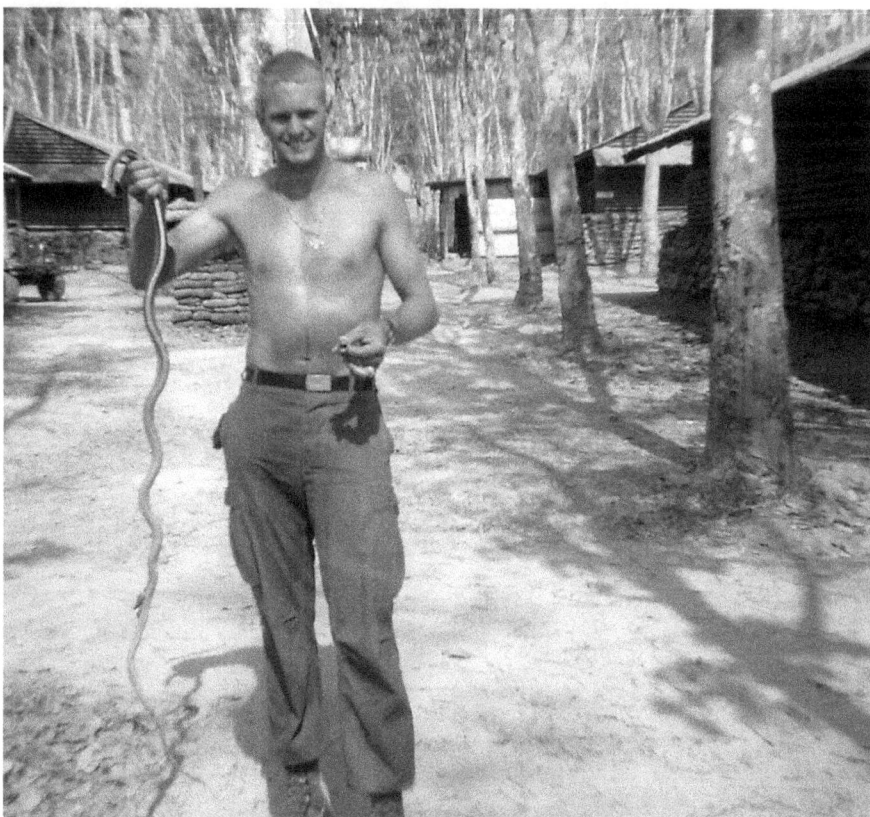

Chris Backman and "friend," Phuoc Vinh Base Camp, March 1968.

160

Herman L Trent,
in helicopter enroute to a
combat assault, April 1968.

Tiger Bravo soldiers receiving Combat Infantryman Badge (Colonel Mowery on left in
soft cap & Rick St John with rucksack & rifle on right), Rocket Belt, April 1968.

Left (Tom McClear, Dak To, May 1968). Below: Artillery Forward Observer Team (front Joe Palagyi, 2nd in line with sun glasses Chuck Limer), Rocket Belt, April 1968.

*Doc Franks
(standing on left)
and Joe Palagyi
(seated), top of
Ngok Peng Dieng
vicinity of Dak
Pek, May 1968.*

*Dan Bernard (on left) and Dave "Sparrow" Spencer (right), top of Ngok Peng Dieng
vicinity of Dak Pek, May 1968.*

Bill "Pop" Plemons, Dak To,
May 1968.

Ron Albertson, Fire Support Base Concord,
Rocket Belt, May 1968.

2/506 battalion commander (David E Grange in center with helmet) and Tiger Bravo
leaders (front row 2nd from left Dan Bernard, 4th from left Rick St John with sun
glasses, on Grange's left side Herman Trent with Bill Tellis next to Trent), Rocket Belt,
May 1968.

Harry Brown with captured ChiCom (Chinese Communist) CkC50 machine gun, Fire Support Base 25, west of Dak To, June 1968.

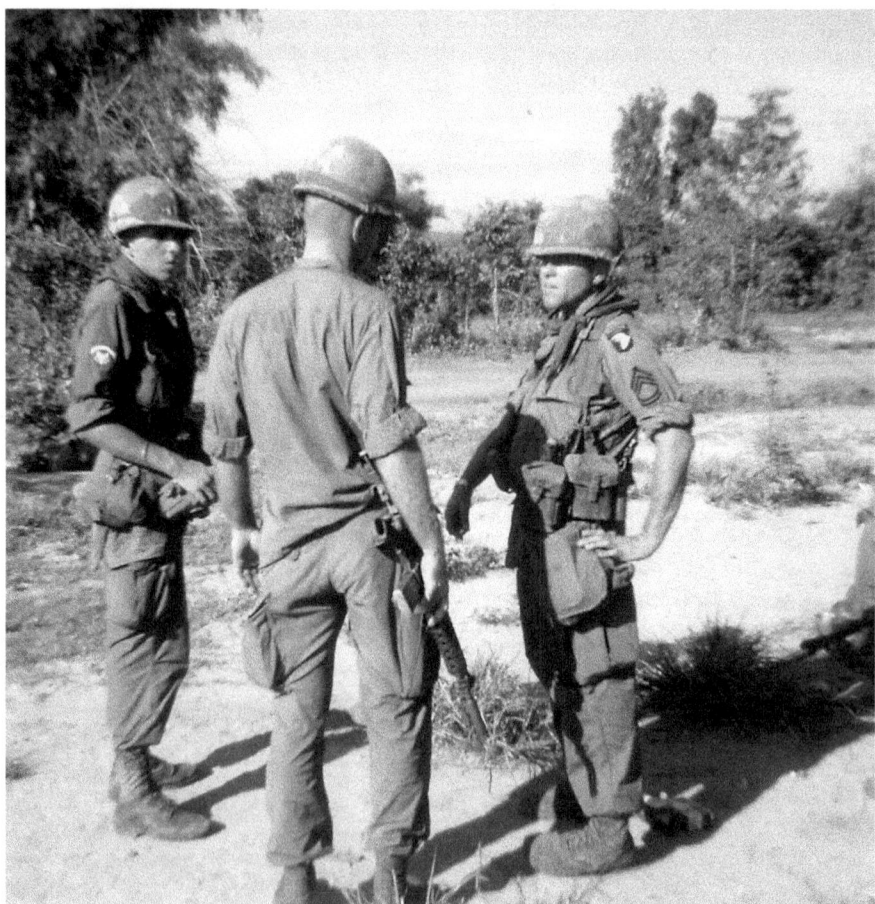

Andy "Mad Dog" Matosky (on right), vicinity of Trang Bang, June 1968.

Left: Chuck Hanson, Song Be Bridge, July 1968. Above: Paul Nabors, Airborne School, date unknown.

Terry Van Meter (left) and unknown soldier (right), vicinity Trang Bang, July 1968.

166

Medics Johnny Walker (left) and Johnnie Greeson (right), vicinity Trang Bang, July 1968.

Song Be Bridge, Highway 1A south of Phuoc Vinh, July 1968.

Chuck Kudla (on drums) and Joe Hillman (seated on right), Song Be Bridge, date unknown. Below: 2/506 Memorial Service, Phuoc Vinh Base Camp, September 1968.

Warren Kilehua,
east of
Camp Evans,
November 1968.

Tiger Bravo at the beach, Joe Adams (seated on right) and Pop Plemons (seated on left),
Eagle Beach, November 1968.

Larry Burton, Camp Evans, November 1968.

Tiger Bravo soldiers (front row left to right Basil Rivera, Ted Tilson & Bill Tellis, 2nd row center Dan Soto) location and date unknown.

Part III

A Currahee Summer

9

Numb and Falling to Pieces

Vicinity of Trang Bang
(Hau Nghia Province)
July 22–25, 1968

SOMETHING BAD

It was midafternoon on July 22. With no one around, I quietly ambled into and out of the rubber trees at Phuoc Vinh Base Bamp, headed to the Tiger Bravo Company area. There was no urgency to my step, no purpose other than to drop my kit bag and settle in after my R & R in Hawaii. It had been a blissful week in the "real world" with my infant son, who had been only six days old the last time I cradled him. My time on the line was officially up, and waiting for me was a staff job as the battalion's S2, intelligence officer.

A messenger's voice ended my reverie: "Captain St John! Captain St John! The brigade commander wants you ASAP at Cu Chi. Something is going on with Bravo Company. They're in contact, a big one." He knew little else about the situation or why I was summoned, only that a helicopter was on its way to the assault pad to pick me up. Something bad had happened; I could sense it.

Waiting for the chopper to take me to Cu Chi, a battle boiled inside me. I could sense that I would be called to take command of the company again. In *Rumors of War*, Philip Caputo wrote that "nine-tenths of the war is waiting for the other one-tenth to happen."[469] While I was waiting at the helipad, the other tenth was happening to Tiger Bravo. I felt a soldier's love for the men I had stood with under so many harrowing circumstances, but I knew that leaving Tiger Bravo afforded me the best chance for survival so I could hold my infant son again. It was tearing me apart. The father I had become battled head-to-head with the soldier in me. When the chopper landed, the father smiled, sighed, and stepped aside, while the soldier I had become squared his shoulders, threw his ruck on the chopper, and climbed aboard.

Arriving at Cu Chi, I went to see Captain Bob Sturdivant, who was my first company commander when I reported to A Company 2/506 right after graduation from West Point. He was now a staff officer in brigade operations who spent his days and many nights monitoring combat operations across the brigade. Gone was the friendly, jovial sort I remembered; this time there was no smile. Instead he grabbed both my elbows, stared straight into my eyes, and said, "B Company is in a lot of shit. The brigade commander wants to see you. You're getting Tiger Bravo back."

HOTBED OF VC

During the summer of 1968, the Allies feared an imminent, all-out assault on Saigon by enemy forces using traditional VC base areas in the district of Cu Chi as staging areas. At this point in the war, control of the Cu Chi district by Allied forces was still very much in doubt. The Viet Cong base areas had been scenes of bloody battles since 1965. One account called the district the most bombed, shelled, gassed, defoliated, and generally devastated area in the war. For years, most of the district was classified as a "free fire zone," where artillery H & I (harassing and interdicting) fires went on all night long and where pilots dropped unexpended bombs before returning to base. It has been called the "Iron Land" or "Land of Fire" in Communist accounts of the war, with most villages glorified as "heroic villages" for their part in supporting the Viet Cong and the NVA.[470]

The strategic significance of the district was self-evident: It straddled the main land and river routes into Saigon. Mai Chi, a Viet Cong political commissar in Saigon during the war, described Cu Chi "as a springboard for attacking Saigon." The troops and supplies necessary for the attacks on Saigon during the Tet Offensive were assembled in Cu Chi district and hidden in the vast tunnel system that crisscrossed the area.[471]" The underground tunnels of Cu Chi were a complex network of multilayer tunnels that stretched from the gates of

Saigon to the border with Cambodia. There were hundreds of kilometers of tunnels, connecting villages, districts, and even provinces. At the height of the war, they held living areas, storage depots, ordnance factories, hospitals, headquarters, and almost every other facility or function that could be burrowed into the earth.

These staging areas offered concealed locations, among a VC-dominated populace, where enemy commanders could concentrate far-flung units and prepare for the summer offensive in relative safety. Or so they thought. Allied intelligence reports began tracking the buildup and soon had a grim picture of a sizable force assembling near Trang Bang and in and around two long-time VC strongholds—Bo Loi Woods and Ho Bo Woods. Trang Bang was one of only three small towns in Hua Nghia Province. Many of its buildings were brick with corrugated metal or Spanish-style red tile roofs. The surrounding villages or hamlets were nothing more than a haphazard collection of farmhouses made of mud walls with roofs of palm and rice stalks. The streets and roads in the area were unpaved, made from the red earth that was found in the province.[472]

All of the enemy units and staging areas fell within the 25th Infantry Division's area of operations, which was already stretched thin across a wide swath of territory protecting Saigon. To give the 25th Infantry Division a fast-reacting, offensive punch, the Wandering Warriors were once again thrown into the breach. With its airmobile capabilities, the 3rd Brigade, 101st Airborne Division was called upon to:

"INTERDICT THESE FORCES AND DENY THEM ACCESS TO TRADITIONAL AVENUES OF APPROACH INTO SAIGON THROUGH THE CONDUCT OF BOTH DAY AND NIGHT HELIBORNE ASSAULTS INTO AREAS OF SUSPECTED ENEMY CONCENTRATION."[473]

In late July, Tiger Bravo was operating under the control of the 1/506 while conducting operations in the vicinity of Trang Bang. It went on a full week of search and destroy missions, that curious euphemism used by the army to make the infantry's primary role in South Vietnam—to hunt down the enemy and kill him, by whatever means possible—more palatable to Mom and Dad back home. For the troops on the ground, it simply meant another day of scouring a hostile land, replete with explosive mines, booby traps, and deadly snipers, in search of an enemy that was seemingly invisible until it decided to show itself. Every few days during combat operations around Cu Chi a soldier would fall into a well. They were never clearly marked, sometimes abandoned when the village was destroyed and seemed to always be directly in the path of a line of American soldiers. Lieutenant Jim Roach was particularly prone to this mishap, or at least that's what the battalion newspaper reported. The tongue-in-cheek article read in part "that it does not happen every day—but it does to Lt. Roach, who incidentally is 21 and fell in 21 different wells in the Cu Chi area, achieving a well a day.[474]

Unlike the Rocket Belt, with days and nights of little enemy contact, the area around Trang Bang was a hotbed of VC activity and presented one of the most challenging combat environments ever faced by Tiger Bravo. In addition to fighting "in the round," with the possibility of attack coming from all directions—front, flanks, and rear—the infantry commander also faced a three-dimensional threat. There were snipers overhead in trees, and VC machine gunners shooting at any American helicopters flying within range. On the ground, contacts ranged from chance encounters with a few VC to set-piece battles with multibattalion VC/NVA forces. Finally, there was the subterranean aspect to the fighting. Through an extensive network of tunnels and spider holes, the enemy could be in front, behind, or in your midst at any time, shooting or throwing hand grenades, then disappearing in seconds.[475] This was a test for even the most seasoned commander. For a new, untested company commander, such as Tiger Bravo's, it could be overwhelming.

During one five-day period during July 1968, 3rd Brigade units were in constant contact, day and night. Brigade units killed 209 VC and another 56 NVA soldiers. There were more than twenty significant contacts or attacks on NDPs, six sniper attacks, seven mortar attacks, thirteen separate booby trap detonations, numerous hot LZs and PZs, and seven helicopters hit by ground fire. Every night sightings were made of large groups of VC on the move, one with the audacity to light their way with candles. Another night one company reported a convoy of eighteen trucks in the distance, all with their headlights on. A quick check of American and ARVN units showed no friendlies in the area; it had to be enemy.[476]

Within a 10-kilometer radius of Trang Bang, 3rd Brigade companies also encountered an enemy that was firmly entrenched, with a robust, clandestine infrastructure. The units found caches of gasoline, 39 tons of rice in 100-pound bags, arms and ammunition, and homemade devices, such as one discovey of 175 hand grenades made from tin cans and Coke bottles. Every day the companies came across fighting positions and shelters for battalions of VC, all freshly made and in use. In one instance, they found 150 spider holes along a wood line; in another, it was a battalion-size bunker complex along a riverbank. There were trench lines in virtually every hedgerow bordering open areas or around fallow rice paddies. Tunnels crisscrossed the area, most going undiscovered. An underwater bridge across a major river was found while still under construction. Fully operational hidden VC classrooms and kitchens were discovered as well, including one with 100 rice bowls set out for the next meal.[477]

Tiger Bravo's first week in the field had started quietly with a CA into a cold LZ. But the pace quickly picked up and never slowed down. Bravo's new commander was experiencing this for the first time. The rapid succession of enemy contacts, all presenting different scenarios, would be a daunting time for even an experienced hand. A week was not enough time to know the habits and quirks of the enemy, to see likely enemy ambush sites through the same lens as the VC commander, or to sense when the company was being trailed by enemy scouts and counter with a change in direction or a stay-behind ambush. It was not a matter of rank or age, but of time on the line. In that sense, Bravo was no different from any other infantry company in South Vietnam. All would eventually be in the same precarious position, with a new commander trying to learn his trade in an unforgiving classroom.

Nor, was it enough time to know that Tiger Bravo's command group and senior leaders were as proficient as any in the division, and that each had distinctive skill sets that he could depend on. First Sergeant Trent was a better company-level tactician than most captains in the battalion. Joe Palagyi was a master at coordinating the army's vast array of weapons systems, akin to an air traffic controller handling dozens of aircraft in the same limited airspace.

Joe Hillman, back in command of a platoon after a stint as the C Company executive officer, was the one to lead the most difficult assault or to position at the most critical point on the battlefield. Mad Dog Matosky had an uncanny sense of direction, in even the blackest of nights. In particularly difficult cross-country movement, his platoon should lead. Dan Bernard was a second set of ears on the battalion radio net, who could summarize the battalion's battle plans and actions better than any staff officer. When Doc Franks said a soldier was critical and needed to be medevac'd now, he meant now! Not heeding his advice meant filling up another body bag. Seemingly impervious to enemy fire, Dave Spencer always had the commander's radio within arm's reach. Harold Sykes was a rock in every battle, under all circumstances. Tell him once and it is done. Bill Tellis was another stalwart to be counted on, and one of only a few nonairborne NCOs in the battalion. The troops quickly gave him the nickname of Sergeant NAP. Any nonairborne assigned to an airborne unit had the designation NAP, which stood for nonairborne personnel, by their name on the unit roster. Usually it was a derogatory term in the airborne community, often used interchangeably with "leg." But this wasn't the case with Sergeant Tellis. The troops liked and respected him, so "NAP" became more of a good-natured ribbing.

But, the situation for Tiger Bravo was not as one-sided as with most infantry companies receiving a new commander. While the new commander, Captain Terry Van Meter, had never been assigned to a line unit in combat, he had been with the 2/506 from the beginning of its tour in Vietnam. More importantly, he had already proved his mettle under fire. On 4 April 1968, while flying in a helicopter over a battle between C Company 2/506 and a VC company, Captain Van Meter was part of a team that attempted to deliver medical aid in the midst of the battle. C Company had a number of wounded soldiers, including every medic in the company, who needed treatment. Because of the dense jungle, landing a helicopter to evacuate the wounded was impossible. As the helicopter hovered over the jungle canopy, a medic was lowered to the battle site under intense enemy fire. Halfway to the ground the rope slipped and the medic plummeted towards the ground. Leaning out of the helicopter, and while totally exposed to the enemy fire, Van Meter grabbed the rope, stopped the medic's fall, and safely lowered him to the ground, saving the medics life.

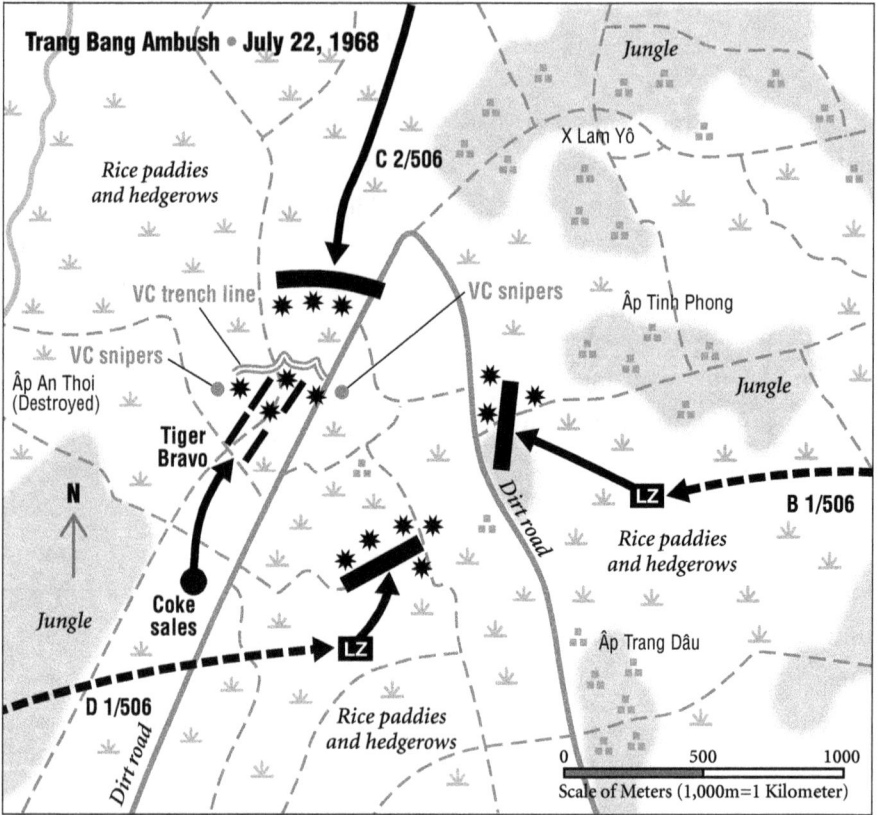

For this action, he was awarded an Air Medal with V for valor.[478] He was a tall, strapping, 1966 graduate of Norwich University who was unafraid to make tough decisions, nor one to shy away from placing himself in the line of fire to get the job done.

COMBAT IN AN OVEN

Tiger Bravo's ordeal started at midmorning on July 22, 1968, and lasted all day. It was a brutally hot summer day in this part of Vietnam. Climate records show that when the first shots were fired the temperature was well on its way toward a record heat index of 134 degrees Fahrenheit.[479] No one could escape the heat; it radiated from the ground up through your boots, slammed you in the face on what little breeze there was. So oppressive was the heat, it was hard to predict which you might succumb to first, the enemy or the heat. For the company's medics, it meant treating traumatic gunshot and shrapnel wounds while, at the same time, countering the debilitating effects of dehydration and heat exhaustion.

What initially was reported as a chance encounter with a squad of VC quickly escalated into an all-out melee between two companies—one US, caught in the open and pinned down, and one VC, dug-in, protected, and expertly camouflaged. "They had us set up good that day," recalled Larry Burton. "They probably let half the company get out into the open before they opened up."[480] By the time the battle ended, the 1/506 commander had committed three more companies into the fray, brought in air strikes and gunships to pound the VC bunkers, dropped CS gas onto the hedgerows, and fired hundreds of artillery rounds. But early on, in the most crucial part of the battle, it was only Tiger Bravo in the fight.

A 3rd Brigade battlefield report sent to the 25th Infantry Division commander summarized the US actions and the lopsided casualty count:

ON 22 JULY AT 1056 HOURS, B COMPANY 2D BATTALION, UNDER THE OPERATIONAL CONTROL OF THE 1ST BATTALION, WAS ENGAGED BY AN ESTIMATED VC PLATOON AT XT 522206. LATER, THE ESTIMATED SIZE OF THE ENEMY FORCE WAS INCREASED TO A COMPANY. VC WERE IN BUNKERS AND SPIDER HOLES AND ARMED WITH AT LEAST TWO MACHINE GUNS, SMALL ARMS, AUTOMATIC WEAPONS, RPG'S AND MORTARS. B COMPANY RETURNED FIRE WITH ALL ORGANIC WEAPONS AND WAS SUPPORTED BY LIGHT FIRE TEAMS, ARTILLERY AND AIRSTRIKES. D COMPANY 1ST BATTALION WAS INSERTED BY AIR TO THE SOUTHWEST OF THE CONTACT IN EARLY AFTERNOON. B COMPANY WAS INSERTED LATER IN THE AFTERNOON TO COMPLETE THE SEAL OF THE CONTACT AREA, AS C COMPANY 2D BATTALION, UNDER CONTROL OF THE 1ST BATTALION, HAD MOVED TO THE NORTH OF THE CONTACT AND ESTABLISHED A BLOCKING POSITION. CONTACT CONTINUED UNTIL 1955 HOURS WITH THE 1ST BATTALION BEING SUPPORTED BY ARTILLERY, LIGHT FIRE TEAMS, CS DROPS, AND 6 AIRSTRIKES. THERE WERE 11 US PERSONNEL KILLED AND 23 WOUNDED, ALL OF WHICH WERE EVACUATED. ENEMY LOSSES WERE 6 KILLED BY BODY COUNT.[481]

CS was a riot control agent widely used in the Vietnam War to disrupt enemy attacks; flush them out of tunnels and bunkers; and, in powder form, to keep the enemy from reoccupying cleared areas. Far more powerful than its weak cousin tear gas, it caused a nearly instantaneous reaction when contacted—irritation of the eyes, tearing, sneezing, coughing, mucous secretions, and a strong irritation of any exposed skin. Many soldiers carried CS grenades along with the standard fragmentation grenades, or "frags." Its use came under fire by some who considered it a chemical weapon, even though riot control agents were not technically part of the Geneva Convention ban on the use of chemical weapons in combat. When questioned at a news conference about this opposition, President Johnson was quoted as saying, "I just wish they were as concerned with our soldiers who are dying as they are with someone's eyes who watered just a little bit."[482] None of us on the ground ever considered the legalities of its use, trusting that the army would not provide us with banned weapons. Our criterion for its employment was simple: if it saves lives, then do it!

The 2/506 Daily Staff Journal entries chronicled what happened next:

DAY 1 JULY 22

- 0900 HOURS TIGER BRAVO STOPS JUST OUTSIDE A VILLAGE, ON THE EDGE OF AN OPEN AREA WITH HEDGEROWS ON ALL SIDES. TWO VIETNAMESE WOMEN APPEAR TO SELL COKES TO THE TROOPS. ORDER RECEIVED TO CHANGE DIRECTION AND CROSS THE OPEN AREA.
- 1000–1030 HOURS (APPROXIMATE) BRAVO STARTS TO MOVE ACROSS THE OPEN AREA, 3RD PLATOON (JOE HILLMAN) IN THE LEAD.
- 1056 HOURS TIGER BRAVO REPORTS CONTACT WITH AN ESTIMATED VC SQUAD ARMED WITH 2 MACHINE GUNS, SMALL ARMS AND RPG'S.
- 1120 HOURS AIR STRIKES COMMENCE. WILL CONTINUE FOR THE NEXT 6 HOURS.
- 1230 HOURS BRAVO REPORTS 9 WIA, CONCENTRATED IN THE LEAD TWO PLATOONS WHICH ARE PINNED DOWN IN THE OPEN.

One eyewitness remembered, "It was a duck shoot. Half the company was caught in the open. There was only one tree in the middle of the rice paddy. All they had for cover was their rucksack. Anytime someone moved, they were dead."[483]

- 1310 HOURS ENEMY FORCE NOW AT COMPANY-SIZE FIRING FROM BUNKERS AND A TRENCH LINE IN THE HEDGEROW ON FAR SIDE OF THE OPEN AREA. BRAVO REPORTS 1 KIA, AND SOME OF THE WIA WERE RESULT OF A FRIENDLY, ERRANT, WP (WHITE PHOSPHOROUS) MARKING ROUND FIRED BY A USAF FORWARD AIR CONTROLLER FLYING OVERHEAD TO DESIGNATE TARGETS FOR THE INBOUND CLOSE AIR SUPPORT AIRCRAFT. 1 WIA EVACUATED, OTHER 8 STILL PINNED DOWN IN THE OPEN AREA.
- 1433 HOURS BRAVO RECEIVES EMERGENCY RESUPPLY; 1 WIA LITTER CASE EVACUATED. 1/506 BEGINS PILE ON SUPPORTING COMPANIES AROUND CONTACT AREA. D 1/506 INSERTED S OF CONTACT AREA INTO COLD LZ; 1 SHIP RECEIVES FIRE WHEN PULLING OUT.
- 1530 HOURS GUNSHIPS MAKING STRAFING RUNS.
- 1605 HOURS SPECIAL MEDICAL TEAM LANDS AT TIGER BRAVO LOCATION; 3RD WIA EVACUATED.
- 1640 HOURS 1/506 BATTALION COMMANDER ON THE GROUND; CONTROLLING BATTLE FROM 1/506 RECON PLATOON POSITION JUST SE OF TIGER BRAVO.
- 1657 HOURS D 1/506 IN CONTACT; HAS 8 WIA. 4 MORE WIA EVACUATED FROM TIGER BRAVO.
- 1810 HOURS B 1/506 INSERTED E OF CONTACT AREA INTO HOT LZ. LINKS UP WITH D 1/506.

After dark, I flew with the brigade commander to the battle site to take the company back, but the company was still in contact and everything flying was being targeted by machine-gun fire, so, despite my urging to drop me off several hundred meters outside the perimeter so I could snake my way back to the company, the ship returned to Cu Chi. On the ground, the fight continued.

- 1910 HOURS B 1/506 RECEIVES SNIPER FIRE.
- 1915 HOURS TIGER BRAVO & B 1/506 STILL RECEIVING SPORADIC FIRE. 1/506 UPDATES CASUALTY REPORT; TIGER BRAVO NOW HAS 9 KIA, 10 WIA & 1 MIA. D 1/506 8 WIA (6 BY FRIENDLY AIR STRIKE).
- 1935 HOURS B 1/506 REPORTS 4 WIA (3 FROM ENEMY FIRE & 1 FROM FRIENDLY M79). C 2/506 MOVES S FROM DAY POSITION TO BLOCK N OF CONTACT AREA. D 1/506 HIT BY SNIPER FIRE.
- 1955 HOURS BATTLE IS OVER. NO FURTHER FIRING REPORTED IN CONTACT AREA.

There was no sleep that night for Tiger Bravo. The company had been decimated. Two platoons were combat-ineffective, and dead and wounded still lay in the kill zone. What was left of the company set up a defensive perimeter in the hedgerow just outside the kill zone, waiting for the VC attack that would surely come. I wrapped up in my poncho liner in a barracks next to the Brigade TOC and waited for a predawn flight to rejoin Tiger Bravo—or what was left of it.

DAY II JULY 23

- 0000–0500 HOURS SURROUNDING COMPANIES REPORT SPORADIC MOVEMENT, BUT NO CONTACT.
- 0944 HOURS ALL COMPANIES RESUPPLIED. TIGER BRAVO REPORTS ONE ADDITIONAL WIA LEFT OFF DAY I CASUALTY LIST.
- 1035–1210 HOURS B, D AND RECON PLATOON 1/506 SWEEP CONTACT AREA. NO ENEMY FOUND. BUNKERS DISCOVERED TO BE CONCRETE, ABLE TO WITHSTAND AIR STRIKES. NEARBY VILLAGE HAS 19 CONCRETE BUNKERS INSIDE HUTS, CONNECTED BY TUNNELS.
- 1120 HOURS TIGER BRAVO DECLARED COMBAT INEFFECTIVE DUE TO CASUALTIES.
- 1255–1400 HOURS B 1/506 IN CONTACT AGAIN.

- 1300 HOURS TIGER BRAVO MOVES ONLINE INTO THE KILL ZONE AND
  RECOVERS ITS KIA'S AND THE 1 MIA (LT HILLMAN), WHOSE STATUS IS
  CHANGED TO KIA.
- 1820–1837 HOURS TIGER BRAVO MOVED BY HELICOPTERS TO CU CHI
  FOR STANDDOWN TO RECOVER.
- 2035–2135 HOURS B COMPANY & RECON PLATOON 1/506 REPORT MORE
  CONTACT.
- 2340 HOURS CO 1/506TH SUMMARIZES CASUALTIES: TIGER BRAVO 10 KIA,
  11 WIA & 2 NON-BATTLE CASUALTIES, B 1/506TH 1 KIA & 4 WIA,
  D 1/506TH 8 WIA.

VIEW FROM A VC BUNKER

I have been unable to uncover any VC accounts of this battle, but assuming that standard NVA/VC tactics and techniques were employed, it is not hard to envision the battle from the enemy's perspective. Following standard VC tactics, the VC commander had Tiger Bravo under observation for several days. Bravo's NDP was identified, its strength counted down to the last man, and in the morning its direction of movement was tracked by scouts or local villagers enlisted by the VC. The fact that Tiger Bravo stopped in broad daylight the day before to set up its NDP was an intelligence windfall for the VC. Larry Burton, who walked point the day before, knew this was a security breach. "The way [St. John] had always operated before was to find a place to setup the perimeter in the daytime, then go past it," he recalled. "About dark, we would backtrack to that point, dig our foxholes, and set up LPs and ambushes. That way they didn't know where we were. But that day [the new commander] stopped us in the middle of the afternoon and set up a defense."[484]

Based on a steady stream of information from scouts and informants, and relying on the fact that his forces were on familiar terrain and that they had been preparing for months for just such an excursion by the Americans, the commander decided that he could meet the basic considerations for any NVA/VC ambush to be set in motion. First, he could use prepared battle plans and standard operating procedures that his forces practiced over and over, just waiting for the right opportunity. Second, he could choose the terrain, where the Americans were exposed and his forces had cover and concealment from American artillery and airpower. Last, his forces could fight from prepared positions, in the bunkers and trench lines that infested the area.

His first move was to locate observation posts forward of the ambush site to detect Bravo as early as possible, so he could have his forces deployed and ready when the company entered the kill zone. In fact, there was a high probability that one or more of the women selling Cokes the morning of the ambush was a VC plant, part of the local network of VC supporters. Through runners or radio, the VC commander knew when Bravo stopped just short of the open area that he had carefully prepared as his kill zone. By this time the ambush force had slipped into the trench line and bunkers in the hedgerow on the far side of the open area, quietly waiting for the signal. His snipers were deployed in the trench line and in concealed positions on each flank to not only fire into the kill zone, but also to delay any American reinforcements trying to outflank the ambush.

Experience has shown that the VC frequently employed snipers in three-man teams, using mutually supporting positions in a triangular configuration, with approximately fifty meters to a side. They normally would fire a series of well-aimed shots, then cease fire and wait for more targets or move to another location.[485] Larry Burton remembered the sniper fire coming from different directions. "I'm not sure they had that many snipers," he recalled. "They had a trench dug and moved from one end of it to the other."[486] Knowing that American doctrine called for any units not pinned down to maneuver around the flanks, the VC commander also had mines and booby traps placed in front of the sniper screen in the wooded areas. Finally, he established his command post in a central location on terrain that afforded him a vantage point overlooking the ambush site, most probably in one of the bunkers hidden in the hedgerow.

One could also surmise that he briefed his subordinates to be patient, to wait until the Americans were almost up to their positions before firing. Having fought the Americans for years, he knew that the only way to negate the US advantage in artillery and air support was to engage at very close ranges, often ten to twenty meters, because the Americans were reluctant to call in their devastating firepower so close to their own troops. This way, the fight would be between two light infantry units, with the advantage going to the ones dug in and concealed. As Larry Burton recalled, "One guy got within feet of the hedgerow when they opened up."[487]

Snipers were trained to initially shoot at American leaders and radio operators. Then they would hold fire, stay concealed, and wait for other soldiers to come to the aid of the wounded—then take them out as well. The VC knew that this characteristic of American infantrymen, while a noble warrior trait, meant that more targets would be moving into the kill zone. Basil Rivera, squad leader in the 1st Platoon, which was the trail platoon about a hundred meters back, remembered it starting with a sniper. "All of a sudden, it was only a few shots. But this guy is smart. He gets one. A guy goes to his rescue, he hits him. Another goes to the rescue and he hits him."[488]

The VC commander knew that the American reaction would be swift, and would include helicopter assaults around his position to try to box him in, so his plan to end the engagement was equally detailed. He could identify likely landing sites for helicopters as well as any American staff officer, so he placed more scouts to keep them under observation with orders to engage the helicopters with harassing fire. This would slow down the American reinforcements and give his main body time to withdraw. In keeping with the principle of one slow and four quick steps to any offensive operation, his plan ended with a quick policing of the battlefield, carrying his killed and wounded and their weapons to safety. On his order, the main ambush force would make a rapid withdrawal along prearranged and concealed routes to designated rendezvous points, where it would break down into smaller groups to continue their dispersal. At this point he would leave the field of battle as well, leaving a small stay-behind force to confuse the Americans and delay any pursuit.[489]

BRAVO FIGHTS BACK

Even though the company had stalled, with the two lead platoons pinned down and cut to pieces, it was not all one-sided. Tiger Bravo soldiers fought back, although in fits and starts. All with only one purpose: save their buddies. By this time in the tour, the bond of friendship had grown far beyond the ability of anyone to explain it to those who had not experienced it themselves. As Matti Friedman explained in *Pumpkin Flowers: A Soldier's Story*, "Armies planned it this way, knowing the strength of this bond is what will keep men together and functioning in the lawless netherworld of war and, when the time comes, cause them to commit the unreasonable act of following each other not away from enemy fire but into it."[490]

Setting the example, as he had done so many times before, was Lieutenant Joe Hillman. "He was right smack in the middle of it," Dan Bernard recalled. "He was up on his knees throwing hand grenades, then he caught one in the head. Not a pretty sight."[491] Pigheaded but fair, he was a favorite among Tiger Bravo's noncommissioned officers. "I loved Lieutenant Hillman," remarked Basil Rivera, who had been an NCO in Hillman's platoon from the beginning. "One thing he would do, every time we would go out on an ambush or operation; he would always tell us who we were fighting . . . this or that famous outfit." Rivera remembered giving Hillman some cherries that someone had scrounged just before the firing started. "You know," said Rivera, "when I touched his hand I get this shock, like something is going to happen. I didn't think about it then, but now I do. He didn't say anything, just a little chuckle. That little laugh he always did."[492]

Basil Rivera and the 1st Platoon were in the rear of the company formation and protected by a hedgerow from the enemy fire. "I hear the shots and take off to the right," Rivera recounted. "All of 1st Platoon was fine, so we ran to the right parallel to the hedgerow." Instinctively, he knew not to rush into the kill zone. But there was no cover or concealment on the right flank, just more of the same flat, open area covered with low grass—not enough to conceal even a squad. As usual, when given time to plan and prepare, the enemy had selected near perfect terrain for the ambush. Nevertheless, Rivera tried—the first of three attempts for him that day. "I had the whole damn squad," Rivera recalled, "I could have gone around the right end. But there was no cover." So, Rivera left his squad in the shelter of the hedgerow and crawled forward, slowly inching across the open ground. "I stopped out in the open and tried the helmet trick. I put my helmet on a stick and held it up just above my head. POW!" The bullet struck a branch just inches above the helmet. "It was one shot. They almost got me."[493] This stopped Rivera cold.

Wriggling back to his squad, he met up with his platoon leader, Sergeant First Class Bill Tellis, who had been told to get an emergency supply of smoke grenades to the troops pinned down in the open. A plan had been put in motion by the 1/506 commander to blanket the far hedgerow with a smoke screen, then call in close air support to make repeated bombing runs. This would create a window of time for the troops in the open to pull back while the VC were blinded and sheltered in their bunkers. For the plan to work, without bombing our own soldiers, the troops in the front had to mark their locations with smoke grenades. But they had used up the last of their smoke grenades and the aircraft were already inbound, only minutes away. After explaining the situation, Rivera remembered Tellis asking, "Can you get all these smoke grenades up front? Take one man with you." Rivera remembered thinking, "Oh my gosh! I've got to go back through that area again and I don't have any cover."

But there was a way. Farther to the right was a fallow rice paddy with waist-high grass that just might conceal two men slithering along on their stomachs. "There were four or five men from my squad with Tellis. When he said to take one man, everybody turned their back on me. It made me so mad! But there was one man. He was a brave one. A new guy named Robert Rohn. All the others turned away but he said, 'I'll go, Sarge.'" Rivera and Rohn made it in time and delivered two bags of smoke grenades.[494] The plan was only partially successful. A few able-bodied, trapped soldiers made it out, but the seriously wounded remained.

Another one taking the fight to the NVA was First Sergeant Trent. The old warhorse instinctively moved to the sounds of battle. All through the afternoon he waged his own fight, at one point actually occupying one of the enemy's bunkers. The citation for the Distinguished Service Cross awarded to Trent for his actions that day tells the story:

AS HIS UNIT WAS CROSSING A RICE PADDY, IT CAME UNDER HEAVY FIRE FROM WELL ENTRENCHED NORTH VIETNAMESE TROOPS. SERGEANT TRENT IMMEDIATELY ORGANIZED A PLATOON AND BEGAN RETURNING FIRE ON THE

ENEMY POSITIONS. REALIZING THAT THE MACHINE GUN POSITIONS WOULD
HAVE TO BE DESTROYED BEFORE THEY COULD ADVANCE, HE ORDERED THE
PLATOON TO FALL BACK WITH THE WOUNDED AND REGROUP WITH THE MAIN
BODY OF THE COMPANY. REMAINING BEHIND, SERGEANT TRENT THEN MOVED
THROUGH THE FUSILLADE WITH THREE OTHER MEN AND ANNIHILATED SEVERAL
ENEMY POSITIONS WITH HAND GRENADES. USING HIS RADIO, HE CALLED IN
AIR STRIKES FROM A SITE LESS THAN FIFTY METERS FROM THE TARGETS. HE
THEN ENTERED A DESTROYED HOSTILE BUNKER AND REMAINED IN IT FOR
SIX HOURS, DIRECTING THE ORDNANCE NEARLY ON TOP OF HIS POSITION.
WHEN HE NOTICED THAT SOME OF THE BETTER CAMOUFLAGED EMPLACEMENTS
REMAINED UNTOUCHED BY THE AIR STRIKES, HE CRAWLED THROUGH A NEARBY
HEDGEROW AND DOWN THE LINE OF ENEMY BUNKERS KILLING THREE SNIPERS.
AS DARKNESS FELL THE NORTH VIETNAMESE FIRE CEASED. RETURNING TO
THE RICE PADDY, HE DISCOVERED A MEMBER OF THE COMPANY WHO WAS
SERIOUSLY WOUNDED AND CARRIED HIM MORE THAN FOUR HUNDRED METERS
TO THE UNIT'S NIGHT DEFENSIVE POSITION.[495]

Midway through the afternoon, Trent crawled back to the soldiers crouched behind the hedgerow just on the edge of the kill zone and "called for volunteers to go out and carry ammo to the guys still in the rice paddy. They were running out of bullets and couldn't back out."[496] Larry Burton recalled hearing the screams and yelling of his buddies still pinned down. "I just couldn't take it anymore, so I volunteered. Three of us volunteered." But they didn't go far. There was no cover or concealment and snipers were shooting anything that moved. "We got only twenty feet past the hedgerow and the guy behind me got shot above his knee. So, I put a tourniquet on him and carried him back, fireman's carry." The other soldier refused to go back out, but not Burton. His buddies were still in trouble, so he once again crawled into the kill zone. "I didn't make it more than a hundred feet before they got me. I was throwing M-16 clips when I was hit. I went down, got up, and they shot me again in the leg and arm. I passed out for the rest of the day."[497] Burton lay in the kill zone until dark, when a rescue party crawled into the kill zone and pulled him to safety.

By this time three more companies had been thrown into the fight. B Company 1/506 came from the south, D Company 1/506 from the east, and C Company 2/506 moved in from the north. Back at Phuoc Vinh, the Tiger Bravo rear detachment was hatching its own rescue scheme. Joe Adams had just been released from the hospital and was at Phuoc Vinh, awaiting a helicopter ride out to the company, when the ambush was sprung. "We had the radio on," he remembered. "We were listening, so three or four of us grabbed our gear." They had no real plan in mind, just to find a helicopter that would fly them to the battle site so they could help their buddies. "When we got to the helipad, someone came over in a jeep and asked where we thought we were going. He told us that we couldn't go and sent us back."[498]

With darkness came opportunity, a lull in the fighting. Now, small teams could crawl into the kill zone and look for dead and wounded. While there had been no firing since 1955 hours, there were no signs that the enemy had pulled back. Since no American unit had been able to clear the bunkers and trench line in the far hedgerow, it was assumed that the enemy was still there, waiting for the Americans to do just as they planned—go into the kill zone and recover casualties.

Rivera led one of those teams. It would be his third foray into the kill zone. "Sergeant Tellis asked me if I would go. He didn't tell me, he asked me. I am tired. I am exhausted, but I say, 'Yeah, I'll go.' So, I lightened my load, took 11 clips in bandoleers and my grenades," Rivera recalled. "I go up to the CP and there's the CO (Captain Van Meter) and Brown, who's telling us where the bodies are and that someone had left a 90mm recoilless rifle in the kill zone too. He knows exactly where two of the bodies are. That made me wonder why he didn't bring them back when he came out. Now the CO is a brave guy; he's a brave guy. I can't see fear in his face, and his voice isn't shaky. Then here comes a lieutenant colonel (1/506 battalion commander). He says to me, 'I want you to find that 90.'" This doesn't sit well with Rivera, and the others standing in the group. They wanted to find their buddies, not worry about missing equipment or weapons. "So, I say something real stupid," Rivera recalled. "No, sir, I'm going to find those people who are wounded and dead." No one remembers what the lieutenant colonel says, "but he walked away real dejected."

The company commander, Rivera, and the others crawled into the darkness looking for bodies. "I'm glad we had the CO along. He's a big guy. No way I could carry anyone back. I was exhausted." The team was able to find two bodies, and drag them back to the company perimeter.[499] Terry Van Meter remembered that foray into the kill zone as well. "We were hallway across the open area when I received a radio message to come back, that the 1/506 battalion commander wanted to see me. So, we went all the way back. I was thinking he was going to tell me that he was turning off the illumination that he was firing over the area so we would not be seen. Instead, he just told me to make sure I got all the bodies. Then, he just walked off." Van Meter went on to say that "the only reason we didn't get killed out there was because Charlie was busy doing the same thing — collecting his own bodies."[500]

PICKING UP PIECES

I flew in at dawn on July 23. I expected the worst, and that was exactly what I found. In a letter home, I described Tiger Bravo as "numb and falling to pieces from exhaustion and shock."[501] The ground was littered with the debris of a major battle—bloody field dressings, discarded IV lines, a radio riddled with bullet holes, a pile of ripped jungle fatigues cut off the wounded, and a stack of discarded equipment and rifles. For me, the morning was surreal. Just two days before I had been holding new life in my arms, my infant son. Now I held a piece of paper filled with names of the dead and maimed, twenty-four in all.

It was the bloodiest day yet for Bravo Company. In this one battle, Tiger Bravo had suffered 30 percent of the US Army's killed in action on that day, across all of the hundreds of battlefields in the Republic of South Vietnam.[502] First Lieutenant Joe Hillman III, twenty-three years old from Piedmont, Alabama, holder of the Silver Star, embodiment of the warrior ethos—gone. They came from large cities. Specialist Four Eugene "Thump" Davis, twenty years old, one of a thousand young men from Chicago killed in the war. They came from small towns as well. Private First Class Ron Albertson, twenty-one years old, one of only three killed in the entire war from Dimondale, Michigan, a town of less than eight hundred inhabitants, and Sergeant Allan Hamsmith, twenty-two years old, one of six from Fairmont, Minnesota.[503]

Those three enlisted men were original members of the company, and survivors of every battle fought by Tiger Bravo before that day. "Davis was a hell of a fighter. One of the best. He died trying to save someone else. That's how he did things,"[504] recalled Basil Rivera. The same with Albertson, who also did not go down without a fight. Albertson earned a Bronze Star with V (Valor) in the first few minutes of the fight. His citation reads:

> WITH COMPLETE DISREGARD FOR HIS PERSONAL SAFETY, PFC ALBERTSON
> CHARGED AN ENEMY POSITION UNDER HEAVY FIRE TO RENDER AID TO ONE OF
> HIS WOUNDED COMRADES. AFTER RENDERING FIRST AID, HE CONTINUED ON TO
> DESTROY THE ENEMY POSITION, BUT WAS MORTALLY WOUNDED.[505]

Albertson was well liked by everyone, and a valued friend. "His warm smiles and good humor kept me going in rough times," wrote Chuck Limer. "He never left my side when I was shot . . . although ordered to. He told his sergeant he wasn't going to leave his buddy."[506] He was remembered by his sister as "not much more than a boy, bent on getting into a bit of trouble, dating and driving too fast, when he left our midst. But, he was also well-liked, hard-working, dependable and a family kind of guy. Ron lives on through photos and sharing of stories, as well as with those who bear a part of his name in honor of him."[507]

Albertson's passage from a 1966 graduate of Holt High School in Michigan to the killing fields of South Vietnam began on a snowy night in November 1966. Ron had just enlisted in the army and was on his way to catch the "inductee bus" in the nearby town of Charlotte, along with twenty other teenage boys from Eaton County. His younger sister Dianne—there were four boys and six girls in the family—remembered that Ron had to dig out the family car while the snow was still falling. "What a ride that was! Eight of us crammed in a vehicle, out on the road," Dianne recalled. "But we were going to have our good-byes."[508] In a send-off reminiscent of small-town farewells to their sons in World War II, the local Chamber of Commerce, along with ladies from the American Legion and VFW Auxiliaries, plus the Red Cross, Gideons, Salvation Army, and Interchurch Council, presented the boys with small gifts and served refreshments before they headed off to basic training.[509] It was a time when going off to war was still expected of young men and brought great honor to their communities, just as their fathers had done a generation before. A time when the war was still new and exotic; when hatred of the war and loathing for the warrior were still a distant rumbling.

News of Albertson's death came quickly to his hometown, Dimondale, Michigan. Just three days after the battle, fifteen-year-old Dianne saw a strange car carrying men in uniform park in front of their house. "I felt like a bucket of ice water had been dumped on me when they knocked on the door. Sgt. Snoody asked to speak to my Mom; I told him she was at work and gave him directions. Then, being as brave as I could, I called my Mom and told her to expect them. It seemed like forever before she could talk . . . then all she could say was, 'When did they leave?'" All across America families were answering similar knocks on their doors, and having their lives changed forever.

Newcomers fell alongside veterans in the July 22 battle. Sergeant Jackie Ray Poling, twenty years old from Scott, Ohio, and with the company less than two months—now dead. Not all fell from enemy fire. Larry Burton recalled, "One was hit by an air strike of napalm that blew his legs off."[510] Two medics were also among the dead: Specialist Four John P. Murphy, twenty-three years old, from Omaha, Nebraska, a tall, soft-spoken, Irish Catholic; and Specialist Five Johnny Greeson, eighteen years old, from Melbourne, Florida, one of the youngest medics in the 101st Airborne Division. Johnny was only seventeen years old when he deployed to South Vietnam. The fact that he was under-age, and technically nondeployable, wasn't discovered until he landed in Bien Hoa. The army quickly sent him back to Fort Campbell, where he celebrated his eighteenth birthday on December 7, 1967. Two weeks later, he came back to Vietnam to rejoin the battalion. Johnny Walker, the Tiger Bravo senior medic after Doc Franks, remembered, "He was like a little brother to all of us older medics. He looked up to us, wanted to learn from us, and had as much or more courage than any of us."[511] High praise coming from a member of an elite fraternity of combat medics, where courage was as common as the aid bag they carried into combat. Asked what the bravest thing he ever saw in Vietnam was, one Tiger Bravo soldier replied, "All the things Docs [medics] did under fire."[512] Greeson's brother, David, also fell in Vietnam.[513]

The other dead included Private First Class Joe Davis, twenty years old and another soldier from Chicago; Sergeant Macklin "Mack" Hughes, twenty-two years old, from Pisgah, Alabama; and Corporal David M. Maymon, nineteen years old, from Fairfield, Illinois.[514]

The list included the wounded as well. Dave "Sparrow" Spencer, a gentle soul, with the sharpest wit in the company, was seriously wounded. He was all of 5 feet, 6 inches tall, weighing almost as much as his rucksack and radio, but with the attitude of a 6-foot, 3-inch, 260-pound linebacker. He was "shot in the stomach trying to crawl back to one of the rear platoons to get a radio"—alive, but paralyzed for life. The list seemed endless; each soldier's name evoked a memory, an image of youthful vigor, a sense of incalculable loss. "I knew them all."[515]

By the time I arrived, most of the dead and wounded had already been evacuated, but the bodies of some of the killed and the missing soldier were still unaccounted for.[516] "You came in and took over," remembered Lieutenant Joe Palagyi, artillery forward observer, speaking to me many years later. "That's exactly what the company needed. Everyone had so much confidence in you because of what we had been through in the past. Everyone was relieved."[517] I was met by the 1/506 Battalion commander, himself a new and untested combat leader, who told me that "our boys have been out there too long, go get them." Putting First Sergeant Trent in charge of the company defensive position, I assembled a platoon and headed for the ambush site.

It was quiet; no firing and nothing moved. I had been told that the hedgerow had not been cleared, so the VC could still be there. It was doubtful, but they had been known to leave stay- behind forces to catch rescue parties in the same trap. I had two options. The first was to walk straight ahead into the open area until we found the bodies, and take my chances that the VC had pulled out. Or take the platoon along the wood line on the left until it reached the hedgerow. If the VC were still in their bunkers and trench line we would attack from the flank, then look for the missing soldiers. My gut told me to go to the left, so I gave the order and we moved out.

Just as we began to move, I heard, "Where are you going, Captain? I told you to get out there and find the MIAs." I explained my plan to the lieutenant colonel, adding that approaching from the left flank was the smartest approach and that walking into the open made no tactical sense. It was obvious he wasn't used to being questioned by a lowly captain, especially one who was explaining basic tactics to him. He promptly ordered me to get out in the open and find those men, and I remembered thinking, what a dumb ass; he was only going to get more of my men killed. So, to keep my options open, I sent a squad through the wood line to follow the original plan, while I put the rest of the platoon on line and led them into the open area. After twenty paces or so, I stopped, turned, and asked, "You coming, sir?" To his credit, he did walk with me. The VC were gone, so thankfully we were able to cross the open area without triggering another ambush.

Good-bye, Joe

The morning was so still and quiet, even artillery firing in the distance paused as if out of respect for the fallen. The dead were easy to find. There were crumpled heaps of olive drab scattered around, with red streaks and splotches, obscene, jagged holes in flesh where they shouldn't be, and pieces of bone strewn about. One was still clutching his rifle. Greeson was the first, then another, and another, ending with Lieutenant Joe Hillman laying three feet from the far hedgerow. He was on his back, right arm extended as if he had just thrown a grenade, a single shot to the head. I just stood over my friend. I didn't grieve. I didn't weep. I don't even remember being sad.

I knew what to do. I had done it too many times in the past few months. I took the grief and pushed it deep into a bleak, cheerless space in my gut, to revisit in a quiet moment when I could mourn his loss. Before the war, if I had come upon the dead body of a friend, it would have immobilized me. But this was Vietnam, and I had a company to command. No time for sentiment or grief. I knelt by his side, but only to check for a pulse with one hand, the other clutching my M-16. I didn't move or cover his body until he was first checked for booby traps. Hiding a grenade under a body was a favorite VC trick. My eyes flickered across his body but never left the trench line and a firing port in a bunker just twenty feet away. I did speak, but not to Joe. I simply turned to my radio operator and said, "Tell Battalion we found the MIA." War can be cruel in many ways, and this was one—my friend Joe had become a nameless MIA. He deserved more respect in death. They all did.

In the late afternoon Tiger Bravo was flown by helicopters to Cu Chi to regroup and replenish. Just a week before, twenty slicks had carried the company into combat—heroes and would-be heroes, all. The company was rested, and as it lifted off from the Cu Chi pad there was an abiding sense that somehow Tiger Bravo would prevail; after all, it was one of the best in the division. If guidons were allowed on combat operations, Bravo's would have been held high and snapping in the wind. The return flight was poles apart. Groups of six soldiers, backs bent, plodded to their assigned chopper and climbed aboard. There was little talking, none of the usual banter about going to the "steam and cream" or where to find a beer in the opulence of the Cu Chi Base Camp. These were the same heroes and would-be heroes from the week before, only this time they didn't feel that way, and their numbers were fewer.

There were casualty lists for the killed and wounded, but none for units. If there were such a list, Tiger Bravo would be on it. Its élan was gone, left on the bloody ground in front of a hedgerow, near a town with a strange name. For the return flight to Cu Chi, a decimated Tiger Bravo needed only eighteen slicks—fifteen for the surviving troops, two for shot-up equipment and weapons, and one filled with bodies.[518]

INCIDENT REPORT

That night in Cu Chi Base Camp, with the company having a hot meal for the first time in a week and resting in barracks without beds or electricity, a brigade staff officer arrived to investigate the "incident" of July 22. Leave it to the army's bureaucratic bent to label a life-and- death battle with such an impersonal and dismissive term. This was, in every sense, a quintessential battle in the Vietnam War—an American infantry company searching for the enemy, suddenly engaged with an enemy force fighting from well-concealed fighting positions. A mess hall brawl, a barracks theft, or a paratrooper caught urinating in public were incidents. What Tiger Bravo went through was an ambush, close combat, an old-fashioned shoot-out, a deadly clash, a meeting engagement, or, as the troops would classify it, a "shitstorm." It was anything but an incident.

A week later the incident report was completed and on the brigade commander's desk. I never knew the outcome of the investigation, nor have I been able to find a copy in my research at the National Archives in Maryland. But I have been able to piece together my own assessment of what happened based on interviews with survivors, letters home, and other sources.

According to one eye-witness account, it all began on the morning of July 22 with Captain Van Meter ordering the company to advance across an open area to its front, as he had been ordered to do, and get on with the mission. First Sergeant Trent and Second Lieutenant Joe Palagyi both saw the danger in walking in the open, straight toward a hedgerow that could easily hide a VC ambush. "We were arguing with him," Palagyi remembered. "We told him, 'Do not advance!' The sun was in our eyes and we could not see what was in or beyond the hedgerow."[519] One report has Trent directly confronting the company commander, saying, "We're not going to do that."[520] Palagyi wanted to fire in artillery on the hedgerow to cover the company as it moved across the open area. "He said no," Palagyi recalled. "He told me, you're just a lieutenant, what do you know. I am the company commander." It was at this point, according to Palagyi, that Trent joined in. "You've got to listen to Palagyi. He knows what he is doing. He's not going to let us down." But the argument was over, rank prevailed, and the decision to move out and not to fire artillery stood. "He was just not going to be told what to do," said Palagyi.[521]

Van Meter does not remember that exchange. "It doesn't make sense that I would fire artillery," he recalled. "My frago (operations order) from the day before was to show ourselves and go through the village. We were supposed to show what 'good guys' we were. So, there would be no reason for me to fire artillery."[522]

Immediately upon contact, Van Meter moved forward into the kill zone. Instinctively, he was doing what had been drilled into all of us: Move to the sounds of battle! Share hardships and dangers! Lead by example! But an experienced commander knows that blindly moving to the point of contact in an ambush is not always in the best interest of the company. A rifle company commander in Vietnam made hard choices every day and night. One of the hardest was to choose his location once contact was made, to be in the thick of the fighting and lead from the front or take a position where he could best maneuver the company and manage the violence that was at his disposal. Often these were one and the same. But not in this case. As the battle ensued, Van Meter would remain trapped in the kill zone and, as he recalled "have a couple of radios blown away."[523]

Presumably, the investigation answered only the obvious questions, the ones that the army's investigative process was designed to answer: What happened? Who was at fault? It failed to address the more fundamental problem of why it happened in the first place. On one level, what happened wasn't anyone's fault, certainly not Van Meter's. It wasn't a display of negligence or incompetence that day. It was simply a matter of a commander making a split-second decision to move forward where the fighting was the heaviest, and becoming pinned down alongside his soldiers. All infantry commanders in Vietnam, myself included, have made similar decisions with the same consequences. On that day, Van Meter proved to be a decent, brave, motivated, and capable officer thrust into a complex, terrifying, and fluid battlefield with devastating results.

In retrospect, no one could have foretold the consequences of the commander's decisions or tactics. Commanders "lived in the moment," making one rapid decision after another in a maelstrom of cracking bullets, deafening explosions, and situational information that ranged from limited to wrong to nonexistent. In most cases, the commander on the ground faced constant demands and directions from another commander one echelon higher—usually a thousand feet overhead, in a helicopter—and an enemy situation that could shift from textbook tactics to the inexplicable, and back again in minutes. Also, all too often, when faced with orders from a battalion commander, especially a new one who was adamant as to how he wanted to manage the battle, a commander on the ground did not have the luxury to say no.

In reality, the underlying cause was an unintended consequence of the army's policy to replace experienced commanders after six months on the line, to give a novice a chance at a coveted command position. To make the situation even more one-sided, the enemy commander on the other side of the ambush did not have this disadvantage, as most VC/NVA commanders had years of battlefield experience.

But the report had reached the conclusions expected. The new commander was out, and I was back in command.

BACK IN THE FRAY

After less than forty-eight hours to rearm, replace shot-up equipment and weapons, assimilate replacements, and reconstitute its chain of command, Tiger Bravo returned to the fray. It was barely enough time to prepare the company for combat, and certainly not enough time for the soldiers to process the horrific experience they had endured, nor grieve for lost friends. With amazing stoicism that belied their youth, they shouldered their rucksacks and climbed aboard waiting helicopters. By the afternoon of July 25, Bravo was back in the killing fields of Trang Bang on a search and destroy mission. The company made sporadic, light contact all day long. At 1510 hours, during one of those contacts, it engaged several VC armed with a machine gun and AK-47s, resulting in one casualty. Lieutenant Jim Cress was shot in the leg.[524]

At the same time, and less than ten kilometers away, C Company, 2/506, bumped into a VC battalion that was firmly entrenched in a complex of concrete and rock bunkers.[525] Sensing a rare opportunity to completely surround such a large enemy force, the 1/506 Battalion quickly assumed control of the battle and began piling on every available company in the brigade, fit to fight or not. The plan was to air assault the companies into a series of blocking positions around the enemy, then pummel anything trapped inside the cordon with massed artillery and air support. First in was B Company, 3/187, to reinforce C Company. Then C Company, 1/506, went in to block the enemy's possible withdrawal route. Next, up, to complete the cordon around the flanks of the enemy, A Company, 2/506, and Tiger Bravo[526] were alerted for immediate pickup from the field and insertion into their respective blocking positions. The cryptic radio call to the CP would send Tiger Bravo on yet another harrowing and deadly mission.

SECURE PZ FOR PICKUP. OPCON TO CARGO (1/506)
IMMEDIATELY.[527]

This time it would be part of a rescue attempt to save two A Company, 2/506 platoons, which had been part of the first wave into that company's blocking position but had landed on top of the VC battalion instead. Tiger Bravo's long, bloody Currahee summer was far from over.

10

"Save Us!" . . . "I Will"

Vicinity of Trang Bang
(Hau Nghia Province)
July 25-26, 1968

LOST PLATOONS

A Company made its initial combat assault with only two platoons,[528] since there were not enough helicopters to carry the entire company in one lift. The slicks carrying the two ill-fated platoons were thirty seconds out from landing when the firing started. First a single round, then scattered shots, then the machine guns opened up and the sound of individual weapons melded into one long, continuous roar. By the time the troops jumped off the choppers onto the open ground of the LZ the firing was coming in sheets, dropping the men as quickly as their boots hit the ground. The platoons never had a chance. They had landed on top of a sizable portion of the VC battalion that had remained hidden in a battle that was still taking shape. In less than five minutes six were dead, including a platoon leader, Second Lieutenant Rich Tolette, and another ten wounded.[529] It was a slaughter. The platoons were in shambles. They needed to be rescued before they were completely overrun by an enemy that sensed it had the upper hand.

On the fly, a plan was put in motion to divert Tiger Bravo from its original blocking position mission, to make a combat assault as close to the surrounded platoons as possible, then fight its way to the survivors. The remaining two platoons from A Company received the same orders: conduct a CA as close as possible to the beleaguered platoons, then move cross-country to save them.

A 3rd Brigade report left little doubt as to the audacity of the plan:

"B 2/506 AND THE REMAINING TWO PLATOONS FROM A 2/506 MADE DARING AIR ASSAULTS INTO PITCH BLACK LANDING ZONES, RECEIVING HEAVY HOSTILE FIRE AS THEY ASSAULTED, IN AN EFFORT TO CORDON THE ENEMY'S POSITIONS AND RESCUE THEIR COMRADES, NOT 50 METERS FROM THE ENEMY STRONGHOLD."[530]

It was well after dark when the rescue started. Bravo Company's two lead platoons, plus the company headquarters, landed under intense fire from an enemy dug into a hedgerow just fifty meters from where the helicopters touched down. The enemy fire started on the final approach while the helicopters were still in the air and intensified as the troops jumped off. Machine-gun fire, small-arms fire, and rocket-propelled grenades (RPGs) raked the landing zone, which offered little cover.

Before Bravo Company could begin the rescue mission it had to get off the landing zone. The only option was to go straight ahead in the dark and take the hedgerow from the VC. In minutes the company was on line and moving into the attack. All anyone could see were tracers crisscrossing in the night, flickering light from overhead artillery illumination rounds that created the illusion of movement when it wasn't there, and rocket-propelled grenades appearing as small fireballs hurtling toward a shadowy line of soldiers closing in on the hedgerow. One RPG fired from the hedgerow, now only fifty feet away, came straight at the command group, then in the last seconds curved and exploded on the ground, sending shrapnel into the stomach of a sergeant, running at full stride.

In a final assault, shooting at muzzle flashes and throwing grenades, Bravo's two platoons cleared the hedgerow. It had taken every bit of firepower the platoons could muster to overwhelm the entrenched enemy. Anything less than an overwhelming volume of fire and the attack would have failed, with disastrous consequences. Unlike our fathers' generation, where less than half of soldiers didn't fire their weapons during a battle in either World War II or Ko-

Rescue Mission • July 25–26, 1968

rea, in Vietnam the majority of soldiers engaged the enemy. A follow-up to the acclaimed *Men Against Fire Study* conducted by S. L. A. Marshall on World War II rates of return fire showed that "in WWII no more than 25% of American fighting men engaged the enemy during the course of a battle, while in the Korean War this value had increased to only 50%. However, in Vietnam over 83% of the soldiers equipped with individual weapons and over 86% of those manning crew-served weapons engaged the enemy." It went on to conclude that, while the fact that more than 80 percent fired weapons was important, "what is critical is that . . . the very tactics which the American Army trains on and fights, rely on riflemen putting out an effective volume of fire."[531]

Meanwhile, the remainder of Tiger Bravo landed 150 meters to the west and quickly moved to join the company. On one of the last choppers in, Basil Rivera recalled seeing the firing. "I'm about a half mile out when I see this jet come in and strafe with machine guns right in the middle of the whole thing.

What's remarkable is there's this lone NVA shooting at the jet. I can see his bullets going right up to the jet. You know, he never made another pass. I said, my God! This is hot."[532]

Now, with the company all together, it was time to find the A Company platoons. Little was known of the situation other than that the decimated platoons were due south of Tiger Bravo's position, and that the company would have to fight its way through an entrenched enemy before it could mount a rescue. In the next four hours, Bravo Company would make two attempts to reach the A Company survivors, losing good men in the process.[533]

RESCUE MISSION

Tiger Bravo's first rescue attempt started with Sergeant First Class Bill Tellis and his 1st Platoon taking the lead. At thirty-seven years old, Sergeant Tellis was one of the oldest members of Tiger Bravo; he was well liked, competent, and had the same quiet and calm demeanor under fire as he had relaxing back at base camp. "He was a good man. I liked him," remarked Basil Rivera, who was one of Tellis's squad leaders. "He was a gentle person."[534]

The last of the flares fired by supporting artillery had flickered out when the 1st Platoon's point man, Private First Class Danny Soto, stood up and stepped into the inky darkness. He edged slowly toward a second hedgerow, approximately fifty meters to his front. Soto had seen it in the last snatches of light from the flares. Sergeant Tellis had pointed it out across an open patch and told him to "take point and move toward the next hedgerow." But now the night had closed back in. Soto had inched forward only a few feet when the far hedgerow erupted with small-arms and machine-gun fire. "I saw the flashes and bullets started hitting the ground all around me. So, I got down and started shooting back," Soto recalled. Thinking that Soto had been hit, Tellis ran to where Soto had gone down, but was shot before he could say anything. "Tellis came to get me. He got to me and grabbed me," Soto went on. "When he grabbed me, he got hit. It happened so fast. Sounded like when a deer gets shot. All I know is, he got shot trying to save me."[535]

Alongside Tellis, and caught in the same deadly hail of bullets, was Private First Class Steven A. Frink, a medic attached from Headquarters Company.

Frink had been with Tiger Bravo for a little over three months and was only twenty when he died.[536] Ironically, at the time of his death, his father had just retired, unscathed from twenty-five years of service in the air force, but PFC Frink didn't even make it through his first year in the army before he was killed. As part of a service family he had lived all over during his short life but called Vancouver, Washington, his home, where he attended Clark College before enlisting.[537] It took twenty minutes to recover the bodies of Tellis and Frink and tend to the wounded. Then it was time to continue with the rescue mission.

Unlike the first attempt, this time we had a better understanding of the ill-fated platoons location. They had not made it off the LZ. The survivors were fighting amidst their wounded and dead comrades, right where the choppers had dropped them. I was in radio contact with one of the survivors. He was scared, unsure of what to do and kept repeating, "Save us! Save us! We're going to die!" I told him "I will," that the 101st Airborne Division doesn't leave anyone behind. At one point, I told him to keep fighting and "collect ammunition from those killed or wounded and pass it out to the survivors still able to fight," which he did.

That night I learned that courage comes in many forms. It was present in both the A Company platoon leader who led the charge, the first to meet the foe, and the young, frightened soldier lying beside him, who stayed on the radio as his chances of surviving dwindled with each passing minute. I never found out his identity, nor anything about him, only that he passed the one, true test of courage for a soldier: doing your duty despite your fears, as real and terrifying as they may appear.

Now I would take the lead as we charged forward. There was no time for elaborate planning, as the lives of American soldiers were at stake. We picked the most direct route and moved out. Approximately fifty meters to the right of where Tellis and Frink had been killed was a dark patch of hedgerow that appeared unoccupied by the enemy, an assumption based solely on not seeing any muzzle flashes there. There also appeared to be a narrow dirt road splitting the hedgerow and going in the general direction of the A Company platoons. No thought was given to aborting the mission, nor was I about to order someone else to walk point. It was the only time in my tour when I knew I was going to die. In my mind, I had no chance of making it all the way; my fate was to be the same as Tellis and Frink. It had an odd, calming effect on me that I had never felt before, or since. For the first time, I was completely free of stress and the fear of dying. It was strangely liberating to no longer play a part in my own demise; it was out of my hands and into the hands of whoever was out there in the darkness.

Creeping down the dirt road, we came upon the A Company remnants, at approximately the same time as the other A Company platoons and a lone helicopter dispatched to pick up the dead and wounded. The situation, however, was far from under control. The VC were still attacking, and a hysterical A Company soldier was trying to climb on the lone chopper, already filled with wounded soldiers. I remembered him screaming, "I can't fight! I gotta get out of here!" Leaning out of the left side window of the chopper, the pilot screamed, to no one in particular, "Someone grab him! We're at max load. Can't take anyone else." I yanked the soldier away from the chopper and shouted at him to calm down. Now it was complete pandemonium—chopper blades turning, soldier pleading, pilot screaming, and me shouting.

The soldier and I were face to face. He screamed that he couldn't fight, so I grabbed him and shouted back, "This is Vietnam! You fight or you die!" In a much quieter voice, almost a whisper, he said, "I don't have any ammo." I gave him the bandoleer of ammunition I carried across my chest and put him in the line of survivors that we would take back with us. Between the two rescue forces, Tiger Bravo and the remnants of A Company, thirty soldiers' lives were saved that night. Thirteen of the survivors walked out with Tiger Bravo.

BAYONETS IN THE MOONLIGHT

Four hours after landing on a hot LZ and participating in the rescue of the A Company platoons, Bravo Company was back in its blocking position on the west side of the cordon that encircled the VC battalion. In the last few hours the enemy had attempted breakouts in two other sectors of the cordon without success, now it was Tiger Bravo's turn.

It is easy to imagine what was going inside the cordon of American infantry companies, less than a hundred meters beyond Tiger Bravo's lines. From

all across the trapped battalion, the VC fighters came into the moonlight —
climbing out of tunnels and bunkers, one or two at a time, extinguishing the
candles and small lanterns that had been their only light while they were underground weathering the American bombardments. Keeping to the shadows
of hedgerows and hidden paths, they quickly assembled under the hushed
orders of their commanders. Most were able-bodied with plenty of fight still
in them, but for these desperate times any wounded who could still fight were
pressed into the breakout force as well. What started as an uncoordinated
collection of small groups, no more than shadows with fixed bayonets in the
moonlight, soon grew into a strong, yet desperate, attack force: silent men
assembling in the dark to attack other silent men waiting for them.

It would be a hasty, frontal assault, with no time for the detailed reconnaissance and rehearsals called for by enemy doctrine. Hitting the American
lines hard, with an overwhelming force that would come screaming out of
the jungle offered the only hope of a breakout. For the VC, every minute
counted. They had to attack quickly before the force was discovered and
another round of artillery and air strikes would drop out of the night and
destroy this one last chance to escape. They knew where to attack. Tiger
Bravo's thin line of foxholes would have been located by the few scouts left
alive above ground.

On the American side, Tiger Bravo had gone to ground, everyone crouching in foxholes waiting for the attack. Based on an intelligence report from
battalion warning of an imminent attack, a message was passed from foxhole
to foxhole that put everyone on alert. "An attack is coming to our front at any
minute. Get ready. Engage anything that moves." For the troops in the line of
foxholes facing the enemy the order left no doubt as to whether to fire or not.
It was simple — when you saw movement, point and shoot, point and shoot,
point and shoot, until the movement stopped or you had to reload.

The attack started slowly with several lone attackers popping up directly
in front of the foxholes. They were gunned down immediately. Next, small
groups of 2-3 enemy fighters came running out of the darkness firing from the
hip, like small holes in a dike spurting water just before it burst. Then the entire woodline seemed to move forward, a dark mass coming closer and closer,
quickly changing from indistinct shapes to a line of VC fighters.

But, Tiger Bravo was not alone in this fight. Supporting the cordon of American companies was a Spooky gunship. It was quickly rerouted to orbit over Tiger Bravo. Immediately, radio contact was established and a plan hatched to mark the forward edge of Tiger Bravo's line of foxholes with trip flares. The M49A1 trip flare was ordinarily used with a trip wire as a defensive measure, exploding into a brilliant fifty thousand candlepower fireball when tripped by the enemy. But, in extreme cases such as this it could be thrown, much like a hand grenade, to mark positions at night. It ignited immediately upon release of the arming lever, but cleared the hand by several feet.[538]

Each time the gunship made a pass, a trip flare was thrown in front of each platoon that was facing the enemy. The sequence was the same each time. On my command, white-hot trip flares were pitched from foxholes to land in front of the advancing enemy force, followed by a radio message to Spooky that "flares are out." Then an earsplitting roar enveloped the company as the AC-47's 7.62mm mini-guns raked the jungle in front of our line of foxholes, exploded in the trees and sent chewed up bodies of VC attackers flying in all directions. Few enemy ground attacks survived such an onslaught from Spooky. This night was no exception. The attack faltered, then stopped; the VC pulled back to try their luck with another company.

ANOTHER GOOD MAN GONE

When daylight came, the company's line of foxholes stretched for almost two hundred meters, with the company command post (CP) in the center. The CP was like any of the other fighting positions along the perimeter—mounds of dirt thrown up around the foxhole, empty C ration cans, spent cartridges, empty ammo boxes, and rucksacks stacked along one side—except for two bodies wrapped in ponchos, laid side by side, waiting for a chopper to pick them up. The firing stopped before dawn, but the company remained alert. There could still be remnants of the VC battalion that had not escaped, or wounded that had been left behind. Even wounded VC were dangerous.

Some soldiers stayed on guard at their foxholes, while others checked the surrounding area for wounded VC and bunkers or tunnel entrances that were nearly invisible during the day and completely undetectable at night. This was a necessary precaution taken each morning after a nighttime action, as Joe Adams discovered one morning in the Ho Bo Woods. "I had been up all night," he recalled. "So sleepy that I rested my chin on my knife point to keep from falling asleep. The next morning, I turned around and looked behind me. The sand pile I thought I was resting against during the night was an NVA rucksack and right behind that was the entrance to a tunnel." Joe knew that he had a close call, but was able to see the humor in it. "Was I scared? Yeah! Had to check my throat to see if it had been slit."[539]

During the search of their platoon area, Staff Sergeant Paul Nabors, twenty-three years old, from Binger, Oklahoma,[540] and Sergeant Dan Bernard came across a blood trail made by a wounded VC. Nabors and Bernard had been assigned to line positions because of a severe shortage of infantry NCOs across the entire brigade. Since the flow of replacements, especially leaders in infantry platoons, could not keep up with mounting losses, NCOs with non-infantry specialties were being reassigned to key infantry platoon positions. In this case, Nabors moved from Bravo Company supply sergeant to squad leader in 1st platoon, then to platoon sergeant when Tellis was killed. Bernard went from carrying the company commander's radio to fire team leader in Nabor's squad, then replaced Nabor's as squad leader. Both had unselfishly answered the call and now found themselves in hazardous jobs, with high casualty rates, in one of the most dangerous provinces in South Vietnam.

Bernard remembered what happened next, "Nabors and I followed the blood trail into a bunker. Nabors decided to check it out. I wanted him to throw a grenade in it first. But Nabors said no, that he didn't want a lot of hoopla that would bring the first sergeant over." The bunker was empty, but they found a tunnel going straight down from the floor of the bunker. Nabors didn't even have a flashlight, so he lit a piece of C-4 and dropped it down the hole. Nabors saw something at the bottom of the shaft, and despite a warning from Bernard to "not go down there," he dropped into the tunnel. At the bottom Nabors found an NVA rucksack, AK-47 ammunition, and some rice, which he passed up to Bernard. "When I was setting it aside," Bernard went on, "I heard a CRACK—CRACK! I hollered for Nabors and no response. He was on his haunches at the bottom of the shaft. One of the little guys, probably the wounded one who left the blood trail, put a bullet in his eye and one in his groin."[541]

Nabors died instantly. His body was pulled from the tunnel, wrapped in a poncho, and placed next to the other two at the company CP. Rather than losing anyone else going after a wounded enemy soldier, I made an improvised satchel charge with two blocks of C-4 and the blasting cap from a claymore mine, dropped it down the tunnel shaft, and set it off. "It had about four or five feet of earth on top of it that all collapsed in the explosion," Bernard recalled. "I remember that well. There was a whole bunch of spider holes coming off that tunnel. People were spreading out saying 'I got a hole, I got a hole.' We had all the exit holes off the tunnel covered."[542] There was no sign of the wounded VC. Escaped? Entombed forever in the tunnel? No one bothered to find out.

Later that day, somewhere at Battalion and Brigade Headquarters, two lists were updated. One recorded the names of soldiers killed in action during the day's fighting. Nabors, Paul H., RA18670456, was added just below the names of Tellis and Frink. The other list tallied the enemy body count and inventoried the enemy equipment/weapons captured or destroyed during the engagement that cost the lives of our soldiers. Just below the tally of seventy-eight[543] enemy soldiers killed in the battle that cost Tellis and Frink their lives, a clerk entered one empty rucksack, half a magazine of AK-47 ammunition, and two cups of rice for the shooting in the tunnel that cost Nabors his life.

SIDE BY SIDE

The loss of Staff Sergeant Nabors was the last contact of the day. The rest of the day was quiet, other than the occasional helicopter landing with supplies. Soon the company was going about its normal, post battle routine. Security was posted. Some soldiers ate C rations or brewed coffee in canteen cups. Others cleaned their weapons. A few read letters from home for the third or fourth time. The first sergeant distributed resupplies of food, water, and ammunition to platoon sergeants. Some slept for the first time in three days. Most remained close to their foxholes, since quiet did not mean safe, just quiet. Casualties to the leadership ranks were replaced from within the company. In this case, Dan Bernard moved up once again, this time becoming acting platoon sergeant of 1st platoon. His trajectory over the last forty-eight hours, moving from team leader to squad leader to platoon sergeant, had been remarkably fast, but not unheard of after a battle.[544]

The bodies of our three soldiers lay next to the company command post's foxholes, waiting to be evacuated to the rear. They were arranged neatly, side by side; in a small gesture of respect no one stepped over them, just as someone avoided stepping on a grave in a cemetery. The collection of the company's dead soldiers at the command post was the first step in a process designed to quickly return the remains—at the CP they were still Tiger Bravo soldiers, once in the system they became remains—to their families. Called the Concurrent Return Program, an expedited process to return America's fallen, it averaged seven to ten days from time of death to their final internment destination. According to the Mortuary Affairs Center at Fort Lee, Virginia:

"THE REMAINS MOVED QUICKLY BY HELICOPTER FROM THE BATTLEFIELD TO COLLECTION POINTS SCATTERED THROUGHOUT VIETNAM, THEN SENT TO ONE OF TWO US ARMY MORTUARIES IN SAIGON OR DANANG. HERE THEY WERE OFFICIALLY IDENTIFIED, EMBALMED AND LOADED ON SPECIAL US AIR FORCE FLIGHTS TO PORT OF ENTRY MORTUARIES AT OAKLAND, CALIFORNIA OR DOVER, DELAWARE. FOLLOWING REPROCESSING, TO INCLUDE COSMETIZING, DRESSING, AND CASKETING, THE REMAINS WERE FORWARDED TO THE PLACE OF FINAL DISPOSITION AS DESIGNATED BY THE APPROPRIATE FAMILY REPRESENTATIVE."[545]

"We treated the remains of the soldiers that were killed with respect," one member of the Saigon mortuary facility said. "It was sad but you had to do the job right and you had to treat them like a member of the family."[546]

Every day, coffins were loaded onto cargo planes for the long trip home, often sharing space with severely wounded soldiers heading back to the United States for treatment or recuperation not available in Vietnam. Steve Lyle, who had been shot during Tet and had significant nerve damage to his leg, remembered coffins being loaded onto his medevac airplane while it was being refueled at Tan Son Nhut Air Base. On his plane were four rows of litters stacked six high. "There was no more than two inches between your nose and the bottom of the litter above," he explained years later. "I was loaded on the tail end stack of litters. The ramp came up, then the pilot announced we were moving to an area to be refueled, then we would be off. At the fueling site, the ramp was lowered and a forklift placed a stack of coffins on the ramp floor." Steve watched as the crew secured the coffins, then raised the ramp one last time. He would never forget that flight home, calling it "one big plane full of misery and hope and prayers."[547]

When the final resupply chopper of the day was inbound with more supplies and replacements, the pilot radioed that he was one minute out and asked if the company had anything to backhaul. When I answered, "Roger, I have three KIAs that need to be evacuated to the rear," the pilot's reply sent me over the edge. "Negative, I just had the chopper cleaned after backhauling other dead. You need to call for another bird" came the response.

It was so unexpected, so disrespectful, and so callous that I exploded. I had seen too many good men die. I stood up, grabbed my M-16, and screamed into the radio's mike, "You either land and pick up my dead or I will blow your ass out of the sky." By this time the helicopter was just seconds from touching down. I walked straight out into the whirling dirt and grass being kicked up by the chopper's blades and took aim at the pilot in the front left of the cockpit. My finger was on the trigger. The standoff lasted only a few seconds, ending with the chopper landing to accept its precious cargo. No other words were said; I lowered my weapon and walked away. The bodies were loaded on board, four men to a body, two at the head and two at the feet, an infantry company's makeshift band of pallbearers.

I have always regretted that outburst. Upon reflection, the pilot and crew were probably the same brave souls who had weathered enemy fire just hours before in the midst of a firefight to pick up the A Company dead and wounded. Would I have pulled the trigger? Probably not, but there was a piece of me that could. No one disrespected Tiger Bravo's dead. No one.

As luck would have it, one of Tiger Bravo's own was waiting to receive the bodies when the helicopter landed at Cu Chi. Chris Backman had just returned from R & R and was trying to get back to the company. "I had heard that the fighting was horrible, and was trying to get back to the company but couldn't. The fighting was just too heavy," Backman recalled. "I was standing at the assault pad when a chopper came in loaded with bodies. All of them wrapped in ponchos except Sergeant Tellis."

No one was around to receive the bodies and start them on their journey home, added Backman. "I took it upon myself to get them to graves registration." Commandeering a passing 2 ½ -ton truck, Backman loaded the bodies and drove to the Graves Registration collection point. "They were a little put out that I had taken it into my own hands. We exchanged some words. But I didn't care." Backman's parting comment to the Graves Registration crew echoed what we all felt.

"Take care of these guys," he said.[548]

GRACE AND GENEROSITY

In many ways Bravo existed in its own world, one defined by anything and everything that was within small-arms or mortar range. This world consumed our attention every minute, day, and night; it devoured our capacity to feel and replaced it with a callousness our mothers would abhor. It pushed aside idyllic thoughts of home, our hopes and dreams and replaced them with one simple aspiration, to stay alive for one more day, then one more after that.

Sometimes, when least expected, the outside world reached in and reminded us that others were in this war as well. Such was the case, several weeks later, when I received a letter from the widow of Bill Tellis. It reached in and tore me away from the harsh, sequestered world of Tiger Bravo.

DEAR SIR,

THIS IS TO INFORM YOU THAT TODAY I MAILED A BOX OF CANNED FOODS THAT I HAD PREVIOUSLY MAILED TO MY DEAR HUSBAND THAT WAS IN YOUR COMPANY, BUT IT WAS RETURNED TO ME ON THE 8TH OF AUGUST BECAUSE MY HUSBAND SGT WILLIAM J. TELLIS (RA16293259) WAS KILLED ON JULY 26, 1968, AT CU CHI, VIETNAM. MY DESIRE IS FOR YOU TO DISTRIBUTE THE CONTENTS OF THE BOX TO THE TROOPS IN HIS BATTALION AS YOU SEE FIT. I HOPE IT WILL MAKE THEM HAPPY AS I KNEW IT WOULD IF MY HUSBAND HAD

RECEIVED IT. WOULD YOU WRITE ME AND GIVE ME THE CAUSE OF HIS DEATH?
IT WAS A SUDDEN SHOCK TO ME AND IT'S REALLY HARD TO UNDERSTAND WHAT
HAPPENED TO HIM, HIS LAST LETTER WAS WRITTEN ON JULY 24, AND HE WAS
KILLED ON JULY 26. I HOPE THE TROOPS ENJOY THE SNACKS.
SINCERELY,
MRS. WILLIAM J. TELLIS[549]

Sitting in the shade of a bamboo thicket during a moment's respite from the day's operation away from the blistering sun, I replied from a place in me that I had not tapped into for months.

DEAR MRS. TELLIS,
I'M VERY SORRY THAT I HAVEN'T WRITTEN SOONER, BUT I COULD NOT WRITE
WITHOUT KNOWING IF YOU HAD BEEN NOTIFIED OF YOUR HUSBAND'S DEATH.
PLEASE ACCEPT MY SYMPATHIES AND THOSE OF THE MEN OF TIGER BRAVO ON
YOUR LOSS. BEING HIS COMPANY COMMANDER DURING THE BATTLE IN WHICH
HE LOST HIS LIFE, I CAN TRULY SAY THAT I AM SORRY FROM THE BOTTOM OF
MY HEART.

YOUR GIFT OF THE BOX OF FOOD WILL BE GREATLY APPRECIATED BY THE
MEN WHO SERVED WITH YOUR HUSBAND. KNOWING HOW FINE A MAN HE WAS, I
AM SURE THAT HE WOULD WANT IT THIS WAY.

THE ACTION IN WHICH YOUR HUSBAND GAVE UP HIS LIFE WAS CENTERED
AROUND A PLATOON FROM A COMPANY THAT HAD BEEN CUT-OFF, SURROUNDED,
AND ALL OF ITS LEADERS KILLED BY THE ENEMY. IT WAS B COMPANY'S MISSION
TO GO TO THE AID OF THE SURROUNDED PLATOON. YOUR HUSBAND WAS KILLED
BY A SNIPER'S BULLET WHILE WE WERE FORCING OUR WAY THROUGH THE
ENEMY SURROUNDING THE PLATOON. YOUR HUSBAND'S ACTIONS PRIOR TO HIS
DEATH WARRANTED ME TO SUBMIT HIS NAME TO BATTALION HEADQUARTERS
FOR AN AWARD FOR VALOR IN ACTION. I FEEL THAT IT WILL HELP YOU TO KNOW
THAT AFTER FOUR HOURS B COMPANY FINALLY BROKE THROUGH AND SAVED

THIRTY AMERICAN LIVES. UNFORTUNATELY, I LOST TWO MORE BRAVE SOLDIERS
IN THE ATTEMPT.

   MRS. TELLIS, I AM NOT QUITE CAPABLE OF EXPRESSING THE DEPTHS OF
MY FEELINGS ON PAPER. AS A COMPANY COMMANDER, I HAVE THE ULTIMATE
RESPONSIBILITY FOR THE LIVES OF MY MEN AND I DIE A LITTLE FOR EACH ONE
I LOSE. I PROMISE YOU ONE THING——I HAVE A SON WHO IS 8 1/2 MONTHS
OLD——I PROMISE THAT HE WILL GROW UP KNOWING THAT MEN SUCH AS YOUR
HUSBAND GAVE UP THEIR LIVES SO THAT HE MIGHT BE FREE. AGAIN, PLEASE
ACCEPT MY SYMPATHIES.

                                             RESPECTFULLY,
                                             RICHARD ST JOHN
                                             COMMANDER[550]

The letter from Mrs. Tellis reminded me that we were not alone, that
what happened in Vietnam could trigger anguish halfway around the globe. It
somehow soothed me to know that grace, generosity, kindness, and compassion still existed in the form of a grieving widow. I needed that letter as much
as she needed mine.

11

Ravage Hau Nghia Province

Vicinity of Trang Bang
(Hau Nghia Province)
July 27–August 26, 1968

INTO AND OUT OF THE MOONLIGHT

It was 0330 hours on July 27 and Tiger Bravo was already out of its NDP and on the move.[551] Radios turned down. Orders passed up and down the column in whispers. There was none of the usual banter between soldiers heard in daylight operations, only an occasional cough or a grunt when a soldier stumbled in the dark. The point team slipped into and out of the moonlight, as it picked its way toward a sleeping village. There was a risk of bumping into an enemy force on the move at night; it had happened before. But it was the chance setting off mines and booby traps that posed the greatest threat. The area around Trang Bang was infamous for these diabolical devices, and had been for years. A day didn't go by without several being discovered or blowing up in our faces. VC-controlled villages and the approaches into them were particularly dangerous. On this operation, it was inevitable; Tiger Bravo would hit booby traps. The only unknowns were when, where, what type, and who would be the unlucky ones. They could be anywhere; take any form; shatter the night's stillness and leave you maimed or dead in a split second.

Tiger Bravo was as much in the dark from an intelligence standpoint as it was trying to discern safe passage through a dark countryside without the aid of artificial illumination. Timely, accurate, and mission-specific information about the size, location, weapons, and intentions of the enemy in a target area rarely reached a company on the ground. Normally, the available intelligence prior to an operation such as this one was limited to unconfirmed reports of "enemy forces in your area," a broad classification of the target being a "VC village," warnings to "expect mines and booby traps," and occasionally an actual "sighting of VC" that could be forty-eight hours old. Little wonder that Tiger Bravo's modus operandi for survival was to expect anything at any time. Trust no one not wearing a Screaming Eagle or US Army patch, and be ready—always.

The previous day the battalion had received the mission to seal a village suspected of harboring VC fighters, then assist the South Vietnamese National Police in rooting them out. It was not a simple operation. For the cordon and search to be successful, three companies (B, C, and D, all from the 2/506 Battalion) had to move cross-country from different directions, find their positions in the pitch dark around the target village and link up with each other's flanks—all before sunrise, without waking sleeping villagers or any VC bedded down for the night. Most importantly, it had to be accomplished without stepping on any mines or booby traps that habitually ringed the villages. By 0441 hours, Tiger Bravo had covered five hundred meters and was in its designated position.[552]

This type of operation in and around a VC village, infested with mines and booby traps, took stress and fear of the unknown to uncharted levels. Philip Caputo explained it best in *The Rumors of War*:

We were fighting an enemy whose principal weapons were mines and booby traps. That kind of warfare had its own peculiar terrors. It turned an infantryman's world upside down. The foot soldier has a special feeling for the ground. He walks on it, fights on it, and sleeps and eats on it; the ground shelters him under fire; he digs his home in it. But, mines and booby traps transform that friendly familiar earth into a thing of menace, a thing to be feared as much as machine guns or mortar shells. The infantryman knows that at any moment the ground he is walking on can erupt and kill him; kill him if he is lucky. If he's unlucky, he will be turned into a blind, deaf, emasculated, legless shell. It was not warfare. It was murder. We could not fight back against the Viet Cong mines or take cover from them or anticipate when they would go off. . . . Waiting for those things to explode, we had begun to feel more like victims than soldiers.[553]

Routinely, the company faced a wide assortment of nasty devices—all masterfully camouflaged. There were:

- **Punji stakes:** Sharpened lengths of bamboo or metal with needle-like tips that had been fire-hardened and set in camouflaged holes. Often, they would be coated with excrement to cause infection.
- **Toe-poppers:** A rifle round buried vertically and resting on a nail or firing pin. Downward pressure from a soldier's step fired the cartridge into the foot of the intended victim.
- **Hand grenades:** Hidden in every conceivable place and detonated by trip wires or movement.
- **Salvaged, dud artillery rounds or bombs:** Targeted at groups of soldiers and detonated by trip wires, pressure devices, or command-detonated.
- **Claymore mines:** An antipersonnel mine that could wipe out a whole squad if they were not spread out.
- **Bouncing Betty:** Triggered by a release of pressure on the firing mechanism. If a soldier stepped on one, then moved his foot, the mine would spring into the air and explode chest high.[554]

By 0645 hours the village was completely sealed. No one could come in or out, at least above- ground. There had to be tunnels under the village and one or two escape routes, but none had been discovered. Just at dawn three shots rang out from a hut on the north end of the village.[555] It was a signal to any VC in the village to hide and head to preplanned escape routes. The search lasted until 1430 hours, with an outcome all too familiar. One or two VC appeared and disappeared, some supplies were confiscated, a few villagers stood by disinterested, and multiple Americans were wounded by mines and booby traps.

Two of the wounded were from Tiger Bravo. One soldier had stepped on a toe-popper that just clipped his foot, traveled straight up within a half inch of his chin, and knocked his helmet off. Right next to him, Platoon Sergeant

Sykes was hit by a Bouncing Betty. It blew his pants completely off and ripped his legs apart with shrapnel. Another stalwart gone. The Battalion's Daily Staff Journal recorded the details:

- 0730 HOURS: B CO REPORTS 2 WIA DUE TO BOOBY TRAPS. 1 SERIOUS. BOOBY TRAPS (BT) NOT COMMAND DETONATED. SET WHERE ONE IS TRIPPED THE OTHER GOES OFF TOO.
- 0801 HOURS: DUST OFF COMPLETED FOR B 2/506. WAS THE 26 LEADER (PSG SYKES).[556]

But that wasn't the end. The piecemeal attrition continued for the rest of that miserable day. One blast, from a 175mm artillery round weighing 147 pounds, was so strong that the victim's body could not be found. The entries for the "Daily Staff Journal" entries for the rest of the day were typical for this type of operation:

- 0810 HOURS: C 2/506 REPORTS VC FIRING AK47 FROM HUT. NO CASUALTIES.
- 0815 HOURS: A 2/506 REPORTS ONE US KIA FROM 175MM ROUND W/TRIP WIRE. ONE NATIONAL POLICE STEPPED ON A 175MM ROUND AND CANNOT BE FOUND.
- 0820 HOURS: 2 NATIONAL POLICE AND 2 US WIA FROM 175MM BOOBY TRAP.
- 0935 HOURS: PERMISSION RECEIVED TO BURN HUTS THAT BOOBY TRAPS HAD BEEN FOUND NEAR.
- 0945 HOURS: A 2/506 FINDS 2400 LBS. OF RICE.
- 0955 HOURS: A 2/506 FINDS 4 PRESSURE TYPE BOOBY TRAPS AND 2 175MM W/ TRIP WIRES.
- 1015 HOURS: ONLY PERSONNEL FOUND IN VILLAGE: 1 OLD MAN, 1 OLD WOMAN, 1 YOUNG BOY—8 OR 9 YEARS OLD.

- 1132 HOURS: BOOBY TRAP 105MM ROUND FOUND IN A HUT. DESTROYED IN PLACE.
- 1140 HOURS: A 2/506 FINDS 50M X 70M MINEFIELD. 5 MINES BLOWN IN PLACE.
- 1343 HOURS: A 2/506 FINDS 7800 LBS. OF RICE.
- 1430 HOURS: SEARCH OF VILLAGE COMPLETE.
- 1500 HOURS: D 2/506 REPORTS ONE MAN WIA. TRIPPED BOOBY TRAP WALKING FROM HIS CP TO HIS POSITION.
- 1845 HOURS: B 2/506 REPORTS 3 WIA FROM BOOBY TRAP.[557]

One battlefield study, conducted by the 25th Infantry Division, explained the enemy's tactical use of mines and booby traps this way:

- LOCATION: SIMPLY PUT, MINES AND BOOBY TRAPS WILL BE PLACED WHERE COMMON SENSE SAYS THEY WILL DO THE MOST DAMAGE. THE ENEMY INTELLIGENCE GATHERERS ARE LOOKING FOR ONLY ONE THING—PATTERNS OF FRIENDLY ACTION.
- MARKING: IN SOME INSTANCES, THE ENEMY HAVE MARKED DANGER AREAS WITH SYMBOLS KNOWN TO THE LOCAL INHABITANTS.
- WHEN: FRIENDLY FORCES CAN EXPECT AN INCREASE IN BOOBY TRAP AND MINING INCIDENTS AS THE ENEMY GROUND FORCES ARE DEFEATED AND DRIVEN AWAY. THIS MAKES THE ENEMY LOSE FACE. THE ENEMY THEN TRIES TO PROVE HIS FIGHTING ABILITY TO THE PEOPLE BY EMPLOYING MINES, BOOBY TRAPS, AND AMBUSHES.
- WHO: SAPPERS AND THEIR TRAINED INTELLIGENCE GATHERERS LOCATE THEMSELVES CLOSE ENOUGH TO THE TARGET TO PROVIDE QUICK ACCESS TO AND CONSTANT SURVEILLANCE OVER THE TARGET AREA. THE INTELLIGENCE GATHERER MAY BE AN OLD WOMAN LIVING ALONG A ROAD TO CHILDREN SELLING SODA. IF AT ALL POSSIBLE, THE SAPPER WHO IS TO SET THE MINE WILL PERSONALLY OBSERVE THE TARGET.[558]

Rifle companies in Vietnam at this stage of the war had little in the way of technology to combat mines and booby traps. Tiger Bravo had just two portable, manpack mine sweepers. In areas known to be heavily mined, the normal practice for the company was to move in two parallel columns, with one mine sweeper at the head of each column. They would clear a narrow path for the soldiers following behind. The point squad and flank security moved unprotected. Their only defense was careful movement, sharp eyes, an awareness of likely spots to be mined, based on hard experience, and a hefty dose of luck. On one occasion the lead mine sweeper discovered five cluster bombs set in a two-foot-by-two-foot pattern along a path just cleared by the point squad. When the leaves covering the bombs were removed there was a fresh boot print from one of the point squad right in the middle of the bombs, yet missing all five by inches. When mines were discovered we usually blew them up in place. When combat engineers weren't available we did it ourselves. Plastic explosives (C4), blasting caps, specialized detonation cord, and detonators were common items carried by companies, and many NCOs and officers were proficient in the basics.

The mine sweepers were useless against Chicom claymore mines, several times larger and more powerful than ours, set off to the sides of trails or in trees. Nor could they detect the ubiquitous trip wire booby traps that dotted the countryside. Again, sharp eyes and a sixth sense for likely areas to be booby-trapped were a soldier's best defenses. All units carried ropes with grappling hooks to detonate booby traps from outside the expected blast area. Luck played into this as well. A forty-foot rope with grappling hook couldn't protect a soldier from the unearthly blast of a booby-trapped 175mm artillery round. It was like tossing a grenade into one corner of a room and taking cover in the opposite corner.

A foolproof defense was to use locals as guides to move quickly through infested areas or to point out booby traps in and around their village. They usually knew where all the booby traps and mines were. Often these same villagers, whether coerced or because they were VC sympathizers, had made the booby traps in the first place. In a letter home, I explained when a soldier was killed or wounded by a booby trap, "The policy now is to use civilians as guides and if any Americans get hurt we burn the nearest hootches."[559]

HURT LIKE HELL

Dan Bernard's luck had run out after eight charmed months on the line, beating the odds first as the company commander's radio operator, then as the leader of an infantry squad. He had survived every Tiger Bravo operation and combat assault since arriving in-country. On July 28, Bernard loaded onto a chopper as part of a flight of ten slicks on a combat assault mission I was leading into an area where VC had been spotted the previous day. The helicopters' approach to the PZ and landings were uneventful, but on liftoff, what had been a cold PZ suddenly turned hot. Right after liftoff, and only fifty feet off the ground, the door gunners on the choppers began firing almost straight down at black-clad VC fighters firing at the underbellies of the choppers. My helicopter made it through the fire unscathed, but the chopper just behind mine, carrying Bernard and five others, was not as lucky. Their chopper flew right into a fusillade of bullets. "The ship got hit nineteen times," Bernard recalled. "One of the door gunners, sitting right behind me, got shot in both femurs. He went into shock. A private named Rohm caught one in his chin, and it came out the back of his neck. He just slumped over and we held on to him so he wouldn't fall out of the helicopter. I got shot in the leg. Right on the shinbone. It hurt like hell."[560]

Basil Rivera, who was on the same helicopter as Bernard, was one of the lucky ones. Rivera recounted:

When we were forty or fifty feet off the ground, I give a thumbs up to Rohn. He was sitting where I usually sat by the open door so I could be the first off. He just smiles back at me, then he gets hit in the neck, and starts falling out of the open door. Then the chopper gets hit and shudders like a heavy animal shudders when it gets hit. All of a sudden, the chopper tilts to a forty-five-degree angle. I look down and see the enemy and they're looking up like "that chopper is going to fall." It was tilting so bad it was just like it was going to flip over. So, Rohn is falling out, I'm grabbing at him and, I can't believe it, but bullets are flying inside the chopper, in one open door and out the other. I can't believe I didn't get hit. Three quarters of Rohn's body is out of the chopper. He's yelling, "somebody help me." I grab him, if he falls out I'm going after him, and jerk him back. At the same time Johnson grabs him too, then our guys grab me and everybody grabs one another.[561]

Rivera threw Rohn onto the floor of the helicopter, now slippery with blood, held a first- aid packet to his neck, and began praying. "I didn't pray much over there, but this time I did. I'm praying real hard. 'Please God, please God, let him live,' I prayed."[562]

The helicopter immediately broke formation and flew straight to the hospital, where the doctors and medics began treatment on the tarmac, before the chopper blades stopped rotating. "Just like in the M*A*S*H movies," Bernard recalled. "Here come all these doctors in their cutoff shorts. They got to the ship and said, 'Who's the worst?' Pointing to Rohn I said, 'If he's still alive, it's him.' I've never seen a tracheotomy performed, but right there they poked a hole in his throat and revived him, right on the spot."

Later, before he was evacuated to Japan to recover, Bernard was told by Rohn's doctor that "he had been without oxygen for a few minutes" and thus "part of his brain had started to decay." But because he was only seventeen years old, "another part of his brain will take over." Bernard also met with the pilot and copilot of the helicopter, who stayed at the hospital to check on their door gunner. They told him that "if it was any consolation all the bad guys who were shooting at us were dead."[563]

On the other side of the battlefield, severely wounded enemy soldiers had no such robust medical system in place. In fact, their chances of survival were limited. There was only so much that could be done for the severely wounded VC or NVA *bo doi* (soldier). Many enemy soldiers died during the evacuation process and were buried in hidden graves. Unless the wounded could be quickly transported overland to a makeshift hospital, often in a tunnel or bunker, pressure dressings and tourniquets were all that were available to keep a man from bleeding to death.

Medical supplies and battlefield dressings were in short supply. Discarded American field dressings and gauze pads were scavenged from the battlefield, washed, and reused. Local VC would also buy and smuggle various types of cloth for use as battlefield dressings. One indicator that a VC or NVA attack would soon be mounted, focused on by US and South Vietnamese intelligence agents, was when large numbers of sanitary napkins were bought in nearby towns. Local VC would buy a supply and smuggle them to the waiting attack force to be used as battlefield dressings. If the wounded lived long enough to be carried to an aid station or hospital, which might require an all-day or overnight trip, blood transfusions might be available from donated blood, but there was a limited supply and it remained usable for only so long. Long-term use of tourniquets often resulted in gangrene, leading to frequent amputations. There was little available for pain other than aspirin and marijuana; morphine was scarce. Anesthesia was simply ether, if available.[564]

Even after reaching medical aid, the mortality rate for VC/NVA fighters was extremely high. Word of an NVA soldier's death or serious wounds was sometimes passed on to families in North Vietnam by comrades who managed to return home from the South, although these were precious few. Parents seeking information about a son from the authorities drew a visit from the police. One simply did not talk about casualties in the South, and many never again heard from their sons. If a family did covertly receive word of a loved one's death, they could not publicly mourn.

The more seriously wounded—amputees, those with brain injuries and other serious physical disabilities—did return to the North after months of arduous travel. They were not discharged or allowed to visit home; instead, they were sent to camps in the highlands and remote coastal areas. It was felt that the constant influx of huge numbers of disabled troops into the civilian population would have a detrimental effect on civilian morale. These men were allowed to write home, often for the first time since they had gone south, but they could not mention their wounds, conditions in South Vietnam, or anything about the military. Their letters were heavily censored, but some were able to smuggle out uncensored messages. As a result, all manner of tragic stories and rumors coursed through the civilian population. This left segments of the North Vietnamese population in a perpetual state of silent mourning.[565]

Dan Bernard returned to Tiger Bravo after three weeks in the hospital, with his leg not completely healed, and went right back on the line as a squad leader. "The gunshot wasn't the problem," Bernard remembered. "It was the infection. They sent me back to Bravo Company and it wasn't even healed. So, I went straight back to around Cu Chi, where we were constantly up to our thighs and knees in mucky water. That wound burst open. It was getting red and oozing green pus. Doc was giving me penicillin, but it wasn't doing any good. I was hobbling around using my rifle as a crutch."[566] Two weeks later, Dan Bernard, one of a dwindling number of original Tiger Bravo members, left the company for good. "That was the end of the war for me," Bernard recalled.[567] He would spend another thirty days in a hospital in Camp Zama, Japan, where doctors finally brought the infection under control. At one point the possibility of amputation was discussed, but never acted upon. By early November, Bernard was out of the army and home to New Hampshire.

It seemed every move on our part brought a reaction from the VC. We no longer planned solely for enemy contact as we landed in a combat assault. Now even the pickup zones were hot. Hot pickup zones became so commonplace that a separate tactic was developed to conduct an extraction under fire. Normally, when contact was not expected, companies would break up into six-man loads and spread out the length of the pickup zone. The helicopters would then land in single file— "in trail"—in the center of the pickup zone and the soldiers would move to the choppers. A colored smoke grenade would be thrown to mark the position where the lead chopper should land. But when contact was expected, or when a unit was in contact but still needed to be extracted, the six-man groups—called "sticks"—would be spread out on the pickup zone in the best defensive, fighting positions. It was up to the choppers to individually fly to their locations, and land as close as possible to the soldiers, who would be firing up to the last seconds before loading. Each separate stick would mark their position with a colored smoke grenade. This type of extraction under fire soon became known as "the Christmas tree" because from the air, the different-colored smoke grenades, set against the green of the jungle or a rice paddy, reminded one of the festive, multicolored lights on a Christmas tree.

A STAGGERING TOLL

Weeks of nonstop missions and daily casualties had taken their toll on every company in the battalion. Halfway through the summer, Tiger Bravo had already amassed a staggering number of casualties for a unit its size—12 killed in action and another 39 wounded. Medevacs were a daily occurrence, which according to Harry Brown "circle us like vultures."[568] From private to captain, the descriptions of Tiger Bravo's decimated ranks were remarkably similar. On July 27, Brown wrote, "Just got back from a two-day operation, we had three more men killed. That makes 13 dead and 20 wounded in six days. . . . Tomorrow we go on a combat assault. . . . Yesterday we had another hot LZ. That makes three in two days."[569] My letters echoed Harry's. On July 28 I wrote, "Last night I took 3 more casualties from a booby trap. . . .. My company is down to 82 people. I used to have 4 platoons, but because I have lost so many I have dropped to 3 platoons, and they're all at half strength."[570]

In August, the pace of operations continued unabated, and casualties continued to mount. Entries of radio calls made by Tiger Bravo in the 2/506 "Daily Staff Journal" for August 2[571] were typical:

- 0420 HOURS: TIGER 2 (AMBUSH #2) SPOTTED 4 VC, DISTANCE 200 METERS. ENGAGING WITH SMALL ARMS AND ARTY.
- 0600 HOURS: TIGER 1 (AMBUSH #1) SPOTTED 1 VC, TOOK HIM UNDER FIRE. RESULTS UNKNOWN.
- 1400 HOURS: 16 ELEMENT (1ST PLATOON) RECEIVED 2 SNIPER ROUNDS. FIRING "4.2" MORTAR IN SUPPORT.
- 2050 HOURS: NDP RECEIVING FIRE. ONE LIGHTLY WOUNDED.
- 2100 HOURS: STILL RECEIVING SPORADIC FIRE. ESTIMATE ENEMY AS TWO SQUADS. 1 SQUAD TO EAST AND 1 SQUAD TO SOUTH.
- 2105 HOURS: RECEIVING MORTAR FIRE FROM SOUTH AT 180 DEGREE AZIMUTH. RANGE 200–300 METERS.
- 2110 HOURS: RECEIVING RPG ROUNDS. HAS 3 OR 4 LIGHTLY WOUNDED FROM MORTARS.

- 2115 HOURS: HAVE 5 WIA AT PRESENT.
- 2140 HOURS: NO ENEMY FIRE FOR 10 MINUTES. HAVE
  8 FRIENDLY WOUNDED. REQUESTED MEDEVAC.
- 2150 HOURS: RECEIVING MORE RPG ROUNDS.
- 2210 HOURS: MEDEVAC COMPLETE—1 LITTER AND
  8 WALKING WOUNDED.
- 2250 HOURS: TOTAL WIA 11. 1 GUNSHOT, 10 FRAGS FROM MORTARS. ALL
  EVACUATED.

One of the wounded was Dan Soto. He had been slightly wounded by shrapnel in his arm on July 22 during one of the attempts to rescue the wounded caught in the kill zone of an ambush. This time it was more serious. "We were starting to dig in," Soto remembered. "I heard this Pop Pop—Pop Pop! It was mortars. I got hit bad, in the legs and nose."[572] Somehow, halfway around the world in Arizona, Soto's mother knew that something had happened to her son. "When I was wounded, my mother felt it," he continued. "She went to the Red Cross and told them—I think my son is hurt." The Red Cross had a network of representatives with all the major combat units in Vietnam that was used by families to check on soldiers in Vietnam, outside of official army channels, and to relay important family events such as births and deaths. "The Red Cross found me and asked if I had told my mother that I was wounded. No, I told them. Well, she knows, someone did, the representative said. So, you better write her."[573]

Four days later, Tiger Bravo was in another battle with special significance to me. I celebrated my twenty-fourth birthday flat on the ground behind a rice paddy dike in the midst of a battle against a VC battalion. It all started with Tiger Bravo and A Company, 2/506, making a combat assault into hot LZs. Machine-gun and small-arms fire raked the helicopters as they landed. RPGs roared and belched smoke as they flew out of the tree line, seeking the helicopters during those vulnerable few seconds when they were on the ground discarding their load of troops. The fight would continue for the next twelve hours. "Well, I'm 24!" I wrote home, "Last night we went into another battle.

What a birthday party. The sky was lit up by flares, tracers everywhere and air strikes. I was up all night and right about midnight the battle was still going on. The gooks were shooting anti-aircraft fire at helicopters, jets, anything. It was quite a show."[574]

A 3rd Brigade operational report for the period summarized my "birthday party" this way:

ON 5 AUGUST THE 1ST BATTALION, WITH OPERATIONAL CONTROL OF A, B, C AND D COMPANIES 2ND BATTALION, CONDUCTED A NIGHT COMBAT ASSAULT . . . IN AN ATTEMPT TO SEAL AN AREA IN WHICH AN ESTIMATED VIET CONG BATTALION WAS REPORTED. A AND B INSERTED INTO HOT LANDING ZONES, RECEIVING SMALL ARMS, AUTOMATIC WEAPONS, AND RPG FIRE ON THE INITIAL INSERTION. THROUGHOUT THE CONTACT, WHICH LASTED FROM 1718 HOURS UNTIL THE FOLLOWING MORNING, THE GROUND UNITS AND FIRE SUPPORT BASE LINDA ALSO RECEIVED MORTAR FIRE. SMALL GROUPS OF VIET CONG ATTEMPTING TO EXFILTRATE WERE ENGAGED THROUGHOUT THE NIGHT. HEAVY ANTI-AIRCRAFT WAS RECEIVED FROM AT LEAST FOURTEEN (14) 50 CALIBER MACHINE GUN POSITIONS . . . THE 3RD BRIGADE COMMAND AND CONTROL AIRCRAFT AND AN OH23 WERE DAMAGED BY GROUND FIRE . . . . US FORCES WERE SUPPORTED BY ARTILLERY, LIGHT FIRE TEAMS, FLARE SHIPS AND 15 AIR STRIKES.[575]

Tiger Bravo's pace continued unabated for the next two weeks—twelve search and destroy missions, sporadic mortar attacks, four combat assaults (one hot LZ), five enemy contacts, two days of fire support base security, plus twenty plus night ambushes, one soldier's leg broken by a charging water buffalo, and three injured by a lightning strike while moving at night during the monsoon rains.[576]

It was also the rainy season and there was no way to dry out. Soldiers were wet day and night. "It is just water, water, and more water," wrote Tom McClear. "Last night we set up an ambush in a rice paddy. Real nice. We spent the whole night in about two feet of water. . ... The other day we were going through rice paddies and I fell in a well. The well was concealed by high water and I took a full 30-inch stride with my right foot smack into the middle of it. My left leg being out and my rucksack are the two things that saved me from going straight to the bottom."[577]

The muddy rice paddies could be lifesavers as well. Twenty-one-year-old Private First Class Mike Scott remembered the night "my squad was heading out on ambush when . . . mortar rounds started to hit in front of us. After seeing two rounds hit, I made a decision to turn around and head back—wrong decision! The third round hit right in front of me and the explosion knocked me down. Good thing it landed in the wet rice paddy. I did not find out I was hit until the next morning. I had a scratch near my eyes and shrapnel in my palm near the thumb. Medic dug it out and told me to keep it dry and clean. I laughed at him and told him we were in Nam. Ended up having the wound lanced every day until it finally healed."[578]

By this time, Tiger Bravo was a shell of the robust, fit-to-fight airborne rifle company that it had been at the start of the summer campaign. The staggering toll of casualties, injuries, and losses from disease, coupled with the inability of the army replacement system to keep up with the carnage, had reduced Tiger Bravo's strength by 50 percent. The field replacement process had become a heart-rending routine. Young, fledgling soldiers flown in on resupply choppers. Broken, damaged, or motionless bodies flown out. Varying day to day, only by the numbers and severity of the casualties. "I'm just not making any headway," I wrote after one contact. "I received 11 brand new replacements yesterday at 1700 hours and at 2300 hours I had 11 wounded evac'd—14 wounded total, only 11 of which needed to be evac'd, and 5 of these 11 were treated and released."[579] Mike Scott, who was on the ground when the eleven nascent soldiers arrived at Tiger Bravo, remembered one in particular. "One of the replacements didn't belong in the field. He was so scared after an enemy attack that he dug a hole into the dike and would not come out, although the hole filled up with water."[580] He soon adjusted to combat, as most did, and became a solid soldier. But Scott's original sense of foreboding proved to be prophetic. The green replacement would be killed in action just three weeks later.

This initial exposure of a new replacement to a line company in the boonies was captured in the monologue *Soldiers* by the guru of small-unit leadership in combat, Mike Malone. Lieutenant Colonel Dandridge "Mike" Malone was the commanding officer of the 3rd Battalion, 8th Infantry on Fire Support Base 25, and the one who sent Tiger Bravo across the valley on June 2, 1968, to find an NVA base camp.

"A fire support base with barbed wire, and sand bags, and artillery pieces, and radio antennae, and holes, and trenches, and bunkers. . . . The chopper with no doors and no seats, waiting on the battalion pad . . . . Door gunners and black machine guns. . . . Frightening speed across the roof of the jungle canopy, with treetops blurring by. . . . Tight, canted circles, and the whop! Whop! Whop! Of rotor blades as the bird eases down an open shaft in the jungle. . . . Troops on the ground, looking up, serious, busy, with longer hair, and beard stubble, and fatigue trousers split open at the rear, and no drawers. . . . Fingernails black and split, sleeves rolled up, and old, nasty, dirty bandages put on by Doc, and patches of swollen, red-brown jungle rot. . . . The powerful, pungent, scrungy, skanky smell of feet and socks too long together. . . . A company commander with old-man eyes, and maturity, and authority, and strength. . . . A radio operator with the quick, alert look of a "college kid." . . . Claymore mines, and machetes chopping brush, and troopers digging, and fresh holes in the ground."[581]

The ratio of veterans to replacements in the company had changed dramatically from the spring, when the majority of soldiers were part of the hard-core nucleus who were veterans of Tet, War Zone D, the Central Highlands, and the Rocket Belt. By the second week in August, only thirty-six soldiers from that original group remained on the company rolls.[582] Among the veterans, there was a growing sense that it was only a matter of time before they too would be part of that grim rotation—new replacements jumping off a helicopter and veterans being loaded on, breathing or not. "I felt that I was a dead man," wrote Ted Tilson. "My only chance of surviving was to do my job as best I could and to trust to luck!"[583]

With luck seemingly playing such a huge role in personal survival, lotteries were not uncommon. Mike Tarpley remembered a 4th Platoon lottery. "Each day, for $1 we put a number from 1 to 24 in a steel pot, our guess as to what hour in the day contact would be made. If we got shot at that day the number closest to that time got the pot," he recalled. "If no contact, the pot rolled over. It never went past $50. Some soldiers complained that it was bad luck to bet on contact, rather than against it. When Platoon Sergeant Edge found out about it, he stopped it."[584] The four line company commanders (A, B, C, and D Companies) in the battalion had a far more macabre arrangement. Each promised $1,000 to the widow or family of the first of their number killed in action; accidental death did not count. The payout was a princely sum, almost $15,000 in today's dollars. Luckily, no one collected.

Everyone carried around daily reminders that casualties were an intrinsic part of an infantryman's fate in Vietnam. Around our necks we wore dog tags stamped with our name, serial number, religion, and blood type, for battlefield identification when we were killed. One would be left on our body, and the other collected for notification to higher headquarters. Our dog tags also hung from our rifles in memorial services to commemorate our passing. An innocuous olive drab (OD) neckerchief tied around our necks served a dual purpose, used to wipe away sweat or soak with water to counter the blistering heat and as a sling or tourniquet. Many carried their own supply of albumin (a blood expander used to treat blood loss). Standard operating procedure called for it to be carried only by medics, but experience showed that it quickly ran out during a heavy contact. Last, soldiers also carried ponchos as rain gear, or to use as a makeshift sleeping bag or construct a hasty shelter. But it was understood that in the end, that same poncho was a soldier's body bag, his own personal shroud.

For many of us, what had begun with our landing in Bien Hoa in December 1967, as somewhat of an adventurous expedition, had turned into an exhausting war of attrition in which we fought for no cause other than our own survival.[585]

For Tiger Bravo, life was not all battles and booby traps, nor snipers and sappers. There was time for life's moments, brief respites from combat, and giving life rather than taking it. In one of the nameless hamlets around Trang Bang, while preparing to set up an NDP, a nearby cluster of thatched huts was checked out by Mad Dog Matosky and his platoon. Upon entering one of the huts, to the surprise of everyone, a Vietnamese woman was about to give birth. A quick radio call to Dr. (Captain) Barrett, the battalion surgeon who was close by at FSB Pershing, provided PSG Matosky and Specialist

Five Cotton, a medic in Tiger Bravo, with an abbreviated, two-minute obstetrics class. With Cotton taking the lead and the gruff, hard-core Matosky assisting, a baby was born.[586]

Exceptionally high casualty rates among the company's officer and NCO ranks also meant there simply were not enough soldiers and leaders to fill the roster of a 4th Platoon. In addition, the platoon leadership positions of the other three platoons, normally manned by officers and senior, seasoned platoon sergeants (E7), were mostly filled by junior sergeants. The roster of platoon leaders and platoon sergeants on August 11, 1968, showed the extent of the problem:[587]

- **1st Platoon Leader:** Lieutenant Vandertuin
  - **1st Platoon Sergeant:** Sergeant (E5) Nordyke
- **2nd Platoon Leader:** Lieutenant Winters
  - **2nd Platoon Sergeant:** Sergeant (E5) Vick
- **3rd Platoon Leader:** Sergeant (E5) McKinstry
  - **3rd Platoon Sergeant:** Sergeant (E5) Santos

Less visible was the effect of the war on our psychological state, especially with the veterans who had been under constant strain for months. We were spent emotionally as well as physically. Some of us worried as much about losing ourselves, as we did about losing our lives; recognizing that who we were becoming was not who we wanted to be. In a letter to his sister, Staff Sergeant Thomas "Biffer" Lamb expressed his feelings this way:

My time in this place will be over soon. I want to go home. I am tired of this war. I'm fed up with killing people, seeing my buddies shot, and almost being killed. This is a nasty, filthy experience, and goes against everything I was raised to think and believe. I want to wash my soul, but I'm not sure I have one anymore. . ... I want to go to college, reclaim my innocence, and forget everything I know and have experienced here.[588]

Staff Sergeant Lamb would never get his wish. Shortly after writing this letter, he would be killed in action. In a cruel twist of fate, due to the difficulties of mail delivery into and out of the war zone, the letter did not arrive at his sister's home until after the family had been notified of his death. It would be the last words the family would ever receive.

TURF WAR

In the last several weeks the battlefield dynamics had shifted once again, from the mid-August booby traps and sniper attacks, "nickel and dime" stuff that the VC were good at,[589] to daily intelligence reports of NVA formations on the move again. Clearly, the enemy was maneuvering its forces to contest the Americans for control of the area around Trang Bang and all of Hau Nghia Province. This was substantiated by the interception of a secret communiqué, from the Central Communist Party to the National Liberation Front (NLF) headquarters in South Vietnam, ordering another offensive:

LAUNCH THE THU DONG (SUMMER AND AUTUMN) CAMPAIGN BEGINNING ON 19 AUGUST 1968 AND ENDING 2 SEPTEMBER 1968. AFTER 2 SEPTEMBER 1968, THE 3RD GENERAL ATTACK CAMPAIGN WILL BEGIN AND LAST TO NOVEMBER 1968.[590]

Within the body of the secret directive was a chilling warning for the 101st Airborne units arrayed across Hau Nghia Province. It listed as one of the three principal objectives of the campaign to:

RAVAGE HAU NGHIA PROVINCE IN ORDER TO SET UP A WAY FOR TROOPS TO ENTER SAIGON.[591]

It wasn't long before this new campaign was in full swing. Reports began flooding in of major enemy troop movements. The 2/506 "Battalion Staff

Journal" entries in late August were typical of the sightings radioed to the companies in the field:

- AUGUST 22, 1700 HOURS: ENEMY COMPANY, 50–80 MEN, MOVED TO VILLAGE BAC SEN XT433259.
- AUGUST 23, 1815 HOURS: PROBABILITY OF 95TH VC DIVISION IN THE AREA XT4325.
- AUGUST 26, 2130 HOURS: LARGE ENEMY FORCE, 33RD NVA REGIMENT, IN VICINITY OF 101ST UNITS. XT4641. MOVING IN A SECURE MANNER TO REACH FRIENDLY (US) UNITS AND DESTROY ONE.[592]

It was shaping up to be a turf war between two rival gangs, each blind to his adversary's battle plans. The NVA commanders could not predict when and where the Americans would attack. They knew from experience that the 101st Airborne Division's eagle flights could drop out of the sky without warning. The American helicopter forays were unpredictable and set no discernible pattern, as had other American units in the past. Moreover, once in contact, these American soldiers with an eagle patch on their shoulder "piled on" from every direction. Day or night, under heavy fire or not, they kept coming. The 101st planners were at a similar disadvantage. While warnings of enemy activity in the area heightened everyone's vigilance, they did little to pinpoint exact NVA locations or intentions. The enemy was still a master at camouflage. Given several days in one location, a battalion could burrow into the ground and not be discovered until stumbled upon.

NEW "OLD MAN"

I would not be with Tiger Bravo as it faced its next battlefield tests. "Well, Tiger B is no longer mine," I wrote on August 26, 1968.[593] It was time. Time for me to finally hand over command of B Company. Captain Terry Van Meter, who had commanded the company for a week in July, had taken command the day before. Nine long months on the line had taken its toll. I was tired both mentally and physically, and emotionally drained. Since being wounded in June, everything was harder. I still did my job, especially under fire. But I no longer carried the false mantle of immortality. Contrary to the laws of probability I learned at West Point, I came to believe that with every passing day in the field the probability of me being hit and killed increased. One day soon, the probability would reach one hundred percent and I would leave the battlefield wrapped in my own poncho.

Part of me was happy to be off the line, with all of its hardships, dangers, and the immense responsibility of command. Yet part of me wanted to stay. "There was just something about getting on the chopper . . . and leaving Bravo Company. I've been through a lot with Tiger B. The things that I've done and seen will stay with me for a long time." I couldn't know it at the time, but I faced a lifetime of daily, intrusive thoughts of the Vietnam War that would bubble up from my subconscious without warning. "There is something about sharing hardships, dangers, and sorrows that forms a tie that is hard to break." Now, close to fifty years later, as I write this book, I know that the bond with those I served with is not hard to break; rather, it is impossible to break. My letter on August 27 summed up my thoughts:

> I was their commander, the "old man," who saw them laugh, cry, die, bleed, and even heard some screams. I know which men are brave and which have too much fear in them to ever control. I know the good and the bad. . . . I've given orders and made decisions knowing that someone might not come back; but, I know in my heart that I did everything within my power to do my job as the company commander . . . . I've had the privilege of leading them into battle. I'm proud, and always will be, of Tiger Bravo.[594]

12

Surround and Pound

Vicinity of Trang Bang
(Hau Nghia Province)
August 27–September 23, 1968

RIVAL BATTALIONS

O n the morning of August 27, 1968, two rival battalions—one US, one
NVA—were separated by only six kilometers. Both laid claim to the same
storied, hard-won unit designation of their respective armies. A battalion of
the 101st Airborne Division, the 2/506, occupied Fire Support Base Judy,
while encamped in a jungle and wooded area just kilometers away was a battalion of the NVA 101st Regiment. Each planned on annihilating the other.
Within hours, these two forces would be locked in a violent clash, leaving dead
and wounded from both sides scattered across the battlefield.

The morning began quietly, with Tiger Bravo and D Company 3/187
defending Fire Support Base Judy, and with A, C, and D Companies, 2/506,
prowling the battalion's area of responsibility. It was a well-established strata-
gem for the 101st Airborne Division, whenever a large enemy force was known
to be somewhere close by, to set up a hasty fire support base composed of a
battery of 105mm howitzers and the battalion's mortars, surround it with one
or more infantry companies for security, then fan out the remainder of the
battalion's companies to find the enemy. It was a simple plan, and exploited
two significant battlefield advantages that the Americans had over any VC or
NVA unit—air mobility and the capability to concentrate overwhelming and
devastating firepower. The first company making contact hunkered down and

served as the base. The remaining companies quickly "piled on" via combat assaults to surround the enemy. Then the trapped enemy would be pounded with artillery, air strikes and gunships.

At 1022 hours, A Company made the first contact and immediately stopped in place and set up a defense. With the base established, the rest of the battalion started to pile on. D Company moved overland to linkup with A Company; then Tiger Bravo (1335 hours) and C Company (1508 hours) made separate combat assaults to complete the cordon around what was estimated to be at most three hundred NVA fighters.[595] Tiger Bravo remained in contact the rest of the day, at one point capturing an NVA lieutenant.[596] According to the 2/506 Daily Staff journal:

ALL COMPANIES REMAINED IN CONTACT THROUGHOUT THE AFTERNOON. AS NIGHT CLOSED IN, THE 4 COMPANIES WENT INTO A TIGHT CORDON, ENCIRCLING THE ENEMY FORCES.[597]

The NVA battalion was trapped inside a box, measuring roughly four hundred meters by eight hundred meters, with no easy way out. The enemy commander had only two choices: attack, or find a gap between the American airborne companies and flow through it. Remaining inside the cordon meant total destruction. Every few minutes the ground would suddenly erupt from incessant artillery barrages, stopping only long enough for the tactical aircraft and helicopter gunships loitering in the airspace outside the box to dart in and add to the violence. The enemy commander's decision was soon evident. All through the night he mounted multiple assaults on the ring of companies surrounding his battalion, often supported by mortar fire from outside the battle area. A few scattered groups managed to bust through the cordon but, on the whole, each was repulsed after heavy fighting. In classic military understatement, the 2/506 Daily Staff Journal summarized a night of intense combat in less than three lines:

Tiger Bravo beat off its share of attackers as well. Harry Brown, who was in a foxhole on the flank by A Company when it was hit by 50 NVA soldiers, wrote, "They broke through A Company knocking out 4 positions and scaring off the two on my left. Our hole started receiving every type of fire they had— AK, MG, RPG, German Mausers, grenades, etc. We only had a chance to fire 2 shots because they came at us from all sides. Then a gun ship came in, killing most of them." Like the rest of Tiger Bravo, Harry "never slept after that."[599]

BLIND TRUST

When direct attacks failed, small groups tried to slip away quietly through the night without making contact. It was impossible in the terrain and pitch-black darkness to completely seal the enemy inside the cordon. Gaps existed between the understrength companies and platoons. The NVA would creep silently up to the American lines, wait until the flares floating above had flickered out, then pass like ghosts through gaps between fighting positions or, by chance, stumble into one of our foxholes in the dark. A successful escape in this manner was as much luck as skill, as both sides were unable to differentiate form from shadow in the darkness. That is, unless the defenders employed a "starlight scope"; then the exfiltrating NVA were at a decided disadvantage. Tiger Bravo carried several first-generation AN/PVS-2 night vision sights, or starlight scopes. It was an image intensifier device, weighing six pounds complete with battery, that took ambient reflected moonlight, starlight, or a sky glow on the horizon and intensified it up to thirty thousand times. Unlike earlier active sights that used an infrared light source that could be detected by the enemy, this technology allowed us to observe and engage an enemy who would be completely unaware of our presence.[600] Under the right conditions, and if the soldier had a steady hand to keep in focus, moving figures could be detected several hundred meters away.

One such encounter took place along Tiger Bravo's line of foxholes. Mike Scott, who had his 90mm recoilless rifle loaded with an antipersonnel beehive round, was on the Tiger Bravo line of foxholes facing the encircled enemy battalion but couldn't see a thing. Next to him was Private First Class Jimmy Pizzano, staring through one of the company's scarce starlight scopes.[601] Scott and the 90mm recoilless rifle were an unlikely pair; one of the smallest soldiers in the company carried the largest and heaviest weapon in the company's arsenal. The M-67 recoilless rifle was an antitank weapon used in Vietnam as a bunker buster and, in the defense, as an antipersonnel weapon. At 37.5 pounds and 54 inches in length, it was a nightmare to carry in the stifling heat and thick underbrush of South Vietnam. With a single trigger pull, the 6.7-pound M590 antipersonnel beehive round sends 2,400 steel flechettes screaming out of the barrel at 1,250 feet per second, mowing down anything in its path out to 300 meters.[602]

Pizzano was "scanning through a starlight scope and spotted an enemy mortar squad coming through the hedgerow toward our hole," remembered Scott.[603] Unable to see in the darkness, Scott told Pizzano to "tap my helmet when they were all out of the hedgerow,"[604] so he would know when to fire. Unlike Scott, who had been on dozens of combat assaults and learned combat skills the hard way during the company's months in War Zone D, the Central Highlands, and the Rocket Belt, Pizzano's initiation to combat had been crammed into August in Hau Nghia Province. But he had learned fast and had proved himself as one who could be counted on during a fight. Proving to be steady under fire was not always the case, as Tiger Bravo was like any other rifle company in Vietnam—a mix of heroes and cowards, good and bad, strong and weak.

This spur-of-the-moment, microtactical plan by Scott and Pizzano was an example of how battles were actually fought in the Vietnam War. A sweeping strategy positioned the 3rd Brigade of the 101st Airborne Division astride the invasion routes used by the NVA to reach Saigon. The 3rd Brigade commander's operational plan sent the 2/506 to the Trang Bang area, and the battalion commander's "pile-on tactics" set up the battle. But it was the actions of the individual soldier and his buddies that actually killed the enemy. Every battle was a disparate collection of individual or small unit firefights—spontaneous, short-range, violent, and fleeting. Without soldiers taking the fight to the enemy, every grand tactical strategy would fail.

A minute later, according to Scott, "Tap, then BLAM!" The 90mm roared. Flechettes buzzed across the open rice paddy, spreading out as they flew. The back blast blew a cloud of dirt and shredded foliage to the rear, halfway to the platoon command post. "Six killed with one shot. Pizzano couldn't believe it until he went out and looked at the bodies."[605] The next morning the heap of dead NVA bodies, riddled with hundreds of tiny darts, were searched; stripped of their weapons, ammunition, and equipment, including an 82mm mortar complete with base plate;[606] and left where they lay, to bloat and rot in the tropical heat. What had been worthy adversaries the night before, soldiers in their own right doing their duty and just minutes away from escaping to fight another day, were now nothing more than battlefield litter—something to walk by and forget. As the battle went into

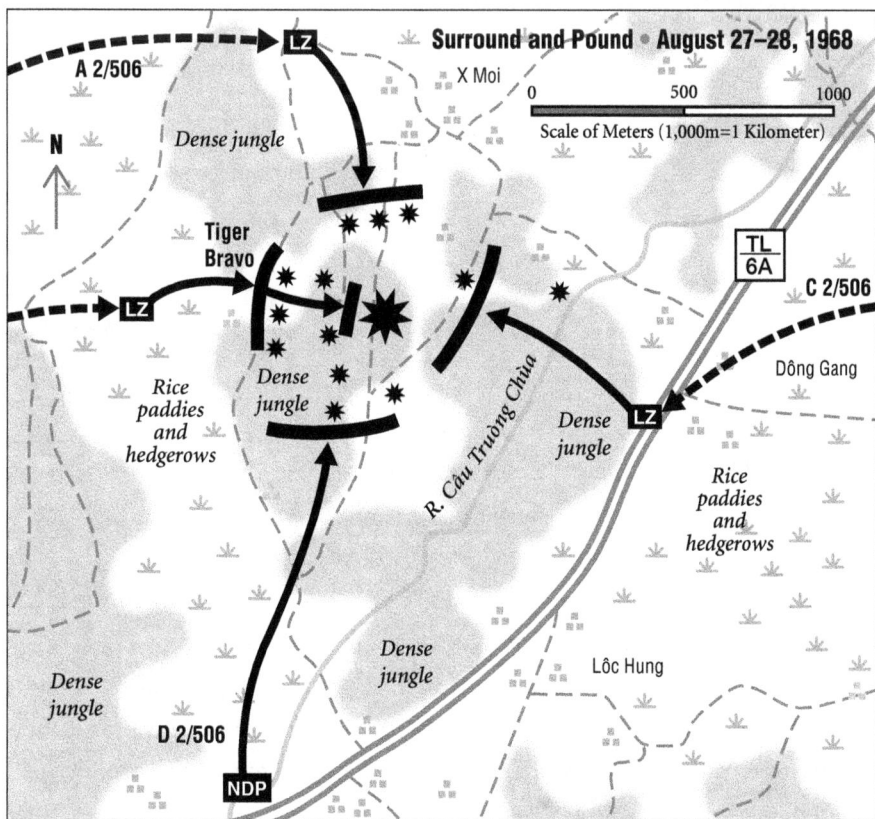

Surround and Pound • August 27–28, 1968

A 2/506

X Moi

0   500   1000

Scale of Meters (1,000m=1 Kilometer)

Dense jungle

N

Tiger Bravo

LZ

Rice paddies and hedgerows

Dense jungle

R. Cầu Trường Chùa

Dense jungle

LZ

TL 6A

C 2/506

Dông Gang

Rice paddies and hedgerows

Dense jungle

Lôc Hung

Dense jungle

D 2/506

NDP

its second day, the two foxhole mates would have other roles to play. In just a few hours, one of these two would go on to earn a Silver Star; the other would be dead.

HEART OF THE ENEMY

It was dawn of the second day and every sign indicated that the NVA battalion, or what was left of it, was still trapped. At 0650 hours C Company captured one enemy soldier who reported that three of his comrades were still inside the box. Additionally, both A and C Companies reported that small groups of ten to fifteen at a time would attack out of the jungle, then retreat after a vicious firefight.[607] Tiger Bravo had two contacts in less than an hour, killing two in the first contact and killing one and capturing one enemy in the second.

By midmorning, Lieutenant Colonel Keesling, the 2/506 Battalion Commander, decided to take the initiative and push straight into the box. His plan: hold A, C, and D Companies in place, then send Tiger Bravo from its blocking position on the west side of the cordon, straight through the middle of the box to split what was left of the enemy force, then destroy each piece individually. The order went out at 0948 hours over the battalion command radio net:

TO B 2/506: BE PREPARED TO MAKE A SWEEP WEST TO EAST IN 10 MINUTES.[608]

At 1022 hours, Captain Van Meter gave the order. Tiger Bravo got up out of its foxholes and moved into the heart of the enemy.[609] Slowly. Cautiously. Deliberately. Knowing that somewhere ahead lay a weakened, cornered NVA force that hadn't capitulated, and wasn't about to. Individual enemy soldiers may surrender, but VC and NVA units fought to the bitter end.

The first part of the sweep went well. Within the first hundred meters Tiger Bravo's point team had flushed out and captured two NVA and their 60mm mortar, then three more fighters close by.[610] Signs of the artillery barrages, napalm strikes, and US Air Force bombing runs were everywhere. Trees shattered. Scorched earth and blasted hedgerows surrounded them. Craters were scattered along the company's route. It was hard to imagine how a viable fighting force could have survived in any shape to fight. But the battle all around Tiger Bravo was far from over. The battalion command radio net was peppered with reports of enemy sightings just outside the cordon. One unconfirmed report sent over the battalion command net had up to two thousand VC in the general area of the 2/506. A constant rotation of gunships overhead darted in and out at targets identified by the other companies in the cordon. Even our helicopters were in the line of fire. The battalion's artillery fire coordination officer was shot in the foot while orbiting above Tiger Bravo.[611]

By 1100 hours Tiger Bravo was almost halfway through its sweep—dead center inside the box—with little to show for it other than the prisoners taken earlier. Up to this point, Tiger Bravo was battling the heat more than the NVA. With a heat index of 124 degrees Fahrenheit,[612] it felt like walking through a brushfire without the flames. But in another hundred meters, it all changed. Bravo ran into resistance, starting with a lone machine gun that wounded three. The contact quickly escalated to more machine-gun fire, RPGs coming out of the hedgerows, and automatic weapons from NVA in trench lines and bunkers, wounding another Tiger Bravo soldier.

In the next forty-five minutes, five more Tiger Bravo soldiers would be hit. Bravo Company had found what was left of the NVA battalion, and it still could fight! This was Tiger Bravo's battle, and it couldn't expect help from the other companies. Reports of new enemy contacts flooded the battalion command radio net—D Company in contact and capturing more NVA and weapons, A and C Companies in contact and being mortared. A new report indicated another sighting of another NVA battalion just three kilometers north of the cordon.[613] The battle was in its second day and showed no signs of ending.

Anyone listening to the battalion command radio net would have heard a seasoned team of commanders and artillery officers artfully choreographing the airspace over the battle. First would come artillery and mortar barrages fired from FSB Patton. Minutes later, the area would be cleared for on-station gunships to strafe with rockets and machine guns. It was a tightly controlled symphony of violence. But on the ground, it melded into one continuous cacophony of sound from rockets, artillery, mortars, automatic weapons, grenades, rifles, and RPGs, interspersed with cries of "Medic! Medic!" For the next two hours Tiger Bravo fought its way forward, foot by foot, suffering six more wounded. The first real sign that the NVA were in dire straits was when Tiger Bravo captured an 82mm mortar tube, then a complete 60mm mortar system.[614] Only desperate VC would discard such valuable weapons to lighten their load.

By this time Captain Van Meter was with his lead soldiers. In this terrain, even fifty feet back in the formation left a commander blind, relying on information over the radio and the occasional report relayed from soldier to soldier that never gave a true picture. Then at 1720 hours this message was recorded in the battalion's "Daily Staff Journal":

CPT VAN METER WOUNDED IN THE NECK AND SHOULDER. VIC GRID 5127.[615]

Another Tiger Bravo company commander had fallen. Reflecting on his decision to move forward, Van Meter recalled, "We couldn't knock out one of their positions, so I had to move to the front. I called the lieutenant back, we came up with another plan and I led everyone back. Turns out, they were a lot closer than had been reported. Caught me by surprise. I got clobbered." Van Meter had been shot at point-blank range — the first shot straight into his throat, another passing through his torso and the last into his arm.[616] He was then peppered with hand grenades as he lay helpless, just feet from the enemy. "I have thought about that decision over the years," Van Meter continued, "and still think that it was the right thing to do."[617]

Now the battle turned from an attack to a rescue of the company commander. The 1st Platoon advanced and came under intense fire, immediately suffering two killed and another soldier wounded. One of those killed was Staff Sergeant Thomas "Biffer" Lamb, the 1st Platoon sergeant. The platoon was moving forward, everyone crouching as low as they could get and darting from tree to tree, with Lamb in the center, when he was shot. Biffer Lamb was gone, that quickly. "He was someone who always moved to the sounds of battle."[618] Basil Rivera, who was one of Lamb's squad leaders, remembered.

"He was one of the best fighting and courageous leaders that Bravo Company had."[619]

The fight continued unabated. At one point, a well-camouflaged machine gun bunker had stopped the advance cold. Without any hesitation Mike Scott and Specialist Four McMillan crawled just short of the bunker and blasted it with a 90mm round. As Mike remembered, "I got the 90mm and a beehive round. McMillan had another. As we got close we could see the fortification. We bore sighted the 90mm and fired. Reloaded and moved. Good thing. Bullets are chewing up the area we just vacated. Fired second shot, firing slowed down. We headed back. Walked upright like we were behind a concrete wall. False sense of security. Bullets are still common from the bushes from other enemy positions."[620]

For his bravery Mike Scott was awarded the Silver star. The citation reads:

UNHESITATINGLY PRIVATE FIRST CLASS SCOTT VOLUNTEERED TO BRAVE THE ANNIHILATING HAIL OF ENEMY FIRE AND ATTACK A FORTIFIED POSITION WITH A 90MM RECOILLESS RIFLE. CREEPING AS CLOSE AS POSSIBLE, AND MIRACULOUSLY AVOIDING THE ENEMY FUSILLADE, HE FIRED UPON AND DESTROYED TWO FORTIFIED BUNKERS. REJOINING HIS PLATOON, HE BEGAN TO HELP WITH THE EVACUATION OF THE WOUNDED PERSONNEL. EXHIBITING EXCEPTIONAL BRAVERY, HE CRAWLED BETWEEN THEM AND THE ENEMY THROWING SMOKE GRENADES TO COVER HIS COMRADES WHO WERE TREATING AND EVACUATING WOUNDED. AT THE SAME TIME HE PLACED EFFECTIVE FIRE ON THE ENEMY POSITIONS ENABLING THE WOUNDED TO BE REMOVED FROM THE FORWARD COMBAT AREA.[621]

An hour later Van Meter had been rescued. But another eight Bravo soldiers had been hit and were missing, presumably dead. Harry Brown was in the 1st Platoon that had taken the brunt of the casualties and was part of the squad trying to recover the bodies of those killed. "My squad went in. We were just to the bodies, then they fired. We lost 3 more dead," Brown wrote. "One sniper had a Winchester 30 and the other a machine gun. Me, Kennedy and my squad leader played dead. Then [the enemy] started peppering the bodies" (of Tiger Bravo's fallen).[622] Then smoke grenades were thrown from the woods, creating a smoke screen for Brown and the other survivors to pull back. "We got out under heavy fire," Brown recalled. "A round went under my stomach, hitting me hard. I thought I was hit. Then three bullets went through a tree one inch above my head and the bark hit me in the head." Brown made it out safely and unhurt, never understanding how. "Never in my life will I know how I made it out. Nobody will, except God who I prayed to while [the sniper] shot up my friend into little pieces just four feet beside us."[623] It was a scene that Brown would never forget. "He was completely torn up," Brown continued. "Just like my new friend from California whose head split in half, and later his face fell off while picking him up."[624]

Mike Scott was in the rear of the 1st Platoon when the lead squads were taking the casualties. He remembered vividly what happened. "A radio call came, requesting smoke to cover a retreat. I volunteer to go, with a bag full of smoke grenades. Saw Harry Brown and Joe Adams flat on the ground. Pizzano, Plourde, Thomas, Washington, and Navarro are killed. I yell for them to get the hell out of there. Started popping smoke and firing. Empty a magazine, reload, fire, and move. Made it out without a scratch."[625]

Twenty minutes later Van Meter was on a medevac helicopter heading to a mobile army surgical hospital (MASH) in Cu Chi. He remembered waving to the soldiers lifting him on the helicopter, and one soldier saying, "the Old Man is going to make it, he just smiled." In one of those strange ironic twists that happen in combat, his medevac helicopter landed at the nearby fire base and picked up two critically wounded NVA soldiers.[626] It was standard practice to evacuate enemy wounded in supply choppers or on a space available basis via a medical helicopter, then treat them at an American or South Vietnamese hospital. Weeks later, I was relieved to receive a letter from Captain Van Meter, who wrote from his hospital bed in Japan, "All of my tubes I had hanging out of me are gone, except one. Here in Japan all that they are doing is trying to make me strong enough to withstand the flight home." Then, being the strong man and soldier that he was, in words without any hint of pity or fear of the future, he broke the harrowing news that "I still do not have a feeling from just above my waist down to my toes. My spinal column was not cut and this is why they believe I should recover completely, but it will take a long, long time."[627] However, Terry Van Meter would never fully recover from his wounds. He did not know it at the time, but he would spend the rest of his days in a wheelchair.

Captain Jim McMonigle, a staff officer at battalion headquarters, was quickly flown in to take command of Tiger Bravo. After dark Tiger Bravo set up a night defensive position, still missing the eight soldiers. Plans were made to move back into the contact area the next morning to recover the bodies. Then at 2200 hours the night was rocked by enemy 75mm recoilless rifle rounds, screaming into the perimeter. Luckily, no one was hit. The rest of the night was relatively quiet for Tiger Bravo, but for the second night, no one slept.[628] The next morning air strikes and artillery were called in to soften the area of yesterday's contact. Then Tiger Bravo moved in with two platoons to find its MIAs, killing two more NVA along the way. By noon all eight missing soldiers had been found and confirmed as killed in action. Tiger Bravo eliminated the last pocket of resistance within the cordoned area at 1230 hours.[629] The battle was finally over.

In three days the 2/506 had surrounded and decimated a battalion of the NVA's 101st Regiment, and held off a second NVA battalion trying to reinforce from the north. Known enemy losses were 106 killed, 19 taken prisoner, and another 3 detained. Actual losses were probably much higher, since the NVA and the VC took special efforts to clear the battlefield of their casualties. Battalion losses were heavy—16 killed and another 28 soldiers wounded. Ten of the US soldiers killed in action and seven of the wounded were from Tiger Bravo.[630] Lost in the battle were:

- Private First Class James J. Criswell, twenty years old, from Red Lion, Pennsylvania
- Staff Sergeant Thomas R. Lamb, twenty-four years old, from Los Angeles, California
- Sergeant Michael J. Mitchell, twenty-one years old, from San Esteban, California
- Private First Class Armondo S. Navarro, twenty-one years old, from San Antonio, Texas

- Private First Class James R. Pizzano, a nineteen-year-old green replacement from Everett, Massachusetts, who had been with the company for only a month. At Everett High School (1967) he was remembered as a "great friend, wonderful guy . . . so very smart, with a quick wit, handsome and with a smile that came easily."[631]
- Specialist Four Victor M. Plourde, nineteen years old,
  from Oakland, Maine
- Private First Class Paul M. Stockwell, twenty years old,
  from Corder, Missouri
- Corporal Isiah Thomas, twenty-two years old,
  from Washington, Louisiana
- Corporal Robert J. Washington, eighteen years old,
  from Indianapolis, Indiana, who had written "uncle with the mostest, 23 nephews and nieces" on his helmet cover[632]
- Private First Class Thomas V. Williams, twenty years old,
  from Lakin, Kansas.[633]

It was by any measure a dark day. Back at Phuoc Vinh, the losses hit everyone hard. "Do you think this war will ever end?' one soldier said, quietly, to no one in particular while reading the casualty list. Tiger Bravo had fought for three days and two nights—with no sleep—at ranges as close as ten feet, had lost a company commander and platoon leader, and seen more than 10 percent of its fighting strength taken off the battlefield by medevac helicopters. Spirits were numb; bodies battered and tired. Morale fluctuated between "ready and willing" to "just get me out of here," but for the most part it hovered closer to "I'll do the best I can, I just want it to be over."

This was not a good time for the next mission, but that's exactly what happened. Only one hour after the last Tiger Bravo casualties were evacuated an order came over the battalion command net for Bravo to move by foot approximately 3.5 kilometers southwest to secure an enemy weapons cache reported to hold 75mm recoilless rifles and ammunition. By 1730 hours the company was on the move, across a landscape fraught with large VC and NVA fighting forces, their exact locations and intentions still largely unknown.[634] In the past ten days the 2/506 had been in contact for nine of them. Even when not in contact, no one could relax. Fifty percent alert at all times was the norm. There was no reason to believe that the next ten days would be any different. Another three-day battle for Tiger Bravo could be just past the next hedgerow, or the next.

New Commanders

The end of August brought significant changes in the command structure for Tiger Bravo and the battalion. On August 30, 1968, the executive officer of Delta Company and the senior lieutenant in the 2/506 became Tiger Bravo's company commander, replacing Captain McMonigle, who returned to the battalion staff.[635] Then, at 1810 hours of the same day, the battalion's command and control helicopter crashed, sending Lieutenant Colonel Keesling and his operations officer to the hospital.[636] Their war was over. Major Freddie Boyd, the 2/506 executive officer, assumed command of the battalion. This went almost unnoticed by soldiers in Tiger Bravo, as they rarely saw Lieutenant Colonel Keesling. Unlike Lieutenant Colonel Grange, who would land his chopper under any conditions to lead on the ground, Keesling preferred to orbit above the battle or command from the Tactical Operation Center bunker at the closest fire support base.

Lieutenant Colonel Keesling was replaced by Lieutenant Colonel John O. Childs on September 1, 1968. This was Childs's second tour in Vietnam. In 1963, he served as an adviser to a battalion in the 33rd Vietnamese Infantry Regiment in the Mekong Delta before he became deathly ill with amoebic dysentery and hepatitis. "I was evacuated to Clark Air Base Hospital in the Philippines for treatment," he recalled. After recovery, he returned to Vietnam and was assigned as an adviser to an ARVN training battalion.[637] "I thought the 2/506 was well trained, smart, cocky, and not afraid to fight," was his first impression of the battalion. "They were not afraid of anything. Everybody seemed to know what they were doing. I was just impressed period."[638] In a move that was reminiscent of the popular Lieutenant Colonel Grange, the new commander, who chose the call sign "Sidewinder" to reflect his Texan heritage, immediately visited every company in the field and spent time on the ground, humping alongside his soldiers.

Cu Chi Battalion

For the next four days, Tiger Bravo and C Company secured Fire Support Base Pershing,[639] giving the company a much-needed respite from the craziness of August. While not safe by any means, providing security of a fire support base was considered a relatively benign mission. Each day, Bravo would conduct reconnaissance in force missions in the surrounding area, then return to the fire support base. At night, it would send one to two ambushes out along likely avenues of enemy approach.

On September 5, A Company took Tiger Bravo's spot along the fire support base perimeter. It was back to the field for Tiger Bravo, this time with the new battalion commander tagging along. The day was not yet half over when Lieutenant Colonel Childs received a radio call from the battalion operations officer: a large enemy force had been spotted in the area. "Major Boyd, the operations officer, sent a C&C helicopter to pick me up," Childs recalled. "He also arranged for helicopters to pick up Tiger Bravo for a combat assault into that same area to investigate."[640] The company's combat assault, near a village four kilometers east of Trang Bang, was into a hot LZ. As soon as the helicopters touched down, the company was in a vicious firefight with an estimated platoon-size enemy force.[641] Soon the volume of small arms, automatic weapons, and RPG fire was so great, and coming at Tiger Bravo from such a wide front, that the estimate quickly grew to a VC battalion.[642] Without knowing it, Tiger Bravo had landed on top of the temporary assembly area of the VC main force "Cu Chi Battalion," which was more than three times Bravo's size.

Lieutenant Colonel Childs, orbiting above the contested landing zone, noted "the exodus of hundreds of civilians, from a nearby village, fleeing for their lives." From a thousand feet above the scene it was impossible to be 100 percent sure that they were civilians, and not VC posing as farmers to escape before the Americans surrounded them with more companies. "I was asked whether to bring artillery upon them," Childs recalled. "I thank my God that I did not do that."[643] The next day it was confirmed that they were unarmed civilians. Childs continued, "It was decided to keep pounding the enemy unit with artillery fire, helicopter fire teams, and bring in more infantry units to surround them, rather than attack them straight on."[644] Over the next six hours seven more companies were flung into the battle, surrounding the VC battalion inside a box less than a mile across. A 3rd Brigade report described the operation:

USING THE "PILE ON" TECHNIQUE, B COMPANY WAS IMMEDIATELY REINFORCED WITH A AND D COMPANIES, WITH ATTACK HELICOPTER LIGHT FIRE TEAMS, ARTILLERY AND AIR STRIKES IN SUPPORT. A TROOP 3RD SQUADRON 4TH CAVALRY, A AND C COMPANIES 2ND BATTALION 27TH INFANTRY AND A COMPANY, WITH ELEMENTS OF D COMPANY 3RD BATTALION 187TH INFANTRY WERE INSERTED LATER TO COMPLETE THE CORDON.[645]

The battle continued all day, with repeated and violent enemy probes repulsed by the companies. The Cu Chi battalion commander had almost run

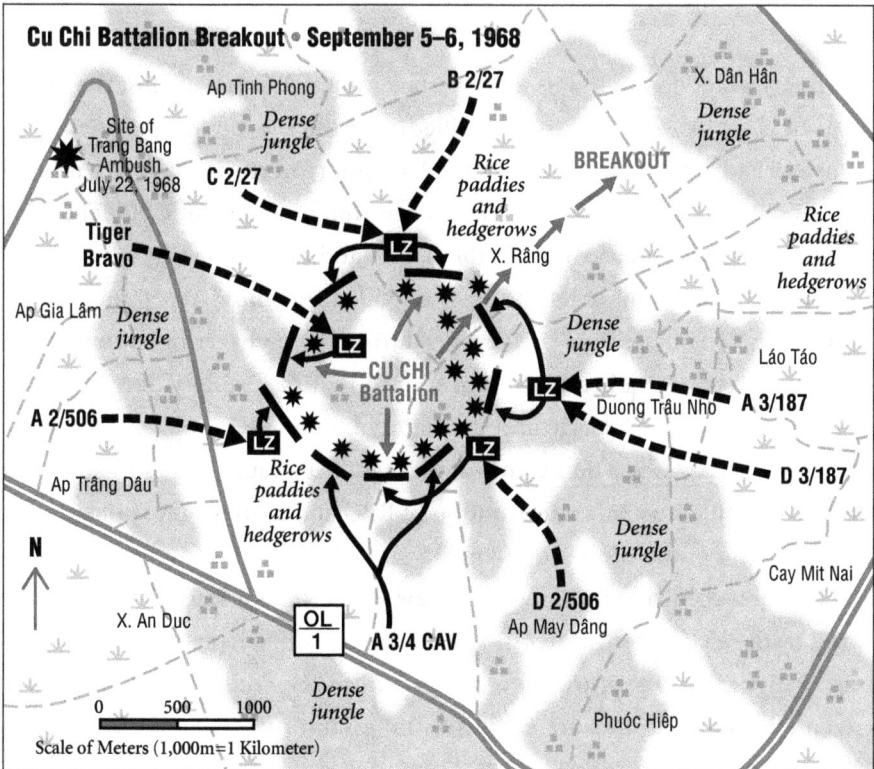

Cu Chi Battalion Breakout • September 5–6, 1968

out of options. His companies couldn't withstand the constant artillery bombardment and air strikes for another day, and the stout cordon of eight American companies was too tight for them to slip through in small groups, even
at night. His only option was to mass his battalion and send an overwhelming
force to punch a hole through the American lines. At 0230 hours he did just
that, bursting through the lines of A Company, 3/187, with wave after wave of
attackers. According to one 3rd Brigade report:

THE ENEMY BATTALION MASSED ITS STRENGTH AND ASSAULTED THE CENTER
OF THE A CO 3/187TH INF SECTOR AFTER INTENSE PREPARATION BY MORTAR
FIRE. A THREE-PHASE "HUMAN WAVE" ATTACK BROKE INTO THE COMPANY
COMMAND POST. FOR APPROXIMATELY 30 MINUTES A CO ENGAGED THE ENEMY
IN SAVAGE HAND TO HAND COMBAT. ENEMY FINALLY BROKE INTO AREA BEHIND
A CO AND RETREATED NORTH.[646]

It was a bloodbath. In less than thirty minutes, A Company lost the equivalent of two platoons, twenty-eight killed, and the remainder wounded. The
company had battled the best they could, but there were too many VC, massed
at the point of attack, to repel. The next morning Lieutenant Colonel Childs
was on the ground to reconstruct the battle, and determine why A Company
suffered such a terribly high number of killed in action. Years later he recalled, "The enemy probed the cordon during the night trying to determine
the weakest spot for a breakout. It appeared that the artillery flares we used
were floating so they clearly and constantly illuminated the command post
of Company A 3/187. The VC duly noted that. At about 0230 hours . . . they
opened the attack with a concentration of RPG and mortar fire, knocking out
the command group of the company."[648]

He then confirmed what we all knew would be our fate if we were captured by the VC, and why the ratio of Americans killed in action to wounded
was so high. "They captured several men from the company and promptly
executed them." From start to finish, the assault on A Company and subsequent escape of the Cu Chi Battalion lasted only thirty minutes. The few

Americans left alive in the center of the A Company perimeter reported seeing a macabre caravan of oxcart after oxcart, piled with VC dead and wounded, emerge from the darkness into the flickering light. The oxcarts quietly trundled past the Americans' foxholes, strewn with bodies, and disappeared into the darkness.[649]

Word of A Company's losses quickly spread across the battalion, along with wild, unsubstantiated rumors about how a debacle of this magnitude could happen. "Having no claymores employed along the perimeter, no listening posts established for early warning, not dug in and everyone asleep when the attack started"[650] were the rumors I included in a letter home. It also mentioned that rumors of how many A Company soldiers were executed ranged from several to "nineteen of the KIAs shot in the head from extremely close range."[651]

In a letter to his parents, Tom McClear wrote,

It was A Co. 187th Infantry . . . that had seen almost no combat since being over here. It was a cordon operation . . . with 500 NVA inside a 1 mile ½ by ½ mile area. [A Company)] were about 300 meters to our right, not dug in and violated several other basic rules. . . . About 1:30 in the morning I heard the firing. There was very little return fire from the American side. The enemy force was believed to be about 100 strong. Many of the GIs were found shot in the head. The theory was that the majority of the company was asleep. The CO and 1SG were killed so there is no one to hold responsible.[652]

Newspapers back home picked up the story as well. One article mentioned that a unit from the 101st Airborne Division was overrun, and included an unconfirmed report that the VC drove women and children in front of them as human shields. The September 6, 1968, edition of the *Honolulu Star-Bulletin* led with a front-page headline that read, "Viet Cong Use Human Shields in Attack on GIs." The story went on to state:

A veteran Viet Cong battalion, herding women and children in front as human shields, smashed into a company of American paratroopers in three waves. . . . About 300 enemy troops charged through rice paddies into one American company 150 meters away. . . . At headquarters the voice of a radio operator came through: "They are coming. They are coming." Then the radio went dead.[653]

Everyone had heard the anecdotal evidence of the VC using villagers to shield their assault forces, and each of us grappled with what our response would be. A strong survival instinct, to shoot before being overrun, battled against a basic tenet of the warrior code to never shoot unarmed women and children. Yet in reality, it would not be so black-and-white. Decision- making in combat comes in every shade of gray imaginable. Imagine after a long, arduous day of fighting and with your nerves supercharged, you barely make out in the flickering overhead illumination a large force coming at you from inside a cordon where you know the enemy to be. You have seconds to decide—shoot or not? Tom McClear spoke for the majority when he wrote, "If I saw someone coming at me at 1:30 in the morning from a cordoned off area which you had been receiving fire from all day, I am afraid I would shoot first and ask questions later."[654] Luckily, according to my own experience and extensive research for this book, Tiger Bravo never faced this dilemma.

On September 7, Tiger Bravo was ordered to a pickup zone for extraction to support another 3/187 company in contact. Once at the PZ and now spread out across an open area waiting for the helicopters, Tiger Bravo came under RPG and small-arms fire from the south. On an order from the battalion commander, it quickly moved into a wood line along the northern portion of the PZ so that artillery could be dropped on the enemy. The exchange of fire ended by 2010 hours, the pickup was canceled, and Tiger Bravo moved into a nearby NDP position, still under 2/506 control.[655] For the next week Tiger Bravo provided security to FSB Pershing and Stuart, conducting daylight sweeps into the surrounding countryside and setting close-in ambushes at night. Only once, on September 11, did Bravo make contact on one of these sweeps outside the wire when its 1st Platoon spotted two VC and gave chase, but came up empty.[656]

Inside the fire support base perimeter was another story. On September 14, it seemed as if every VC gunner in the area wanted to try his hand at dropping a mortar round into the Americans' lap. It was a quiet, hot day, with troops not on guard lolling under whatever shade they could find, until a mortar barrage, eighteen in all, suddenly carpeted the fire support base.[657] Luckily, other than jangled nerves, no one in Tiger Bravo was hit. Then A Company intercepted a partial VC radio transmission outlining a massive ground attack on the fire support base that night. You could carry the tension around in a bucket until the sun came up. Late that night one of Bravo's ambushes had two lightly wounded as a result of a VC contact,[658] but the ground attack never materialized.

DÉJÀ VU, ALL OVER AGAIN

On September 15, I received an unexpected message from the battalion commander. He had removed the Tiger Bravo commander and I was back in command, rejoining the company at Fire Support Base Pershing. I never knew the reasons for the switch, but I had heard that my replacement had problems. A week earlier, Mad Dog Matosky told me that "the CO just wasn't ready for command. He was late, sometimes in the wrong place and does halfway jobs." Other Tiger Bravo platoon leaders had also complained to me that he was "not much on security or good formations when moving."[659] But the suddenness of the announcement did surprise me. Four hours later I led Tiger Bravo on a combat assault into a cold LZ, but one soldier fell into a punji pit[660] and had his leg skewered by an excrement-tipped bamboo spear. I had no problem adjusting. It was as if I had never left.[661]

For the next week Tiger Bravo provided fire base security, first for Pershing, then for the smaller, company-size Fire Support Base Stuart. The routine for soldiers meant days filled with patrolling outside the perimeter, the occasional combat assault, and nights spent pulling guard duty in dank bunkers or ambushing along footpaths and jungle trails, shooting at anything that moved. Then there were the particularly dangerous listening posts at night. A two- or three-man team, often more than a football field's length away from the safety of the perimeter, were used to warn of an approaching enemy. Any sound or movement—man, animal, shadows, a breeze, a whispered voice on the radio, or an occasional phantom that no one else could see—would send adrenaline cascading through the soldier's nervous system, peaking with the dreaded message over the radio from the company commander that there was "movement between your position and the perimeter."

While the Tiger Bravo operations were routine, the results were anything but. The area was still teeming with VC and NVA units, and fire support bases remained juicy targets to be tested, probed, and bombarded with mortars. On the night of the September 15, troopers in a Tiger Bravo listening post outside of FSB Pershing fired on a VC approaching them out of the darkness. The next day another mortar attack on Pershing sent everyone scrambling for cover. Once again, Tiger Bravo's luck held; there were no casualties. September 17 brought a string of contacts, lasting all day and into the night. At 0027 hours and again at 0120 hours Tiger Bravo listening posts engaged small groups of VC probing for a weakness in the defenses. One soldier was wounded.

Before dawn came another mortar attack, this time in conjunction with VC sappers in the wire carrying satchels of explosives to blow up the valuable cannon artillery in the center of the base, resulting in another three wounded in the company. Then, just after first light, a Tiger Bravo patrol made contact with an unknown size force and had one casualty from a hand grenade. And again at 1824 hours, another mortar attack. This time the rounds exploded inside the perimeter before the mad scramble for cover was completed, resulting in another five wounded for Tiger Bravo.[662]

September 18 and 19 brought more mortar attacks on the firebase, this time coupled with movement outside the perimeter and probes of the firebase defenses that put everyone on alert. At one point even the geese, bought by the army and caged inside the barbed wire as an early warning system, worked as the army intended when their honking joined in raising the alarm. An entry in the 2/506 Daily Staff Journal for September 19 read:

2055 HOURS B 2/506 6 (COMMANDER) REPORTED HIS GEESE ARE RAISING HELL ALONG THE PERIMETER. COMPANY ON 100% ALERT.[663]

On September 20, the company was ordered to take over security of Fire Support Base Stuart just outside Trang Bang. A change in scenery didn't seem to help; on its first patrol around the firebase Tiger Bravo made contact, capturing one VC. The next day was a combat assault into a cold LZ and a long, uneventful march back to Stuart. On September 22, it was business as usual for the VC. Another of Tiger Bravo's patrols made contact, and after an exchange of small arms the VC pulled back, leaving no casualties on either side.[664]

BYE-BYE, TRANG BANG

At about this time the rumors flying around the 3rd Brigade about leaving the killing fields of Trang Bang for good and "moving north" to rejoin the 101st Airborne Division came true. All across the 3rd Brigade, a warning order went out to be prepared to stand down from combat operations in the vicinity of Trang Bang and to return to Phuoc Vinh, for further deployment to I Corps. The long, bloody "Currahee Summer" was about to end.

Within hours helicopters were crisscrossing the sky above Trang Bang, ferrying one company after another to Cu Chi for a one-day stand-down before returning to Phuoc Vinh one last time. On the morning of September 23, 1968, there was a changing of the guard at FSB Stuart. An infantry company from the 25th Infantry Division arrived by truck convoy to replace Tiger Bravo. The soldiers had packed their rucksacks at first light without being told to. This was one time when the NCOs didn't have to check on preparations for a move. Escaping the killing fields of Trang Bang was a powerful incentive to be ready on time; no one wanted to miss the helicopters inbound for the short flight to Cu Chi. Unlike the relief company which had three 2.5-ton trucks filled just with personal gear[665] included in the convoy, Tiger Bravo's soldiers carried on their backs everything they needed to fight and survive. At 1045 hours the changeover was complete. Tiger Bravo was in the air, heading to Cu Chi.

The next day was one of complete rest for the company in barracks at Cu Chi. For the first time in weeks, the soldiers could sleep inside and not worry about what "Charlie" would do. For a few, it had been even longer since they had slept under a roof. Signing off on a letter to his parents, Tom McClear wrote, "Well, I am sleeping under a roof tonight for the first time in about 70 days so I think I will make the most of it."[666] The battalion had been scheduled to convoy to Bien Hoa, then load aboard US Air Force C-130s for the short hop to Phuoc Vinh. But the aircraft had been diverted to a more pressing mission, which left a whole day for sleeping, PX runs, letter writing, and the inevitable card games that seemed to appear within seconds whenever two or more soldiers were in the same spot with time on their hands. By September 25 the army and the air force had reworked the plan. A Company flew directly to Phuoc Vinh by CH-47 Chinooks, while Tiger Bravo, and the remainder of the battalion, rode by truck to the airfield at Bien Hoa, then flew to Phuoc Vinh, arriving at 1700 hours.[667] The "Wandering Warriors" had returned to their home base, this time for the last time.

Part IV

Homecoming

13

Ground Zero

Phuoc Vinh
(Binh Duong Province)
September 24–30, 1968

TOXIC DELUGE

K ept secret from all of us at the time was the fact that Phuoc Vinh, the home base of Tiger Bravo, was ground zero for Agent Orange missions during the Vietnam War. None of us was aware that Tiger Bravo soldiers lived, ate, bathed, slept, played, and fought in the epicenter of an enormous, toxic chemical deluge. Of all the bases where Americans launched herbicidal defoliation missions under the name "Ranch Hand" across all of Southeast Asia, Phuoc Vinh ranked number one on the list, in terms of the amounts of toxins flown from its airstrip and sprayed in the general area.

During the years that Phuoc Vinh served as a base camp for US forces, including Tiger Bravo's ten months, a colossal 484,383 gallons of Agent Orange, 146,576 gallons of a secondary herbicide called Agent White, and 12,810 gallons of yet another, named Agent Blue, were launched from the Phuoc Vinh airfield.[668] These missions covered all of War Zone D, and often drenched the landscape well within a ten-mile radius around the base camp. In the calendar year of 1967 alone, sixty-one defoliation missions were flown within five miles of the base camp perimeter wire.[669]

While the location and timing of defoliation missions, in relation to tactical operations, may have been tracked and considered at some level of operational planning, at the battalion and company levels it was not a factor. "As far as Agent Orange is concerned, I do not remember any mention of it except in passing"[670] wrote John Childs, the battalion commander during the last two months of our time at Phuoc Vinh. There were no warnings, or alarms, or briefings on potential hazards; the subject just never came up. "Although we knew the stuff was being sprayed in the area," recalled Steve Lyle, "I don't recall anybody being concerned about it."[671] "I didn't think about Agent Orange while I was in Vietnam," Jim Roach agreed. "I don't think I even considered the topic until the 1980s."[672] Doc Franks recalled the same lack of concern: "I had not seen any sickness or heard any complaints about Agent Orange while in country. I don't think anyone understood it's danger at that time."[673]

It was impossible to avoid areas that had been sprayed with the toxins. A map of III Corps marked with defoliation missions over a period of years looked like the Vietnam equivalent of carpet bombing in World War II. Wide swaths of War Zone D had been hit multiple times. The effect was easy to see. Chris Backman described walking through a defoliated area as if it were hit by a forest fire. "It was bare, everything was dead. There weren't even any leaves on the trees."[674] Occasionally, Tiger Bravo operated in areas with an active defoliation mission in progress. There was never any warning or orders to move out of the path of the planes. The first indication that a Ranch Hand mission was in the area was when the planes appeared and commenced spraying. Doc Franks remembered one incident in War Zone D when the entire company was hit. "We were paralleling a partially defoliated area when the planes dropped their loads. On one of the passes we were sprayed with a heavy concentration of the chemical as a plane passed overhead. I recall its mist filtering down through the jungle canopy."

Like the rest of us, Franks was not concerned. "I wasn't alarmed by coming in contact with it at that point as I never envisioned that the US government would put any of us in peril. How wrong I was!"[675] Ted Tilson remembered a similar experience. "I remember quite vividly," he recalled. "On a RIF (Reconnaissance in Force) in the middle of War Zone D, in the middle of the afternoon, the company broke out of some thick jungle onto an old roadbed or a long clearing. Three silver-sided C-123s flew directly overhead in a V formation spraying a fine mist onto the entire company formation. I distinctly recall my helmet cover being very damp as well as the top of my rucksack." Like everyone else, he wasn't aware of the danger and even looked on the bright side of the incident. "We knew it was defoliant, and thought it was a good idea because we were tired of getting ambushed."[676] Joe Palagyi remembered waking up one morning in War Zone D, drenched with an unknown, sticky substance. "During the night, the agent orange had dripped down through the jungle canopy and covered the ground and us."[677]

The ground in front of the bunkers around Phuoc Vinh base camp was also a favorite target. The bunker line was periodically sprayed to clear the fields of fire for machine guns and eliminate cover and concealment for sapper attacks. Every base camp in South Vietnam sprayed defoliants for the same reason. Additionally, all roads heavily traveled by US supply convoys were routinely sprayed to eliminate potential ambush sites and to give the Military Police guarding the convoys a clear view on either side of the road.

The Agent Orange story had begun long before Tiger Bravo started its combat tour. The series of events that led to the decision to send US Air Force C-123s to South Vietnam to spray herbicides seems to have begun on April 12, 1961, when Walter W. Rostow, a foreign affairs adviser to President Kennedy, forwarded a memo to the president on "gearing up the whole Vietnam operation." Nine specific courses of action were mentioned in his memo. The fifth course of action recommended that a military hardware research and development team go to Vietnam to explore the usefulness of various "techniques and gadgets" then available or under development. Aerial defoliation later became one of these unspecified "techniques and gadgets."[678] Shortly thereafter, defoliation missions began on a limited basis and officially began as Operation Ranch Hand in January 1962. By the time it ended nine years later, some eighteen million gallons had been sprayed on an estimated 20 percent of South Vietnam's jungles, including 36 percent of its mangrove forests.[679]

Agent Orange was the most frequently used herbicide, by far. Named for the color of a stripe girdling the barrels in which it was shipped, it combined two herbicides, one of which turned out to be contaminated with a highly toxic strain of dioxin. No need for alarm, Washington officials professed at the time, and chemical company executives reiterated for years: "Agent Orange did not harm humans."[680] It was not just their opinion. Early reports by the US Air Force made the same claim, backed by its version of scientific fact. One of the first US Air Force scientific studies on Agent Orange, conducted in 1968, made that point perfectly clear in its conclusions: "Direct toxicity hazard to people and animals on the ground is nearly nonexistent. . .. Extensive studies have shown that the risk of human and animal toxicity from the herbicide components of Orange is very, very low."[681]

In its definitive 1982 report on the herbicide missions in Vietnam, the US Air Force avoided the issue entirely. It did, however, defend its choice of the chemical ingredients by using the sort of "everybody else does it" argument every parent has heard from their teenager:

THE CHEMICALS PRESENT IN THE DEFOLIANT MIXES EMPLOYED BY THE US AIR FORCE IN SOUTHEAST ASIA WERE DEVELOPED ORIGINALLY TO CONTROL WEEDS. NONE OF THE HERBICIDES WERE OF A NEW OR EXPERIMENTAL NATURE. THEY WERE FAMILIAR AGRICULTURAL CHEMICALS, AND AERIAL SPRAYING OF THEM WAS COMMON. IN 1961, THE YEAR BEFORE THE RANCH HAND PROGRAM BEGAN, APPROXIMATELY 40 MILLION ACRES PLUS HUNDREDS OF THOUSANDS OF MILES OF ROADSIDES, RAILROADS AND UTILITY RIGHTS OF WAY WERE TREATED WITH HERBICIDES IN THE UNITED STATES. OF THIS TOTAL, MORE THAN TEN MILLION ACRES, AN AREA ABOUT ONE-FOURTH THE SIZE OF SOUTH VIETNAM, RECEIVED AERIAL SPRAY APPLICATIONS.[682]

Notwithstanding the devastating environmental impacts and long-term health problems of our exposed veterans, the tactical benefits of defoliation cannot be denied. In a 1991 speech to a conference on dioxin use in Vietnam, Admiral Elmo Zumwalt, retired chief of naval operations, stated that during operations along the Cambodian border:

THE CASUALTY RATES IN MY FORCES, AS I HAD TO PUT ONE THOUSAND SMALL CRAFT INTO NARROW CANALS AND RIVERS, SOARED TO 6 PERCENT PER MONTH. THIS MEANT THAT THE AVERAGE YOUNG MAN HAD ABOUT A 70 PERCENT

PROBABILITY OF BEING KILLED OR WOUNDED DURING HIS YEAR'S TOUR. WITH NO DATA TO SUGGEST ANYTHING OTHER THAN COMPLETE EFFICIENCY AND NO HARMFUL EFFECTS TO PERSONNEL, I CHOSE TO USE AGENT ORANGE TO DEFOLIATE THE BANKS OF THOSE NARROW RIVERS AND CANALS. OUR CASUALTY RATES PROMPTLY WERE REDUCED TO LESS THAN 1 PERCENT PER MONTH. I THEREFORE CONCLUDE THAT IT IS PROBABLY THE CASE THAT MANY THOUSANDS OF LIVES WERE SAVED, ALTHOUGH HUNDREDS AND HUNDREDS HAVE BEEN AFFECTED THEREAFTER.[683]

Admiral Zumwalt later came to believe that there was a direct link between long-term exposure to Agent Orange in Vietnam and serious medical issues later in life. In an ironic twist of fate, his son, Elmo III, who commanded one of those one thousand small craft that patrolled the defoliated canals and rivers, was later diagnosed with Agent Orange–induced non-Hodgkin's lymphoma and Hodgkin's lymphoma, and fathered two children with birth defects.[684]

Equal in value to the American war effort was the destruction of VC crops. In War Zone D, the VC considered crop damage to large vegetable gardens that they planted, to supplement what were often meager food supplies, to be the severest consequence from defoliation. Once sprayed, an area was considered unsuitable for future planting, primarily because they considered it too dangerous to work in a contaminated area. One captured VC reported that after a spraying operation, the foliage began to droop in about one or two days. Then it began to lose its green color and started turning yellow. After two or three months all foliage and crops, except bamboo, died. The exposure of their jungle camps to aerial observation was secondary, as there were vast tracts of untouched jungle to relocate to, and the jungle would soon return during the next rainy season. They habitually moved their base camps out of areas sprayed by defoliation missions as soon as possible, and usually did not return.

At the same time that the Americans were downplaying the dangers of exposure and not requiring that ground troops take any protective measures, the VC took an opposite approach. Everyone in a base camp was told to run away from spraying aircraft and to immediately bathe if any of the defoliants contacted their skin.For protection, the VC were issued masks, similar to surgical masks. The masks consisted of a 15cm x 9cm piece of transparent plastic with white cotton attached to one side. A blue liquid, smelling similar to ether or camphor, was poured on the cotton side of the mask and placed against the face. Strings held it in place over the mouth and nose. [685]

The dispute over whether there is a direct link between exposure to herbicides such as Agent Orange and the onset of certain diseases has raged for decades. Initially, a veteran had to prove that he was in close proximity to an active defoliation mission to receive benefits. This was impractical at best and impossible to prove for the majority. Since information on locations, patterns, and timing of defoliation missions in an area of operation never reached the company level, the typical veteran in a line unit would never have this vital information, nor the means to research it. Moreover, the even more difficult task was to prove a link between this exposure and the disease. Rarely was the veteran successful. Finally, after years of study and analysis, the Veterans Administration's policy was changed to reflect what veterans, and their families, had been saying all along. It reads in part:

VA PRESUMES THAT VETERANS WERE EXPOSED TO AGENT ORANGE OR OTHER HERBICIDES IF THEY SERVED IN VIETNAM ANYTIME BETWEEN JANUARY 9, 1962, AND MAY 7, 1975.[686]

Even if the individual was in-country for only one day, for purposes of VA benefits, the soldier was assumed to have been exposed.

VA ASSUMES THAT CERTAIN DISEASES CAN BE RELATED TO A VETERAN'S MILITARY SERVICE. FOR EXPOSURE TO AGENT ORANGE, OR OTHER HERBICIDES, THESE "PRESUMPTIVE DISEASES" ARE AL AMYIODOSIS, CHRONIC B-CELL

LEUKEMIA, CHLORACNE, DIABETES MELLITUS TYPE 2, HODGKIN'S DISEASE, ISCHEMIC HEART DISEASE, MULTIPLE MYELOMA, NON-HODGKIN'S LYMPHOMA, PARKINSON'S DISEASE, PERIPHERAL NEUROPATHY (EARLY ONSET), PORPHYRIA CUTANEA TARDA, PROSTATE CANCER, RESPIRATORY CANCERS, AND SOFT TISSUE SARCOMAS.

To this day, stories abound of Phuoc Vinh's toxic history with Agent Orange, from a reputed higher incidence of cancer among Phuoc Vinh veterans to unconfirmed rumors that the bladders used to store and haul Agent Orange also were being used to refill the troops' outdoor shower water tanks. One unsubstantiated report, circulated among veterans, postulated the risk of Agent Orange contamination to the Phuoc Vinh potable water supply. It contends that the soil at Phuoc Vinh, and for twenty-five kilometers in every direction, is classified as Podzologic (characterized by moderate leaching), with one of its properties being that whatever washes into the soil, such as a toxic chemical, begins to move laterally and finds its way to the water table on its way to sea level. Based on an analysis of elevations in the area, the general direction of the subterranean flow of groundwater in the Phuoc Vinh area should be south and west. Unfortunately, the majority of the heaviest defoliant spraying in the area was immediately to the north and east, thus ensuring that any toxins seeping through the soil would have ended up in the groundwater beneath Phuoc Vinh. This potentially contaminated water was pumped daily to the surface at the Phuoc Vinh water point, treated, and then disbursed throughout the camp.[687]

While the VA does not have a published scale to identify those veterans at most risk due to Agent Orange exposure, by any measure—official US Air Force reports, eyewitness battlefield accounts, anecdotal stories, or circumstantial science—Tiger Bravo and other members of the 3rd Brigade, 101st Airborne Division would surely rank near the top.

INTERLUDES

The downtime at Phuoc Vinh, while packing up for the move north, meant a welcome break from what had been a steady diet of "Meals Combat, Individual," universally called C rations or C rats by the soldiers. The company mess halls put out a steady stream of hot meals. Alcohol was also readily available. After a visit to the Phuoc Vinh Base Camp NCO Club, Harry Brown wrote, "Had two beers and seven mixed drinks. First time I had hard liquor in four months. I feel relaxed and safe. No more shooting at me or hiding from the enemy." Simple pleasures, taken for granted in civilian life, added to Brown's bliss. "Got my boots off, have clean clothes, cool drinks, radio and much time to sleep. . . . I am enjoying every minute."[688]

Among the few events that Tiger Bravo marched to in company formation while in Vietnam were the memorial services at battalion headquarters. All the companies in the battalion would stand in solemn ranks while the names of those killed in action, or in accidents, were read aloud for the last time. It was that time again on September 27. Time for prayer, and speeches, and quiet grieving.

On that bright, sunny morning the troops assembled in the company street without the accoutrements of war—no weapons or LBE, wearing floppy brim, boonie hats instead of steel pots. They were a motley crew. Pasty white replacements, in dark OD uniforms still displaying the sheen of newness, and carrying the ten to fifteen pounds they would soon lose humping in the torrid tropical heat, marched to pay tribute to men they didn't know. Striding alongside the replacements were bronzed veterans, in sun-bleached jungle fatigues that displayed a locally produced Screaming Eagle patch, more akin to an angry chicken than a proud bald eagle. Those vets knew the names of the dead only too well. Down the dusty, dirt road and through the rubber plantation they tramped, the first sergeant calling cadence and making corrections. "Get in step! Quit bunching up! I don't know where in the hell you learned to march!"

Waiting at the end of the road were sixty-eight helmets arrayed in a circle, resting on the butts of upright M-16s, their bayonets thrust deep into the dried ground, one for each of the battalion's soldiers lost during the Currahee Summer. Twenty-four of the helmets symbolized Tiger Bravo's losses, both those assigned and those attached medics from the headquarters company. John Childs, the 2/506 battalion commander, remembered it well: "I spoke at the memorial service, but felt so inadequate because I had not been in the battalion very long and did not know most of the men. Also, I was stunned at the number of men who had been killed in action."[689]

Part of the service was a memorial prayer read by the battalion chaplain: "Eternal God, our heavenly Father, whose love for us we do not understand as we would like. We ask for Thy help that this day and those of our future be richer in meaning and purpose. We ask for greater understanding and a deeper appreciation for the lives we love and those about us today. With hearts that have heaviness and awe we turn to Thee for Thy help to know eternal things. Forgive us all that we talk too much and think too little. Forgive us all that we worry so often and pray so seldom. Speak to us through the inner ears of our hearts some may find greater faith and have our own assurances of Thy light and peace in our life. Give to us and the families of these men grace to seek Thee with the whole heart, that seeking Thee and finding Thee may love Thee, and loving Thee may keep Thy Commandments and do Thy will through Jesus Christ we pray. Amen."[690]

OPERATION GOLDEN SWORD

On October 2, 1968, rain fell in a soft drizzle as a C-130 aircraft came to a halt at the Hue–Phu Bai airfield and the Screaming Eagle Band struck up "Rendezvous with Destiny," the official song of the 101st Airborne Division (Airmobile). The ramp came down and a tall, slender colonel led the command element of the 3rd Brigade, 101st Airborne Division (Airmobile) straight to Major General Melvin Zais, commanding general, 101st Airborne Division (Airmobile), halted the formation, saluted, and reported, "Sir, the 3rd Brigade reports for duty." Colonel Joseph B. Conmy, commander, 3rd Brigade, had brought the "Wandering Warriors" home.[691] Over the past seven months the brigade had fought across the length and breadth of South Vietnam—in all four tactical zones, in all types of terrain and against the most elite, formidable VC/NVA units on the field of battle. But now its odyssey was over. The fighting would continue under the banner of the 101st Airborne, but the "gun for hire" missions had come to an end.

While the 3rd Brigade was fighting in the South, its vacancy in the 101st Airborne Division's organization had been ably filled by another airborne brigade: the 3rd Brigade, 82nd Airborne Division. However, in September 1968 it was decided to exchange these two relatively similar brigades, sending the 3rd Brigade, 82nd Airborne Division south to III Corps and bringing the 3rd Brigade, 101st Airborne Division back under the control of its parent division in I Corps. The move was more of a realignment of command and control of combat units rather than an operational necessity brought on by any immediate, significant enemy threat. The timing to make the switch in the fall seemed propitious, since there had been a deceasing number of enemy contacts during July and August on the coastal lowlands surrounding Hue. Intelligence reports portrayed an enemy who was refurbishing his materials and infrastructure, resting and retraining his units, preserving his strength, and avoiding contact with 101st Airborne Division units at all cost.[692]

Operation Golden Sword was the code name given to the movement of the 3rd Brigade, 101st Airborne Division (Airmobile) from III Corps, and the end of its "fire brigade" missions. The brigade advanced party left Phuoc Vinh in early September, with the last of the brigade units reaching Camp Eagle on October 24, 1968. A total of 289 US Air Force C-130 aircraft and 1 US Navy LST were used to move 3,000 tons of equipment and vehicles and all 3,452 brigade personnel. Tiger Bravo was part of Task Force Bravo, which was composed primarily of the 2/506 and its supporting units that flew north on September 30, 1968.[693] The move began at 0900, with 115 soldiers from A Company climbing aboard a C-130, and ended ten hours later, with the last plane of 2nd Battalion personnel landing at Hue–Phu Bai Air Base.[694] Finally, there was an end to Tiger Bravo's nomadic, gun-for-hire role. The Wandering Warriors had come home.

North, where the People's Republic of [North] Vietnam and the Republic of [South] Vietnam clashed along a demilitarized zone (DMZ). The DMZ lay near the seventeenth parallel and generally followed the Ben Hai River, which ran east to west. A no-man's-land extended five kilometers on each side of the river. To stem the flow of men and supplies coming across the DMZ from the

North the Allies built a series of fire support bases about six miles south of and parallel to the DMZ. These bases became known as the "McNamara Line," named for Secretary of Defense Robert S. McNamara.[695]

North, where the marines at Khe Sanh had held off the combined might of two NVA divisions.

North, where the NVA controlled the strategic A Shau Valley, with its vast supply depots and staging areas.

North, where the fighting during Tet in Hue's Citadel was added to the lexicon of great American battles.

North, where the Republic of [South] Vietnam had committed its elite divisions to join with the 101st Airborne and the US Marines to fight off full-strength NVA units fresh off the Ho Chi Minh Trail.

North, where Tiger Bravo would be tested again—and bloodied again—before its epic first year in combat was over.

The move north also ended my connection with Tiger Bravo, this time permanently. There would be no returning, as I had done twice before. Prior to leaving Phuoc Vinh, the brigade commander told me that as soon as we landed I would take command of Echo Company, with "Mad Dog" Matosky as my first sergeant. A welcome respite for two soldiers who had earned it many times over.

14

Back to Bastogne

Fire Support Base Bastogne
(Thua Thien Province)
October 1–November 30, 1968

CARVED OUT OF A MOUNTAINTOP

While the deployment of the 3rd Brigade to I Corps was a significant shift in Allied combat power from south to north, for the soldiers of Tiger Bravo it was just another move to another strange place. The routine was all too familiar; it had been through it many times before. Land in an unfamiliar airfield—this one was Phu Bai, close to the 101st Airborne Division headquarters at Camp Eagle. Assemble for several days in buildings—in this case what could loosely be called barracks, with no electricity or running water. Receive a quick briefing on the terrain—this time rugged mountains that formed the western border of South Vietnam instead of muddy rice paddies and swamps. Digest the latest intelligence on the enemy—here it was mostly NVA, with few villages to support VC units or booby trap "factories." Finally, pack everything needed to live and fight into a rucksack, and climb on helicopters for another round of "fighting and surviving" in Vietnam.

While the enemy, terrain, and weather might be different from those of III Corps, the constant threat from mortar and rocket attacks was the same. Rockets greeted the 2/506 on its first night in I Corps. "Camp Eagle got rocketed last night," I wrote home. "I was in a ditch watching them hit farther in. It was pretty spectacular."[696] What had been a frightening, stressful event earlier in the year was now as commonplace as a Fourth of July fireworks display.

Reminders of the bloody summer also appeared unexpectedly as well. While unloading the battalion's supplies and equipment that had made the journey from Phuoc Vinh to Camp Eagle, Joe Adams, who had been plucked from line duty and assigned to the battalion's S4 team, came across a poignant reminder of Tiger Bravo's battle on July 22. "I was unpacking stuff that had been sent to Camp Eagle," Adams recalled, "when there was Hillman's helmet with a hole in it."[697]

It was a short CH-47 flight to just fifteen kilometers east of the A Shau Valley, to Fire Support Base Bastogne, named after the famous town at the center of the Battle of the Bulge in World War II. Bastogne was perched around a hilltop overlooking Highway 547, and within 175mm artillery range of the of the A Shau Valley. It was built in the spring of 1968 when units of the 101st Airborne Division and the army of the Republic of Vietnam launched Operation Delaware, into the heart of the A Shau Valley.[698] It was part of a string of fire support bases along Highway 547, starting with Birmingham to its east and ending with FSB Vehgel, carved out of a mountain- top overlooking the A Shau. If the NVA wanted to push out from the A Shau Valley and split the country from west to east, they would have to run a gauntlet of fire from bases bristling with artillery all along Highway 547. Lieutenant Colonel Childs remembered that "it was a pretty well developed firebase with good bunkers. But it was a bitch to defend. There was a very steep hill that we surrounded. If the enemy ever took that hill, the firebase would fall."[699]

The fire support base was already overrun by large, bold rats that seemed invincible to traps. Joe Adams recalled one soldier making a rat proof bed to keep them from scurrying across his legs while he slept. "My RTO got a couple of engineer stakes, some ammo boxes, and communication wire and made himself a bed, to get off the floor because of the rats."[700]

The move from Camp Eagle to Bastogne would prove to be anything but uneventful. For the first time, the battalion was faced with an organized, group refusal to leave the security of base camp for the relatively dangerous life on a fire support base. "When we moved north we had the mission of taking over FSB Bastogne," remembered Lieutenant Colonel Childs.

So, one morning helicopters came in to take the battalion to the fire support base. Four black soldiers from D Company refused to get on the helicopters. After discussing the situation with 1SG Trent, who had been selected to be the acting Battalion Command Sergeant Major, I gave orders to force them on the helicopters. They would be in the field while we preferred charges. We constructed an enclosure (on Fire Support Base Bastogne), with bunkers of course, for them to await their court-martial. They were all sentenced to six months in the LBJ stockade (Long Binh Jail).[701]

This incident was a harbinger of what lay ahead for the army in the years when the war-fighting responsibilities shifted to ARVN forces. Units across Vietnam would experience serious problems with drug abuse, racial tensions, weakened discipline, and lapses in leadership that would see morale and combat effectiveness plummet. Outright refusals to fight, such as the one experienced by the 2/506 in October 1968, were few in number during the remainder of the Vietnam War. However, incidents of "fragging," which were attacks on officers and noncommissioned officers (NCOs), occurred frequently enough in the years when US forces were being withdrawn to compel commands to institute a host of new security measures.[702] By 1972 some were even arguing that virtually all officers and NCOs had to take into consideration the possibility of fragging before giving an order to the men under him.[703]

Fragging was one of the most disturbing aspects of the unpopularity of the war among soldiers. Disenchanted soldiers used highly lethal fragmentation grenades, universally called frags, or other explosives to threaten or kill officers and NCOs they disliked. The full extent of the problem will never be known, but it increased sharply from 1969 to 1971, when the morale of the troops declined in step with the diminishing American role in the fighting. A total of 730 well-documented cases involving 83 deaths have come to light. There were doubtless others and probably some instances of fragging that were privately motivated acts of anger that had nothing to do with the war. Nonetheless, fragging was symptomatic of an army in turmoil.[704]

For the first time since arriving in-country, Tiger Bravo experienced a month without enemy contact, save for one combat assault on October 22, when the last helicopter to land received a few rounds of small-arms fire.[705] To be sure, the soldiers were still in harm's way every day, and the tension brought about by expecting a booby trap to explode at any time shadowed every step as well. But for once in its tour of duty, Tiger Bravo's days and nights were relatively quiet. "We had more action in Trang Bang than in I Corps," recalled John Childs. "It seemed that anytime you stepped out of FSB Patton, that was located near Trang Bang, that you would run into a firefight. In I Corps we initially were securing areas around the foothills and did not run into many VC or NVA types." But the specter of the A Shau Valley, and its assemblage of NVA combat units, continued to loom just beyond the foothills. "The closer you got to the A Shau Valley," Childs continued, "the more action you got into, and you had more NVA up north in comparison to the VC in Trang Bang."[706]

Being back in the fold of the 101st Airborne Division also meant closer scrutiny by higher headquarters, to include how the battalion looked and the appearance of its area. Childs recalled, "At Cu Chi the battalion called itself the "Jungle Hat Battalion" because they loved wearing the issued jungle hat instead of the steel pot. After we moved north, the division frowned down on the jungle hat moniker." Proximity to division headquarters also brought layered command and control to manage the enemy contacts regardless of their size. "Being at Camp Eagle meant more people looking over us. If you got into a firefight, you could be assured that the Brigade Commander, General Zais and the I Corps Commander would show up in choppers listening in on our radio nets."[707]

On most days Tiger Bravo provided security for FSB Bastogne, each day sending patrols out to sweep around the base of the hill and clear the highway east to FSB Birmingham. Improving the fire support base's bunkers and fortifications kept everyone busy. It rained most days, so the company was constantly fighting collapsing bunkers and not NVA. The only casualties came from accidents. One self-inflicted wound came when a soldier shot himself in the foot as he (allegedly) rolled onto his M-16 in his sleep—yet another "accidental" foot wound. Another soldier was burned by a bunker fire when a lamp fell over. Another soldier contracted malaria, one of the few recorded cases in the company. Yet another was sent to the rear in the middle of the night on a medevac chopper for reasons not recorded in any report.[708]

The mountainous terrain also took its toll. During October, the battalion rotated companies on fire support base security for Vehgel, Bastogne, and Birmingham, with one always in the mountains looking for NVA and their base camps. The terrain was murderous and difficult to traverse with any semblance of a tactical formation. One night, unable to find a suitable flat area to set up an NDP, the soldiers had to lash themselves to trees to keep from falling down the mountainside.[709]

A slip and fall down the mountain usually ended in broken bones. Such was the case on October 23, when two Tiger Bravo soldiers rolled and bounced down a steep slope, resulting in a medevac for both, one with a fracture and the other with a knee pushed out of joint.[710] The jungle canopy was so thick that a helicopter couldn't land. The two soldiers were pulled up through the trees on a jungle penetrator, which sometimes left the patient with a whole new set of cuts and bruises. One of the two, Private First Class Stevenson, was heard to say, "I was all right until they dragged me up through the treetops on that cable."[711]

When October came to a close, so did Fire Support Base Bastogne. The battalion was ordered to dismantle the base and prepare for another move, this time farther north, closer to the DMZ.

ROLLING THUNDER

By the fall of 1968, US Air Force, Navy, and Marine attack aircraft saturated the air over North Vietnam daily. In October 1968 alone, American airpower flew a total of 11,931 sorties against North Vietnamese military, infrastructure, and industrial targets.[712] Nicknamed Operation Rolling Thunder, this air campaign had escalated from a few air strikes in the southern portion of North Vietnam to widespread bombing of North Vietnam's industrial base and its principal seaport, Haiphong Harbor. It had only two strategic objectives:[713]

- To apply steadily increasing pressure against North Vietnam to cause Hanoi to cease its aggression in South Vietnam.
- To make continual support of the Viet Cong insurgency as difficult and costly as possible.

But the end of October also brought a significant shift in the strategic direction of this air campaign. In a televised address to the nation on October 31, 1968, President Johnson announced that on the basis of developments in the Paris peace negotiations, he ordered the complete cessation of all air, naval, and artillery bombardment of North Vietnam.[714] While the cessation of the air campaign was obviously under consideration for some time at the highest levels of government, a US Air Force CHECO (Contemporary Historical Explanation of Current Operations) report alluded to the tactical and operational commanders being caught unawares that their missions would be halted: "Suddenly and dramatically, Rolling Thunder came to an end on 1 November 1968 by Presidential proclamation. Before the stand-down became effective, 346 attack sorties were flown on the final day by US Air Force, Navy, and Marine pilots."[715]

The halt to the air campaign over North Vietnam meant little to the units on the ground in South Vietnam who were fighting the day-to-day battles. This was made quite clear in a message from the commanding general, 101st Airborne Division, that followed just twenty-four hours later: "President Johnson has ordered a bombing halt over North Vietnam. This in no way alters present operations in the 101st's AO."

Knowing that few soldiers on the ground would agree with what they considered to be an egregious "show of weakness," it went on to order that "US and civilian personnel will refrain from commenting to press or public on the bombing halt message. The President's message speaks for itself. Operations in South Vietnam will continue on present basis."[716]

THE LOWLANDS

After a relatively quiet month operating in the rugged mountains east of the A Shau Valley, the battalion moved even farther north, to Camp Evans Base Camp.[717] It had been recently vacated by elements of the 1st Cavalry Division, which had deployed south to III Corps at about the same time as the 3rd Brigade's move north. Camp Evans was in the coastal region of South Vietnam. Called "the Lowlands" by Tiger Bravo's soldiers, it was a fertile, rice-producing area that stretched the entire length of South Vietnam, starting at the Red

River Delta in North Vietnam and terminating at the Mekong River delta in the South.[718]

Keeping its headquarters at Camp Evans, the battalion placed the companies on a rotation between base camp defense of FSB Jack, a one-company fire support base east of Camp Evans, and conducting sweeps of the battalion area of operations, looking for an enemy that seemed to not want to be found. The battalion also manned FSB Jeannie with E Company, which was on the eastern edge of the battalion area of operations. Bob Pagano recalled the Lowlands as fairly quiet. "We didn't have too many engagements," he recalled. "The occasional sniper. It was mainly booby traps."[719]

A 3rd Brigade report for the period characterized the operations as joint missions with South Vietnamese assets, in an area replete with mines and booby traps:

OPERATIONS IN THE AO WERE OFTEN COMBINED IN NATURE, USUALLY CONSISTING OF A PF PLATOON AND 12 NPFF. . .. KIT CARSON SCOUTS HABITUALLY ACCOMPANIED THE MANEUVER ELEMENTS. . .. THE 2-506 WAS CONTINUOUSLY PLAGUED WITH EXTENSIVE ENEMY MINING AND BOOBY-TRAP ACTIVITIES IN THE SOUTHERN PORTION OF THEIR BATTALION AO. MANY OF THE BOOBY-TRAP TECHNIQUES ENCOUNTERED WERE COMMON THROUGHOUT RVN BUT SOME INNOVATIONS DEMANDED THE DEVELOPMENT OF COUNTER-TECHNIQUES AND EXTRA PRECAUTIONS ON THE PART OF THE MANEUVER ELEMENTS.[720]

Tiger Bravo was a different company, when it began its first operation in the Lowlands on November 7, 1968, from the one that had fought in III Corps, in the South. Unlike the summer, when replacements received their baptism by fire within hours of joining the company, many replacements that had arrived since the end of September were still "unblooded." Also, the proportion of veteran paratroopers to untested, newly assigned soldiers was dangerously low. By this time, only 25 percent of troop strength had been with the company from the beginning of its tour of duty.

The ratio of paratroopers to non-airborne soldiers had also shifted dramatically in the past four months. Since July, packets of mostly non-airborne personnel had been the norm. Bob Pagano was in one of those replacement packets. He recalled being in the replacement center at Cam Ranh Bay when unit assignments were made. "We all were slated to go to the 1st Cav [1st Cavalry Division], so we were all surprised that we were being sent to the 101st. We all just looked at each other, as none of us were jump qualified."[721]

Lieutenant Colonel Childs also noticed a change in the battalion from the one he led north from Trang Bang. "Replacing experienced troopers with draftees began to have an impact on the efficiency and toughness of the battalion," he recalled. Childs went on that it was a transformation of the battalion from top to bottom. "1SG Trent's replacement as Battalion Command Sergeant was a 'leg' and he did not fit in very well with our paratroopers and did not do a very good job. He was later relieved after my departure."[722]

The original members of the company were now officially "short-timers," counting the days until their tour was up. Each day closer would be marked with the same phrase, only the number of days would decrease by one as November dragged by. "Twenty days and a wake-up" or "Twenty days and a bowl of cornflakes" would be answers to "How short are you?"

The company was also on its second company commander since my departure. After only a few weeks in command, and for reasons unknown to the rank and file, my replacement was abruptly removed and replaced by Captain Warren Kiilehua,[723]called Captain K by the soldiers, who quickly earned the respect of his new charges. Bob Pagano remembered that one day the old commander was there and the next he was gone: "I don't know what happened to him. He just disappeared."[724]

Captain Kiilehua was on his second tour in Vietnam, spending all of 1967 with the 1st Brigade, 101st Airborne Division during its "Nomad" days of being the II Field Force fire brigade. He had seen it all, but was not prepared for his return to the Screaming Eagle Division. "I arrived at Bastogne the day before Halloween," he recalled. "As I was dropped off at the top of the fire base, I witnessed a Special Court Martial of a soldier who had refused to move out with his unit."[725] Immediately, Captain Kiilehua noticed stark differences from the unit he had left. "In nine months, everything had changed. We were no longer airborne, and now operated as part of the division. Support bases were massive in comparison to 1967. The morale of the officers and men was not as high, nor was the unit cohesion. Finally, when I left country the first time, we all felt that we were winning. That feeling never returned."[726]

Tiger Bravo stayed in the field for the entire month of november, conducting search and destroy operations in the day and ambushing at night. During the entire month only one ambush made contact, and that was a minor action, with three VC unlucky enough to walk into a platoon's kill zone.

THE JUNGLE SHOOK

November was a relatively quiet time until November 12, when the war came crashing down, again, on Tiger Bravo. A booby-trapped 82mm mortar round blew up in the faces of two soldiers, wounding one and killing another. Private First Class Ralph Guarienti from Sacramento, California, who had joined the company in Phuoc Vinh a few days before its deployment north to I Corps, was the unlucky soldier. Wounded in the blast was an army scout dog.[727]

Four days later, on another typical sweep operation, death came calling again; this time it was Private First Class Stanley Wilton, twenty years old and also from Sacramento.[728] He would lose the daily struggle between his own sharp eyes, constant training, a cautious attitude, and the sometimes unavoidable, random, and tragically bad luck of tripping a booby trap. Bob Pagano and Stanley had arrived in Tiger Bravo with the same batch of replacements that arrived in Phuoc Vinh before the company moved north. "We became good friends since we both ended up in the same platoon," Pagano recalled.[729] During a stop in the Lowlands, he and Wilton were sitting close to each other, as buddies tended to do, just leaning back, resting on their rucksacks. It was a scene that had been repeated dozens of times in the past few weeks. "Someone came over and said, 'Hey, we need someone to go out on a RIF.' He looked at me. Looked at Stan. Looked at me again. Looked at Stan. Picked Stan. He wasn't gone more than a minute or two when the jungle shook." Stan Wilton was killed instantly by a booby trap. Bob Pagano couldn't bear to see what had happened to his friend, something that just as easily could have happened to him. "I never saw what happened and I didn't want to see what happened."[730]

The prevention of booby trap casualties had been a priority of the 3rd Brigade commander since landing at Camp Eagle. Every occurrence when a soldier was brought down by a booby trap was treated as an "incident" and investigated accordingly. A 3rd Brigade report of lessons learned from those investigations, listing the main causes of booby trap casualties, laid the blame squarely on the backs of the soldiers and their leaders. The report listed only a lack of training, carelessness and improper movement as the main reasons for casualties.[731]

Left out of the report were other reasons equally to blame. The enemy was a master at concealing booby traps, constantly shifting his tactics to respond to American countermeasures. Also, bone-deep fatigue from days of grueling marches and sleepless nights on ambush dulled the senses of even the sharpest soldier. Finally, it took weeks and months of dealing with the deadly devices and learning how to think like the enemy for leaders to effectively translate the training and lessons learned into tactical actions—a luxury green replacements did not have. "Its policy on booby traps was so true of brigade," Childs remembered. "It was not helpful at any time during my tour."[733]

Patience was often an overlooked virtue for successful combat leaders. We had all learned the hard way that the VC rarely fired randomly, then ran away without an "endgame" in mind. Usually, they were trying to entice the overzealous leader to pursue directly into an ambush, minefield, or series of booby traps.

For the next two weeks, the company worked with the 1st ARVN Division in an operation designed to completely clear the lowlands, east of Camp Evans, of any cover and concealment used by the VC to cover their night-time movement or to conceal booby traps. "We spent weeks providing security for Rome Plows," Kiilehua recalled. "The plan was to plow the entire area, leaving nothing sticking above ground. Since it was hard for infantry to keep up in the thick brush I was assigned a company of ARVN armored personnel carriers (APC) for the company to ride on." And so, it went, day after day, each Rome Plow followed by an ARVN APC bristling with heavily armed Tiger Bravo soldiers.[735]

Midway through the clearing operation, on November 24, Kiilehua was added to the exclusive list of Tiger Bravo company commanders wounded in action. "I was riding on one of the APC's when we uncovered booby traps in our path," remembered Kiilehua. They were in low brush, about to the soldiers' belt buckles. "We couldn't really tell what they were," Kiilehua went on. "Could have been grenades or mortar rounds or even a 250-pound bomb. I rigged one with demolitions to blow in place, and the others were rigged by soldiers I was training in how to destroy a booby trap. The fuses were set and everyone took shelter behind the APC's." The first two demolition charges went off as planned. "When the 3rd one, which was at least fifty feet on the other side of the APC, went off, this rock came flying over the APC and landed at our feet. But, it wasn't a rock, it was a grenade!!!"[736] The subsequent grenade blast wounded thirteen soldiers, including Captain Kiilehua, adding his name to those of Rankin, St John and Van Meter — all Tiger Bravo company commanders wounded in action in 1968. Wounded in the arm, Kiilehua was treated at a medical clearing station and returned to the company several days later.

The next five days, comprised of search and destroy operations for the company, and a bridge security detail for one platoon, were uneventful—no enemy contact and no casualties. The only excitement came one morning when battalion radioed to let Tiger Bravo know that the battleship *New Jersey*, which was sitting just offshore, would be firing its main guns over Tiger Bravo's head against targets well to the south of the company. On November 30 the company headed back to Camp Evans for a forty-eight-hour break from the monotony of patrolling the lowlands.[737]

This would be the last combat mission for those few rotating home, who now had a new slogan. When asked "How short are you?" The response was "INSIN = I'm not short, I'm next!"[738] Many were already in the out-processing system and enjoying the comforts of "civilization" for the first time in a year—showers, cots, hot meals, cold drinks, and not having a weapon within arm's reach.

As for me, my first steak dinner in a year was memorable. "I had two T-bone steaks, french fries, salad and bourbon," I wrote in one of my last letters home. As I recall, the salad was only lettuce and there was no dressing other than oily mayonnaise, but that mattered little to someone starving for anything other than C rations. The next night, I had two more steaks.[739]

15

An Epic Year

Camp Evans
(Thua Thien Province)
December 1-2, 1968

FIRST ANNIVERSARY

December 2, 1968: the first anniversary of Tiger Bravo's arrival in South Vietnam. Exactly a year before, an untested airborne rifle company had landed at Bien Hoa Air Base, Republic of Vietnam. One hundred sixty paratroopers marched off the plane into an uncertain future. Records show that one year later, only eighteen remained in the company from the original troupe of bright faces. Massive casualties, injuries, disease, transfers of veterans to assignments at battalion, and a few early rotations home had taken their toll.

Most line units in the 3rd Brigade faced a similar situation, although in rear service support units the proportion of experienced soldiers leaving was much higher. Across the 101st Airborne Division, more than six thousand officers and enlisted men, who had arrived with the division in December 1967, rotated home from late November to early December 1968. The replacement center at Bien Hoa was operating night and day, reaching a peak of two hundred soldiers out-processed and a similar number in-processed every twenty-four hours.[740]

The day was marked in many different ways. The company, minus its few remaining veterans who were scheduled to rotate home, was on another operation in the Lowlands, this time under the control of the 3rd Battalion, 5th

Cavalry Regiment. It would be an uneventful day, with no enemy contact or casualties. Its ranks included Bob Pagano carrying a radio and Mike Scott still walking point.

First Sergeant Trent, now the battalion command sergeant major, assembled a small group of original members for a beer to say good-bye. As Chris Backman remembered, "There was only four or five of us. Not many left."

Mike Tarpley was on his two-hundredth day in the William Beaumont Medical Center at Fort Bliss, Texas, still recovering from the leg wound suffered in the Rocket Belt and having his jaw reconstructed because of lingering problems from the grenade blast to the face during Tet. It would be another five months before he would be released from the hospital.[741] Steve Lyle was at Beaumont as well, finishing his active duty enlistment as a medic, and working at a base clinic. He would be discharged in January 1969, just in time for the spring semester at college, eventually graduating with a BS in chemistry/biology.[742]

Jim Roach was now a platoon leader in A Company, 2/506 which also had a high casualty rate during its first year. "When it came time to DEROS in December 1968, there was only myself and one other soldier in all of A Company that had originally come over with the battalion a year earlier," he recalled.

Chuck Kudla was already out of the army and preparing to go to college in Minneapolis. Dan Bernard was home in New Hampshire, having recovered from his leg wound.

Chuck Limer also survived his wounds. After spending four months in the hospital, he was back at Fort Campbell, Kentucky, packing parachutes as a rigger.[743]

Doc Franks had left Vietnam and Tiger Bravo in August when his enlistment was up. By December he was working part-time for Bay Cities Ambulance Service in Chula Vista, California, and attending Grossmont College in La Mesa, California.[744]

Dave Spencer was in the US Naval Hospital in San Diego, still recovering from wounds received on July 22, 1968. It would be another six months before he would be released.

Chuck Hanson was enjoying the comforts of Camp Evans, having extended his tour an additional six months to continue as the Tiger Bravo company clerk. Each day he and Bill "Pop" Plemons, who did not extend and was keeping a low profile until his time to fly back to America, would drink a cold beer outside the company's orderly room. "We have a cooler outside the door, which is kept full of beer," Hanson wrote. "Bill Plemons, who works in S-4, comes by once a day and drops off 200 lbs. of ice to keep it cold. It's 7 PM now and most of us are sitting around with a cold beer and reading, writing or bull-shitting."[745]

Basil Rivera was already home in Alvin, Texas, on a thirty-day leave, expecting to be married. However, his fiancée's priest said they had to wait to give him time to get himself together, since "he had been to hell and back." Basil and his fiancée, Lupe, complied and wed six months later, a union still strong after almost fifty years.[746]

Dan Soto was at Camp Evans, spending his last few days in-country working in the communication section and on company work details.[747]

Terry Van Meter was in Valley Forge Veterans Hospital, Valley Forge, Pennsylvania, still recovering from wounds received on August 28, 1968.

As for me, I was already out of the country and stranded in Manila, Philippines. Lieutenant Colonel Grange and I had boarded the same flight to Hawaii the day before, where our families had waited out the year. After refueling in Manila, the plane was just lifting off the tarmac when a tire blew, sending the plane skidding off the runway and Grange and me to a nearby hotel for the night. It was a loud, boisterous, slightly dangerous R & R hotel full of drunk soldiers, marines, and airmen, partying their last hours before heading back to the war zone. I knew it was not the usual tourist destination when we saw a contingent of national police, armed with machine guns, guarding the lobby. This was reinforced upon entering our hotel room and finding a pool of partially congealed blood on the bathroom floor, possibly the remnants of a brawl between the previous tenants. But after a year in Vietnam, nothing as minor as a pool of blood fazed us. Grange went to bed, while I called for maid service to clean it up. Then I headed to the bar.

SACRIFICE AND VALOR, BY THE NUMBERS

In his book *Band of Brothers: E Company, 506th Regiment, 101st Airborne from Normandy to Hitler's Eagle's Nest*, Stephen Ambrose called World War II combat a "topsy-turvy world, where perfect strangers are going to great lengths to kill you. If they succeed, far from being punished for taking a life, they will be rewarded, honored, celebrated." During all of World War II, E Company had been in combat for less than eight months (232 days), stretching from the Normandy invasion to the final push into Germany. By the end of the war, forty-eight members of Easy Company had given their lives for their country and more than a hundred had been wounded.[748]

Twenty-plus years later, Tiger Bravo, of the very same 506th Infantry Regiment, found itself in its own "topsy-turvy world" of ground combat, amassing a staggering 309 days of combat in just its first year. By the end of the twelve months, 30 members of Tiger Bravo had been killed in action, 118 more were wounded, and another 21 were non-battle casualties from injury or disease. This grim tally included four company commanders wounded in action and another twelve platoon leaders/platoon sergeants either killed or wounded. The average field strength, boots on the ground, during the summer of 1968 hovered at about 120 soldiers. At one point, during a period of the heaviest fighting, it dipped to 82 able-bodied soldiers, enough to man only three of four authorized platoons. One unofficial report put the number of able-bodied soldiers as low as 69.[749]

Based on an assigned/attached authorized strength of approximately 164 soldiers (the organization would change several times during the year), the company's combat casualty rate for both battle and non-battle casualties was an astonishing one hundred and three percent. The probability of being killed or wounded, if assigned or attached to Tiger Bravo, during the year was equally daunting. Taking into account the approximately one hundred replacements that passed through its ranks, some lasting only a few weeks before they too became a casualty, the probability of being killed in action if you were in Tiger Bravo was eleven percent, and you had a forty-five percent chance of being wounded in action. The chances of being killed or wounded were even more foreboding - - - being assigned or attached to Tiger Bravo in 1968 meant you had a greater than fifty percent probability of being killed or wounded.

As part of the Wandering Warriors, the company fought local VC guerrillas, main force VC, and conventional NVA units up to regimental size in all four of the country's tactical zones and six of its provinces, including the top two provinces in terms of US casualties—Thua Thien in I Corps and Binh Duong in III Corps. It conducted operations in all of the combat environments found in the Vietnam War: up and down the jungle-covered mountains of the Central Highlands and A Shau Valley; in the triple canopy jungles of War Zone D; through the streets of Bien Hoa; into the tunnels and bunkers of VC-fortified villages of the Ho Bo Woods and other enemy strongholds; along the fringes of the Mekong Delta, where the company would move all day through a flooded landscape and send out patrols just to find dry land for the night; and in booby- trap-infested coastal lowlands. It fought under or alongside of seven major combat units: 1st Infantry Division, 4th Infantry Division, 9th Infantry Division, 25th Infantry Division, 199th Light Infantry Brigade (Separate), 11th Armored Cavalry Regiment, and the ARVN Rangers.

The company's record of missions included search and destroy, reconnaissance in force, patrolling, night ambushes, eagle flights, road clearance operations, attacks on enemy fortified base camps, day and night airmobile combat assaults, cordon and search of heavily bunkered VC villages, and defense of fire support bases from ground attacks. It also included some not taught in tactics classes at Fort Benning, such as a "moving Tiger" operation, where the company moved silently and stealthily through the night attempting to intercept rocket-carrying parties up to four hundred strong. Or, on more than one occasion, fighting its way off a hot pickup zone to make an airmobile combat assault into an equally hot landing zone. Two hundred nights were spent in company night defensive positions, manned by 100 soldiers or less, deep in enemy territory and miles from the nearest American unit. On 277 night ambushes, of squad or platoon size, Tiger Bravo soldiers crouched silently in the pitch dark waiting for an enemy, that often outnumbered them, to walk into their kill zone.

While not facing the awesome firepower and massed armor of the German Army, the men of Tiger Bravo faced their own uniquely dangerous and terrifying enemy. Unlike the campaigns in Europe during World War II, Vietnam was a war without a front, flanks, or rear. Everything outside of fixed installations and American base camps was considered enemy territory. One day the company would fight a formless enemy, who could seemingly booby trap everything and then evaporate after a contact like the morning jungle mist. On the very next day it would find itself toe-to-toe with main force VC or NVA forces in a set-piece, conventional-style battle. Yet at the same time, there were striking similarities between the Vietnam War and the war fought by our fathers—the savageness of close combat, the intensity felt at the heart-wrenching loss of a friend, a degree of comradeship among buddies never to be duplicated, the unabashed yearning for home and who we used to be, and the immutable effect of war on the human spirit.

Researching Tiger Bravo's record of awards for valor and gallantry in action has been a nearly impossible task. But, in my research, I did discover a memorandum from the headquarters of the 2/506 to the awards and decorations section at the 101st Airborne Division that provides a glimpse into Tiger Bravo's courageous actions on the battlefield. The memorandum listed all the individuals in the battalion who had been recommended for awards for bravery during the summer of 1968. For Tiger Bravo, four soldiers had been recommended for the Silver Star, twenty-one for the Bronze Star with V for valor and another six soldiers for the Army Commendation Medal with V for valor[750]. A prodigious number, by any measure.

SOLDIER ON

Through it all, the men of Tiger Bravo soldiered on, taking each mission a day at a time. They looked out for each other and trusted one another. All the while, they were buttressed by a brotherhood that lived the words of one of its soldiers:

"We laughed, we cried, we got drunk, we fought, we died, and we never gave up on one another!"[751]

Epilogue

THE COMPANY

Tiger Bravo remained in Vietnam until 1972, when the 101st Airborne Division (Airmobile) returned to its home base in Fort Campbell, Kentucky, as part of the American withdrawal of forces from South Vietnam. During its time in the war zone the company amassed a total of 1,573 days in combat, participated in twelve major campaigns, and had sixty-five soldiers killed in action.

Along with other units of the 3rd Brigade, 101st Airborne Division, the company received the prestigious Presidential Unit Citation (PUC) for its exceptional combat record during the summer of 1968 in the vicinity of Trang Bang, when elements of the 3rd Brigade were in continuous contact with the enemy for a hundred consecutive days and nights. It was awarded a second PUC in May 1969 for its part in the Battle of Dong Ap Bia, known to history as "Hamburger Hill." To receive a PUC a unit must display such gallantry, determination, and esprit de corps in accomplishing its mission under extremely difficult and hazardous conditions so as to set it apart from and above other units. The degree of heroism required on the unit level is equivalent to that which would warrant the awarding of the Distinguished Service Cross (our nation's second-highest award for gallantry in action) to an individual. To put this prestigious award in the proper perspective, 2/506 companies have received the PUC for only two other combat operations: the invasion of Normandy and the Battle of the Bulge (Bastogne) in World War II.

Twenty years later, the company would return to combat as part of the 2/506, participating in the Iraq National Resolution Campaign (2005–2007) and Afghanistan Resolution Campaigns I (2008–2009) and II (2009–2011).[752]

AFTERMATHS (LIFE AFTER THE WAR)

Unlike our fathers, we didn't come back to a grateful nation. It took a generation and another war, Desert Storm, for America to come back to us.[753] We returned home to a changed world, to a country that was awash in antiwar protest, where anger at the war enveloped the unsuspecting warrior as well. We soon learned that while we were fighting across the length and breadth of South Vietnam, the American soldier had become a scapegoat for our government's involvement in an unpopular war. It was a circumstance alien to any previous returning veteran, in any of our nation's foreign wars, and a level of outrage that we could neither cope with nor understand.

For the most part, we kept our combat experiences and wartime service bottled up, not only to avoid painful memories, but also to sidestep confrontation over the war. We soon learned that support of the war, and especially for its warriors, was not an acceptable topic outside the obligatory embrace of our families or the walls of a veterans' club.[754] We were caught in a trap not of our making. On the one hand, we felt an intense pride for having served in and survived a battlefield as dangerous as any faced by earlier generations of American infantrymen. On the other hand, we were unable to freely and openly display that pride without facing the wrath of the very population we looked to for succor.

To all outward appearances, returning Tiger Bravo veterans showed little effect of our time in Vietnam. It seemed as if we had assimilated back into civilian life without much difficulty. Many of us have lived full and productive lives. We married, had children and grandchildren, went back to school, found jobs, and weathered life's ebbs and flows. After attending five reunions of the 1968 Tiger Bravo veterans, Chuck Hanson described us as "honest, honorable, polite, modest, courteous, self-starters, law abiding, hardworking, and patriotic."[755]

After his tour ended, Joe Adams returned to his hometown of Nacogdoches, Texas, where he proudly lives to this day as a BIN (born in Nacogdoches). He received a degree in criminal justice and worked in law enforcement for thirty-six years as a deputy constable and inspector/investigator for the Fire Marshal's Office. He has one son and two grandchildren.[756]

After Vietnam, Chris Backman finished his enlistment with the 82nd Airborne Division at Fort Bragg, North Carolina, where he shared a house with Tom McClear. In 1969, he returned home to Seattle, Washington where he still resides with his wife, Georgia. They have a son and daughter, and one grandchild. In 1988, Chris started his own paint contracting business, which he is still active in, but looking forward to retirement.

Dan Bernard left the army when his enlistment was up. He worked for six years in a precision sheet metal shop in New Hampshire, then began a twenty-seven-year career working on US Navy nuclear submarines at the Portsmouth Naval Shipyard in Kittery, Maine, retiring as a general foreman (GS-12). He has two daughters. Dan and his wife, Karen, moved to Fort Meyers, Florida, in 2009. He keeps active in the Elks, Shriners, CIB organization, and various 101st Airborne activities.

After he returned from Vietnam, Harry Brown married Virginia Alanzo. They had one son, Shalako. Harry initially found work as a geologist, then became interested in computers, and spent the next twenty-eight years as a senior computer analyst. His other interests included motorcycles and gunsmithing. Harry passed away from cancer in 2012.

Larry Burton settled in Kentucky after the war. He worked for thirty years at General Motors as a sheet metal fabricator. After retiring from GM, he moved onto eight hundred acres in Columbia, Kentucky, with his wife, Julie. They have two children. Larry spends his free time as a judge at Tennessee Walking Horse competitions and assisting other veterans work through the administrative maze of the VA claims process.

John O. Childs remained in the army, retiring as a lieutenant colonel. After retirement, he returned to his native Texas, where he became a banker and an adjunct professor in the School of Business, University of Texas at Tyler. He, and his wife of sixty plus years, Thelma Ruth, live in Tyler, Texas. They have three sons, six grandchildren and one great grandson. One of his grandsons is a Special Forces noncommissioned officer who has served several tours in the middle east.

Everett "Doc" Franks returned to California after his enlistment ended. For thirty years he was employed by the County of San Diego, California—twelve years at the Department of Social Services and eighteen years as an investigator for the District Attorney's Office. He has two children and is married to Cindy, "the girl of my life," who was his next-door neighbor during one of his foster home placements in 1955.

David E Grange Jr. remained in the army, retiring as a lieutenant general. The US Army's "David E. Grange Best Ranger Competition" is named in his honor. He and his wife, Lois, live on a horse farm in Dover, Pennsylvania.

Chuck Hanson returned to Minnesota after his enlistment ended and became a salesman for a Fortune 200 company. Now retired, he does volunteer work in between hunting and fishing trips. Each winter he travels to Arizona to fly gliders "for the pure joy and peacefulness." Chuck lives in Moorhead, Minnesota, and has never married, explaining that he "came close a few times, but never found a woman who could put up with him for more than five minutes."

After his enlistment ended, Chuck Kudla moved to Detroit, where he joined the Detroit Police Department as a patrolman, and later in a dual role as an EMT. After fifteen years in the police department, he had a motorcycle accident, which eventually led to a medical retirement from the force. He lives in Lake Orion, Michigan, and has three daughters — "three wonderful beauties."

Chuck Limer left the army in 1969 and worked as a carpenter for the next thirty years. In 1999, he went to work for the US Postal Service as a carrier, but was forced into retirement in 2004 due to complications from his wounds suffered on June 2, 1968. He has six children and twelve grandchildren. He and his wife, Diane, reside in Arlington, Texas.

Steve Lyle graduated from college in 1972 with a BS in chemistry/biology. He worked for an immune-chemistry diagnostics firm, and later went back to his medical roots and obtained an RN degree. He has a daughter and three grandchildren. Retired, he lives in the house he built on 43 wooded acres in East Texas.

Tom McClear returned to Michigan and graduated from Michigan State University with honors. After receiving his Juris Doctor degree from the University of Detroit in 1975, he joined his father in his law practice in Owosso, Michigan, where he was joined several years later by his wife, Rebecca, also an attorney. Tom spent the last twenty years of his law career as an assistant attorney general of the State of Michigan. He was an avid reader, a gentleman farmer who loved "to watch the corn grow" on his small farm, and was known for his razor-sharp wit and intellectual curiosity. He and Rebecca had two daughters: Sheila, who is a writer in New York City, and Elisabeth, who is an attorney in Columbus, Ohio. He passed away in 2015.

After his tour in Vietnam, Bob Pagano returned to college, but left without completing his degree to join the family restaurant business. Fourteen years later he completed college and obtained a graduate degree, which he used to start a tax preparation business, which he maintains to this day in Harrisburg, Pennsylvania.

From Vietnam, Basil Rivera was assigned to the 82nd Airborne Division
at Fort Bragg, North Carolina as a squad leader in A Company, 4th Battalion,
325th Infantry. After eleven months, he completed his military service as a
Basic Rifle Instructor. He returned to Texas, where he worked in a chemical
plant, became a barber and hairdresser, then retired as a US Postal Service
mail handler. He and his wife, Lupe, live in Friendswood, Texas. They have
a son and a daughter, both teachers. Basil enjoys family, friends, walks in the
park, reading, and movies, and is active in many local charities.

Jim Roach remained in the army, retiring in 1995 as Special Forces colonel. He was one of fifty-five original US trainers in El Salvador, and has spent
the past ten years in Latin America, serving as a US adviser to government
military and police forces. He and his wife, Peggy, live in Oxford, Pennsylvania. They have three children (the oldest, Steve, has served more than
twenty-seven years on active duty, and is a UH-60 Blackhawk Instructor Pilot)
and seven grandchildren (Steve's daughter, Kirsten, was commissioned in
2016 as a navy nurse).

Mike Scott returned to Georgia after he was discharged from the army,
where he and his wife still reside. With an aptitude for anything mechanical,
Mike spent many years riding and working on motorcycles, then became a
maintenance technician until his retirement. His main focus now is his "two
lovely daughters and three grandsons."

Dan Soto returned to Arizona after Vietnam, and worked in the Phoenix
Cement Company for forty-three years, retiring as the control room supervisor.
He and his wife, Thalia, have three sons, three daughters and fifteen grandchildren. In memory of the soldier who died coming to his aid on July 25, 1968 he
keeps a picture of Sergeant First Class Bill Tellis on the wall of his home.

Dave "Sparrow" Spencer was medically retired from the army in 1969.
The wounds he received on July 22, 1968, had left him paralyzed from the
waist down. He spent the rest of his life supporting veterans' causes, once
"wheel-chairing" from Los Angeles to the state Capital in Sacramento to raise
funds for the California Vietnam Memorial. His police escort was reputed to
have clocked him at close to 70 miles per hour on one of the downhill legs. He
also trained scores of service dogs for returning veterans of both Gulf Wars.
The wheelchair never defined who he was and no one ever heard him com-

plain or lapse into self-pity. Whether learning to water-ski, building and selling homes, or traveling to Afghanistan and befriending Russian paratroopers
fighting the Afghans, Dave lived life to its fullest. He passed away in 2016 and
is buried at the Veterans' Cemetery in Miramar, California.

Rick St John remained in the army, retiring in 1993 as a colonel. His assignments included commander, 2nd Battalion, 9th Infantry Regiment (Man-
chus); commander, 29th Infantry Regiment; chief of staff, 25th Infantry Division (Light); and chief of staff, US Army, Pacific. He also earned an MS degree
in industrial engineering from the University of Arizona. After his army service he worked for TSYS, a global credit card processing company, establishing
its overseas operations in Mexico, Japan, and the United Kingdom. He retired
from TSYS in 2012, as a group executive, to pursue lifelong dreams of writing
and teaching. He currently teaches at a small college in Columbus, Georgia.
Rick has two sons and a daughter, plus four grandchildren. He and his wife,
Susan, reside in Midland, Georgia.

Mike "Bird Dog" Tarpley was hospitalized for a full year following his second wound in April 1968. He worked for thirty-five years in the Texas oil fields
until medically retired in 2006, from lingering effects of his wounds. Since that
time, he has devoted his life to helping veterans by creating a memorial to all
returning veterans in his hometown, serving on numerous veterans' commit-
tees, conducting a weekend dove hunt for disabled veterans, and facilitating a
PTSD discussion group at a VA domiciliary in West Texas.

Ted Tilson remained in the army, retiring as a first sergeant in 1992. After
graduating from Ranger School in 1971, he returned to Vietnam for a second
tour, as a hunter/killer team leader in Company P (Ranger), 75th Infantry,
and also served as first sergeant in the 2nd Battalion, 7th Infantry Regiment,
during Operation Desert Storm in Iraq. He resides in Gainesville, Georgia,
with his wife, Cheri, and three children.

Herman L. Trent remained in the army, retiring as a command sergeant
major in 1976 as one of the most highly decorated soldiers of the Vietnam War.
His decorations include the Distinguished Service Cross, the Silver Star, four
Bronze Stars for valor, three Purple Hearts, five Air Medals, and three Army
Commendation Medals for valor. Upon retirement he remained in Clarksville,
Tennessee, where he attended Austin Peay State University and worked for

the next eighteen years in banking. In 1994, he returned to his hometown of Hazard, Kentucky, with his wife, Sharon, and grandson, Jordan. In 2005 the Marksmanship Training Facility in Fort Campbell, Kentucky, was named in his honor. He passed away in 2007.

Johnny Walker returned to Mississippi to complete his college, earning a bachelor of science and a Masters of Business Administration. He continued to serve his country in the Army National Guard until his retirement. He has two sons, and he and his wife reside in Madison, Mississippi.

AFTEREFFECTS (LIVING WITH THE WAR)

The war remains an undercurrent in our lives, mostly silent and out of sight, yet always tugging at us under a thin veneer of normalcy. In Vietnam, we learned lessons about fear, cowardice, suffering, cruelty and comradeship. Most of all, we learned about death at an age when it is common to think of oneself as immortal. Everyone loses that illusion eventually, but in civilian life it is lost in installments over the years. We lost it all at once, in the span of just a few months. That proximity to death severed us from our youth as irrevocably as a surgeon's scissors had once severed us from the womb.[759]

> "I didn't realize it but I had issues. . . . I thought everyone slept only four hours a night and checked the perimeter periodically. . . . I knew it was unusual to sleep with a weapon in my hand, but did it for years afterward."

> "I'm lucky in the sense that it did not affect me as much as other men in the company. I do not dwell in the past and the most important things in my life are my three grandsons."

> "The war still haunts me as I go over and over what I could have done better. At the present, my life is in complete turmoil. What a mess. I see no end in sight . . . and of course my weekly visits to the VA to deal with PTSD, as I am considered an 'at risk' veteran."

"So much has happened between then and now. Enough, you might say, to put things into their proper perspective . . . and generally they are. But there are just times, or things, or moments when those hundred pounds we carried were rather light."

"I can't remember what I had for dinner last night; but I can remember everything about the battles I was in."

"When I graduated from college in '72, I lied on my résumé about those three years in the army out of fear I wouldn't be invited to any job interviews."

"My life has been anything but boring. Divorced, and remarried to a wonderful lady. I belong to more charitable organizations than is legally allowable—I tend to gravitate to those that work with kids and veterans. Life is great!"

"In August '68 I was at home on leave from Fort Riley. I received a letter about the passing of Hillman, Albertson, et al. I don't think I told anyone, but I went out and got drunk. I drank a lot for about seventeen years. . . . I used alcohol as a painkiller and didn't know it."

"I was not greatly affected [by the war]. I chalked it off as a life experience and went on where I left off."

"For many years I didn't tell anyone I was a Vietnam vet. But now I am so proud to be one."

"I knew I could lose my life in Vietnam. I didn't know that I would keep my life, but lose who I am."

"I learned to live one day at a time and there was no guarantee of tomorrow."

"Cronkite said we couldn't win, LBJ quit, the war was a crime against humanity, etc. The message was clear: when discharged, blend in, tell nobody you were in the military or Nam, conceal it, and I remained that way for twenty-one years."

"Not a day that doesn't go by that something rolls through your head. You just never forget. You could be walking in a park, no one around, on a beautiful day or spitting rain and have a little bit of flashback."

"I just thank God I'm alive. I've got memories. It bothers me."

"I carry guilt feelings around. So many times you survived and the guy next to you died. Time after time after time. You never know who was watching over you."

"The world that I had known before my army service was gone. No one outside of a select few seemed to understand or care that I had watched my friends being killed and I, in turn, had killed enemy soldiers."

"My personal life was a wreck for thirty years, but the last twenty years have been great. Five marriages, the fifth one has lasted longer than the other four combined."

"The death and devastation have not caused me any problems in later life. I sleep like a baby with no bad dreams of the war."

"The war affected me and still does. As a Catholic, I was taught not to kill; it took me a long time to be strong again."

"I was accused of not having feelings by all who loved or knew me. I was not a nice person."

"After getting out of the military, no one knew where I was for two years. What a mean thing to do to my family."

"I am not as emotional about death as others in my family. Death is a part of life, and although I don't like it when family members die— death is not a great emotional event for me."

"I will never forget picking up our soldiers and putting them in their ponchos; the smell was so bad."

"I was in the hospital when I received the sad news about Ron (KIA). It broke my heart. Now, all I can do is stop by the Wall in DC and touch his name."

A Hero, Forgotten No Longer

On October 22, 2016, Doc Franks was finally awarded a Silver Star for his heroism on June 2, 1968, in the Central Highlands of South Vietnam. The original recommendation had vanished decades ago in the army's bureaucratic labyrinth. There is no clear answer as to why or how.

Weakened from a debilitating regimen of chemotherapy for a cancer that had plagued his later years, Franks nevertheless stood straight and tall for the awards presentation, his eyes fixed on fellow Tiger Bravo veterans and their families seated in front of a Vietnam War Memorial at the National Infantry Museum in Columbus, Georgia. Colonel Kelly D. Kendrick, a senior commander at Fort Benning, Georgia, who was not even born when the heroics took place, presented the award "to a humble hero who symbolizes the enduring legacy of the 101st Airborne Division, and its warrior ethos; attributes that the current generation of soldiers revere and strive to emulate." When Franks attempted to down play his actions, saying that "anyone would have done the same thing," Colonel Kendrick, himself a veteran of the 101st in Iraq and Afghanistan, replied, "That is probably so, but you are the one who did it."

Sitting in the front row at the ceremony next to Doc's wife, Cindy, and watching his friend of almost fifty years being presented with America's third-highest medal for saving his life, was Chuck Limer. In acknowledging

Limer's presence and recognizing his role in starting the battle on June 2, 1968, Colonel Kendrick observed that "often in combat someone has to be the first one to poke a stick into the hornets' nest. On that day, it was Chuck Limer. But we've all done something similar somewhere along the line."

It was only a brief, simple ceremony, but to Doc Franks and Chuck Limer, it was a golden moment in their lives. It was five decades in the making, and half the world away from the scene of the battle, but a golden moment nonetheless.

Killed in Action

January 16, 1968 (Binh Duong Province), three KIA

Specialist Four Eugene S. Hicks (Arcadia, California), nineteen years old

Private First Class Homer E. Pierce Jr. (Cillicothe, Ohio), eighteen years old

Specialist Four John H. Wrisberg III (Mason City, Iowa), attached medic, twenty years old

February 1, 1968 (Bien Hoa Province), one KIA

Private First Class Gene L. Keahi (Ewa Beach, Hawaii), eighteen years old

July 22, 1968 (Hau Nghia Province), ten KIA

Private First Class Ronald D. Albertson (Dimondale, Michigan), twenty-one years old.

Specialist Four Eugene F. Davis (Chicago), twenty years old

PRIVATE FIRST CLASS JOE M. DAVIS (CHICAGO), TWENTY YEARS OLD

SPECIALIST FIVE JOHN E. GREESON (MELBOURNE, FLORIDA), ATTACHED MEDIC, EIGHTEEN YEARS OLD

SERGEANT ALLAN F. HAMSMITH (FAIRMONT, MINNESOTA), TWENTY-TWO YEARS OLD

FIRST LIEUTENANT JOSEPH HILLMAN III (PIEDMONT, ALABAMA), TWENTY-THREE YEARS OLD

SERGEANT MACKLIN O. HUGHES (PISGAH, ALABAMA), TWENTY-TWO YEARS OLD

CORPORAL DAVID M. MAYMON (FAIRFIELD, ILLINOIS), NINETEEN YEARS OLD

SPECIALIST FOUR JOHN P. MURPHY (OMAHA, NEBRASKA), ATTACHED MEDIC, TWENTY-THREE YEARS OLD

SERGEANT JACKIE R. POLING (SCOTT, OHIO), TWENTY YEARS OLD

July 25, 1968 (Hau Nghia Province), two KIA

PRIVATE FIRST CLASS STEVEN A. FRINK (VANCOUVER, WASHINGTON), ATTACHED MEDIC, TWENTY YEARS OLD

SERGEANT FIRST CLASS WILLIAM J. TELLIS (DETROIT, MICHIGAN), THIRTY-SEVEN YEARS OLD

July 26, 1968 (Hau Nghia Province), one KIA

STAFF SERGEANT PAUL H. NABORS (BINGER, OKLAHOMA)
TWENTY-THREE YEARS OLD

August 28, 1968 (Hau Nghia Province), ten KIA

PRIVATE FIRST CLASS JAMES J. CRISWELL (RED LION, PENNSYLVANIA),
TWENTY YEARS OLD

STAFF SERGEANT THOMAS R. LAMB (LOS ANGELES), TWENTY-FOUR YEARS OLD

SERGEANT MICHAEL J. MITCHELL (SAN ESTEBAN, CALIFORNIA),
TWENTY-ONE YEARS OLD

PRIVATE FIRST CLASS ARMONDO S. NAVARRO (SAN ANTONIO, TEXAS),
TWENTY-ONE YEARS OLD

PRIVATE FIRST CLASS JAMES R. PIZZANO (EVERETT, MASSACHUSETTS),
NINETEEN YEARS OLD

SPECIALIST FOUR VICTOR M. PLOURDE (OAKLAND, MAINE),
NINETEEN YEARS OLD

PRIVATE FIRST CLASS PAUL M. STOCKWELL (CORDER, MISSOURI),
TWENTY YEARS OLD

CORPORAL ISIAH THOMAS (WASHINGTON, LOUISIANA), TWENTY-TWO YEARS OLD

CORPORAL ROBERT J. WASHINGTON (INDIANAPOLIS, INDIANA),
EIGHTEEN YEARS OLD

PRIVATE FIRST CLASS THOMAS V. WILLIAMS JR. (LAKIN, KANSAS),
TWENTY YEARS OLD

September 21, 1968 (Hau Nghia Province), one KIA

SPECIALIST FIVE HOWARD OWENS (MEMPHIS, TENNESSEE), ATTACHED MEDIC, TWENTY YEARS OLD

November 12, 1968 (Thua Thien Province) one KIA

PRIVATE FIRST CLASS RALPH GUARIENTI (SACRAMENTO, CALIFORNIA), TWENTY YEARS OLD

November 16, 1968 (Thua Thien Province), one KIA

PRIVATE FIRST CLASS STANLEY F. WILTON (SACRAMENTO, CALIFORNIA), TWENTY YEARS OLD [760]

Bibliography

BATTLEFIELD JOURNALS, REPORTS, AND STUDIES

"Daily Staff Journals," Headquarters, 2nd Battalion, 506th Infantry, various locations in Republic of Vietnam, 330 journals, 1 January–15 December 1968.

"Daily Staff Journals," Headquarters, 3rd Brigade, 101st Airborne Division, July 1968.

"Operational Report of 3rd Brigade, 101st Airborne Division for Period Ending 30 April 1968," Headquarters, 3rd Brigade, 101st Airborne Division, 16 May 1968.

"Operational Report of 3rd Brigade, 101st Air Cavalry Division for Period Ending 31 July 1968," Headquarters, 3rd Brigade, 101st Air Cavalry Division, 12 August 1968.

"Operational Report of 3rd Brigade, 101st Airborne Division (Airmobile) for Period Ending 31 October 1968," Headquarters, 3rd Brigade, 101st Airborne Division (Airmobile), 15 November 1968.

"Operational Report of 3rd Brigade, 101st Airborne Division for Period Ending 31 January 1969," Headquarters, 3rd Brigade, 101st Airborne Division, 5 February 1969.

"Recommendation for Award of the Presidential Unit Citation," Headquarters, 3rd Brigade, 101st Airborne Division (Airmobile), 25 November 1968.

"Operational Report of 101st Airborne Division Artillery for Period Ending 30 April 1968," Headquarters, 101st Airborne Division Artillery, 10 May 1968.

"Operational Report—Lessons Learned, 1 May 1968–31 July 1968," Headquarters, 101st Air Cavalry Division Artillery, 15 August 1968.

"Operational Report—Lessons Learned, 1 August 1968–31 October 1968," Headquarters, 101st Airborne Division Artillery, 1 November 1968.

"Operational Report for Quarterly Period Ending 31 January 1968," Headquarters, 101st Airborne Division, 31 March 1968.

"Inclosure 1: After Action Report—Premovement and Movement Phases, dated 31 January 1968, Operational Report for Quarterly Period Ending 31 January 1968," Headquarters, 101st Airborne Division, 31 March 1968.

"Operational Report of 101st Airborne Division for Period Ending 30 April 1968," Headquarters, 101st Airborne Division, 24 May 1968.

"Operational Report of 101st Air Cavalry Division for Period Ending 31 July 1968," Headquarters, 101st Air Cavalry Division, 15 August 1968.

"Operational Report—Lessons Learned for Period Ending 31 October 1968," Headquarters, 101st Airborne Division, 11 March 1969.

"After Action Report for Operation Golden Sword," Headquarters, 3rd Brigade, 101st Airborne Division, 11 November 1968.

"Operational Report—Lessons Learned for Period Ending 31 January 1969," Headquarters, 101st Airborne Division, 24 February 1969.

"S3 Daily Staff Journals," !st Brigade, Headquarters, 1st Brigade, 4th Infantry Division, Dak To, 30 –31 May and 1 June 1968.

"Combat Operations After Action Report, Operation Mathews," Headquarters, 4th Infantry Division, 13 June 1968.

"Operational Report—Lessons Learned for Quarterly Period Ending 31 July 1968," Headquarters, 4th Infantry Division, 18 August 1968.

"Chronological Summary of Significant Activities, 1 May–31 July 1968," Headquarters, 4th Infantry Division, n.d.

"Combat Operational After Action Report," Headquarters, 3rd Brigade, 9th Infantry Division, 20 August 1968.

"Operational Report of 9th Infantry Division for Period Ending 30 April 1968," Headquarters, 9th Infantry Division, 12 May 1968.

"Operational Report of 11th Armored Cavalry Regiment for Period Ending 30 April 1968," Headquarters, 11th Armored Cavalry Regiment, 10 May 1968.

"Feeder Report for Operational Report—Lessons Learned," Headquarters, 2nd Brigade, 25th Infantry Division, 6 November 1968.

"Combat Operations After Action Report, TOAN THANG (Phase II)," Headquarters, 2nd Brigade, 25th Infantry Division, 20 March 1969.

"Mines and Booby Traps," Headquarters, 3rd Brigade, 25th Infantry Division, 5 August 1967.

"Operations Report of 25th Infantry Division Artillery for Period Ending 31 October 1968," Headquarters, 25th Infantry Division Artillery, 16 November 1968.

"Combat After Action Report of Battle of Tay Ninh, 17 August–27 September 1968," Headquarters, 25th Infantry Division, 7 February 1969.

"Commander's Combat Note Number 16, Combat Analysis Number 4," Headquarters, 25th Infantry Division, October 1968.

"Operational Report for Quarterly Period Ending 30 April 1968," Headquarters, 25th Infantry Division, n.d.

"Operational Report of 25th Infantry Division for Period Ending 31 July 1968," Headquarters, 25th Infantry Division, 1 August 1968.

"Operational Report for Headquarters, 199th Infantry Brigade for Period Ending 31 July 1968," Headquarters, 199th Infantry Brigade (SEP) (LT), 22 August 1968.

"Operational Report of Lessons Learned for Period Beginning 1 May 1966," Headquarters, 168th Engineer Combat Battalion, 14 August 1966.

James M. Wright, "History of 162nd Assault Helicopter Company 1 April 1968–31 June 1968," Headquarters, 11th Combat Aviation Battalion, n.d.

"Report on Booby Traps," Headquarters, Americal Division, 3 October 1968.

"After Action Report on Ranger Operations in Zone D, Phase III "Operation Holiday," 3 January–9 January 1963," Headquarters, US MAAG Detachment, Phuoc Binh Thanh Special Zone (Song Be), 17 January 1963.

"Vietnamese Ranger Operations in War Zone D, 1962–63," Phuoc-Binh-Thanh Special Zone, August 1998.

"Analysis of Communist Vietnamese Special Operations Forces During the Vietnam War," Headquarters, US Naval Postgraduate School, June 2005.

"Area Analysis Study, 66-16, War Zone D," Combined Intelligence Center Headquarters, US MACV, Office of Assistant Chief of Staff J-2 and Headquarters, Armed Forces of RVN, Office of Joint General Staff J-2, 19 May 1966.

"Human Factors Considerations of Undergrounds in Insurgencies," Special Operations Research Office (SORO), American University, Washington, DC, 1966.

"Memorial Affairs Activities—Republic of Vietnam, Mortuary Affairs Center, Fort Lee, VA, March 2000.

"Project CHECO Southeast Asia Report '"The Defense of Saigon,'" Headquarters, Pacific Air Force, Directorate of Tactical Evaluation, 14 December 1968.

"Project CHECO Southeast Asia Report 'Short Rounds, June 1967–June 1968,'" Headquarters, Pacific Air Force, Directorate of Tactical Evaluation, 23 August 1968.

"Project CHECO Southeast Asia Report 'Rolling Thunder, January 1967– November 1968,'" Headquarters, Pacific Air Force, Directorate of Tactical Evaluation, 1 October 1969.

"Interrogation Report: Defoliation Operations in War Zone D," Combined Military Interrogation Center, 22 June 1967.

"Tet Offensive II Field Force Vietnam: After Action Report 31 January–18 February 1968," Headquarters, II Field Force, n.d.

"Annex A (Intelligence) to Tet Offensive After Action Report," Headquarters, II Field Force, n.d.

"Operational Report—Lessons Learned, Headquarters, II Field Force, Vietnam, for Period Ending 31 January 1968," Headquarters, II Field Force, 21 February 1968.

"Operational Report of Headquarters, II Field Force, Vietnam, for Period Ending 31 October 1968," Headquarters, II Field Force, n.d.

"The Impact of the Sapper on the Vietnam War: A Background Paper," US Mission in Vietnam, October 1969.

"Report of the Seminar on Night Operations in RVN," Headquarters, US Army, Vietnam, 13 April 1968.

Thomas B. Bennett, "Night Jungle Operations," School of Advanced Military Studies, US Army Command and General Staff College, Fort Leavenworth, KS, 1992–93.

"USARV Seminar Report: Attack of Fortified Position in the Jungle," Headquarters, US Army, Vietnam, 2 January 1968.

"A Review of the Herbicide Program in South Vietnam," Scientific Advisory Group, Working Paper No. 10-68, Commander in Chief, Pacific, Scientific Advisory Group, August 1968.

"The Air Force and Herbicides in Southeast Asia 1961–1971," Headquarters, US Air Force, Office of Air Force History, Washington, DC, 1982.

"Services Herbs Tape: A Record of Helicopter and Ground Spraying Missions, Aborts, Leaks, and Incidents," US Army and Joint Services Environmental Support Group, Washington, DC, 12 September 1985.

"Study 67-080: NVA Rocket Artillery Units," Headquarters, Armed Forces of RVN, Office of Joint General Staff, J2, 1 September 1967.

LTC Edwin T. Cooke, "Study on Combat Stress and Its Effect," Medical Field Service School, Fort Sam Houston, TX.

S. L. A. Marshall and David H. Hackworth, "Vietnam Primer: Lessons Learned May 1966–February 1967," Headquarters, Department of the Army, Washington, DC, n.d.

Don Starry, "Vietnam Studies: Mounted Combat in Vietnam," Headquarters, Department of the Army, Washington, DC, 1989.

"Vietnam Studies: Tactical and Material Innovations," CMH Publication 90-21-1, Headquarters, Department of the Army, Washington, DC.

"A Study of Strategic Lessons Learned in Vietnam: Volume VII, Soldier in RVN", BDM Corporation, McLean, VA, April 11, 1980.

"A System Analysis View of the Vietnam War 1965–1972, Volume 8, Casualties and Losses", OASD (SA) Southeast Asia Division, Washington, DC, February 18, 1985.

BOOKS

Ambrose, Stephen E. *Band of Brothers: E Company, 506th Regiment, 101st Airborne from Normandy to Hitler's Eagle's Nest.* New York: Simon & Schuster, 2001.

Army Historical Series: American Military History. Washington, DC: US Army Center of Military History, 1989.

Arnold, James. *Tet Offensive 1968: Turning Point in Vietnam.* Long Island City, NY: Osprey Publishing, 2012.

Atkinson, Rick. *An Army at Dawn: The War in North Africa 1942–1943.* New York: Owl Books, 2002.

———. *The Long Gray Line: The American Journey of West Point's Class of 1966.* Holt McDougal, 2009.

Becton, Julius. *Autobiography of a Soldier and Public Servant.* Annapolis, MD: Naval Institute Press, 2008.

Boyne, Walter J. *Beyond the Wild Blue: A History of the US Air Force 1947–2007.* St Martin's Griffin, New York, NY, 1998.

Caputo, Philip. *A Rumor of War.* New York: Owl Books, 1977.

Donahue, James C. *Blackjack 33: With Special Forces in the Viet Cong Forbidden Zone.* Ballantine Books, New York, NY, 1999.

Farwell, Byron. *Encyclopedia of Nineteenth Century Land Warfare: An Illustrated World View.* R. S. Farwell Living Trust, 2001.

Friedman, Matti. *Pumpkin Flowers: A Soldier's Story.* Chapel Hill, NC: Algonquin Books of Chapel Hill, 2016.

Fussell, Paul. *Wartime: Understanding and Behavior in the Second World War.* New York: Oxford University Press, 1989.

Galloway, Joseph L., and Harold G. Moore. Harold G. *We Were Soldiers Once . . . and Young: Ia Drang—the Battle That Changed the War in Vietnam.* New York: Open Road Integrated Media, 2004.

Gilliam, James T. *Life and Death in the Central Highlands: An American Sergeant in the Vietnam War 1968–1970.* Denton, TX: University of North Texas Press, 2010.

Haslam, Tim. *Stars and Stripes and Shadows: How I Remember Vietnam.* Bloomington, IN: AuthorHouse, 2010.

Junger, Sebastian. *Tribe.* New York: Machete Book Group, 2016.

Infantry in Battle. 2nd ed. Washington, DC: *Infantry Journal,* 1939.

Kelley, Michael P. *Where We Were in Vietnam: A Comprehensive Guide to the Firebases, Military Installations, and Naval Vessels of the Vietnam War.* Ashland, OR: Hellgate Press, 2002.

Lam, Truong Buu. *A Story of Vietnam.* TBL at Smashwords, 2013.

Lung, Hoang Ngoc. *The General Offensives of 1968–69.* Indochina Monograph. US Army Center of Military History, 1981.

Malone, Dandridge M. *Small Unit Leadership: A Common Sense Approach.* Novato, CA: Presidio Press, 1983.

Mangold, Tom, and John Penycate. *The Tunnels of Cu Chi: A Harrowing Account of America's "Tunnel Rats" in the Underground Battlefields of Vietnam.* New York: Random House, 2005.

Menzell, Sewall. *Battle Captain: Cold War Campaigning with the US Army in Vietnam, Cambodia, and Laos, 1967–1971.* Bloomington, IN: AuthorHouse, 2007.

Nolan, Keith. *Battle for Saigon.* New York: Pocket Books, 1996.

Owen, Wilfred. *Wilfred Owen: The War Poems.* Ed. Jon Stallworthy. London: Chatto & Windus, 1994.

Roach, Mary. *Grunt: The Curious Science of Humans at War.* New York: W. W. Norton, 2016.

Rottman, Gordon L. *North Vietnamese Army Soldier 1958–1975.* New York: Osprey Publishing, 2009.

———. *The US Army in the Vietnam War: 1965–1973.* New York: Osprey Publishing, 2008.

———. *Viet Cong and NVA Tunnels and Fortifications of the Vietnam War.* New York: Osprey Publishing, 2006.

_____. *The US Army Infantryman in Vietnam*. New York: Osprey Publishing, 2005.

Shay, Jonathan, MD, PhD. *Achilles in Vietnam: Combat Trauma and the Undoing of Character*. New York: Scribner, 1994.

Sheehan, Neil. *A Bright Shining Lie: John Paul Vann and America in Vietnam*. New York: Modern Library, 2009.

Sorley, Lewis. *A Better War: The Unexamined Victories and Final Tragedy of America's Last Years in Vietnam*. Orlando, FL: Harcourt, 1999.

Spurgeon, Neel, MD. *Medical Support of the US Army in Vietnam 1965–1970*. US Army Medical Department, Office of Medical History, Washington, DC, 1972.

Stanton, Shelby S. *The Rise and Fall of an American Army: US Ground Forces in Vietnam 1965–1973*. Novato, CA: Presidio Press, 1985.

———. *Vietnam Order of Battle: A Complete Illustrated Reference to US Army Combat and Support Forces in Vietnam, 1961–1973*. Mechanicsburg, PA: Stockpile Books, 2003.

Stewart, Richard W., ed. *American Military History, Volume II: The United States Army in a Global Era, 1917–2008*. US Army Center for Military History, 1956.

St John, Rick. *Circle of Helmets: Poetry and Letters of the Vietnam War*. Bloomington, IN: AuthorHouse, 2002.

Summers, Harry G. *On Strategy: A Critical Analysis of the Vietnam War*. New York: Random House, 1982.

Veith, George J. *Code-Name Bright Light*. New York: Dell, 1998.

Wilkins, Warren K. *Grab Their Belts to Fight Them: The Viet Cong's Big Unit War Against the US, 1965–1967*. Annapolis, MD: Naval Institute Press, 2011.

Willbanks, James. *Vietnam War: The Essential Reference Guide*. Santa Barbara, CA: ABC-CLIO, 2013.

INTERVIEWS

Joe Adams, interview by author, January 21, 2015.

Chris Backman, interview by author, July 11, 2016.

Dan Bernard, interview by author, March 3, 2015.

Larry Burton, interview by author, January 2, 2016.

Everett Franks, interview by author, November 11, 2014.

David E. Grange, interviews by author, April 4 and 23, 2014, and September 11, 2014.

Chuck Hanson, interview by author, December 11, 2014.

Warren Kiilehua, interview by author, May 31, 2017.

Chuck Kudla, interviews by author, February 5 and 11, 2014, and April 4, 2015.

Chuck Limer, interviews by author, December 11, 16, and 20, 2014.

Bob Pagano, interview by author, May 4, 2016.

Joe Palagyi, interview by author, February 12, 2015.

Basil Rivera, interviews by author, March 4, 2015, and March 22, 2016.

Jim Roach, interview by unknown source, November 21, 2011.

Dan Soto, interview by author, October 23, 2016.

Dave Spencer, interview by author, September 12, 2014.

Mike Tarpley, interview by author, May 14, 2015.

Ted Tilson, interview by author, April 16, 2016.

Terry Van Meter, interview by author, May 2, 2017.

Johnny Walker, interview by author, November 3, 2015.

MISCELLANEOUS

"Bill Plemons Memorial Service," September 16, 2010.

FM 6-141-1, "Field Artillery Target Analysis and Weapons Employment," Headquarters, Department of the Army, Washington, DC, 15 February 1978.

FM 23-11, "90MM Recoilless Rifle, M-67," Headquarters, Department of the Army, Washington, DC, 6 July 1965.

General Order 10, "Award of the Purple Heart," Headquarters, 6th Convalescent Center, 13 June 1968.

General Order 38, "The Presidential Unit Citation (Army)," Headquarters, Department of the Army, Washington, DC 25 November 1968.

General Order 8630, "Award of the Silver Star," Headquarters, 101st Airborne Division, TCO 320, 30 October 1968.

DA Pam 750-30, "The M-16A1 Rifle: Operation and Preventive Maintenance," Headquarters, Department of the Army, Washington, DC, 1 July 1969.

Army Regulation 600-8-22, "Military Awards," Headquarters, Department of the Army, Washington, DC, 25 June 2015.

Awards Roster, Headquarters Company, 2nd Battalion, 506th Infantry, Phuoc Vinh, Vietnam, 25 June, 1968.

"Tour 365: For Soldiers Going Home," Office of the Commander, US Military Assistance Command, Vietnam, 1971.

DA Pam 360-521, "Handbook for US Forces in Vietnam," Headquarters, US Army, Vietnam, June 1966.

Unit Roster, Headquarters, B Company, 2nd Battalion, 506th Infantry (Airborne), Fort Campbell, KY, 27 November 1967.

"2/506th Unit History," Headquarters, 2nd Battalion, 506th Infantry, n.p., n.d.

"145th Combat Aviation Battalion History, Volume II," Headquarters, 145th Combat Aviation Battalion, n.d.

"Memorial Service, 2nd Battalion, 506th Infantry Airborne, Phuoc Vinh, Republic of Vietnam," n.d.

Special Order 33, "Combat Infantryman Badge (First Award)," Headquarters, 101st Airborne Division, 2 February 1968.

Map Tan Uyen (1:50,000) Sheet 6331 II Series L 7014, Headquarters, 29th Engineer Battalion, 1971.

Map D Go Kram, Vietnam, and Laos (1:50,000) Sheet 6539 III Series L 7014, US Army, 1966.

William Tresder, quote, US Marine Corps, n.d. d

Incident investigation letter, Headquarters, 2nd Battalion, 506th Infantry (Dak To), 3 June 1968.

"Message from the Commanding General, 101st Airborne Division," Headquarters, 2nd Battalion (Airmobile), 506th Airborne Infantry, 1 November 1968.

Russell W. Glenn, "Men Against Fire in Vietnam," School of Advanced Military Studies, US Army Command and General Staff College, Fort Leavenworth, KS, 8 November 1987.

Robert H. Scales, "Life Is like a Box of C Rations," speech at the Truman Library, 12 December 2009.

Memorandum, "Personnel Pending Awards," Headquarters, 2nd Battalion (Airmobile) 506th Infantry, 16 November 1968.

Gregg, Knoltown, "Agent Orange: Phuoc Vinh Ground Zero," n. d.

Brandon Gold and Stanley Horowitz, "History of Combat Pay," Institute for Defense Analyses, Alexandria VA, August 2011.

NEWSPAPERS

"Paratroop Arrivals Put US Buildup Past Korea Peak," *San Francisco Examiner,* December 13, 1967.

Dean Phillips, "Vietnamese Sgt Explains Phuoc Vinh Landmark," *Screaming Eagle,* n. d.

"VC Lose by a Nose: Airborne Finds Supply Cache," *Screaming Eagle,* n.d.

"VC Denied 42 Tons of Rice; Chinooks Lift Out Cache," *Screaming Eagle* 1, no. 5, March 1, 1968.

"Empty Helmets Tell Full Story," *Nashville Tennessean,* February 16, 1968.

"Screaming Eagles Maul Enemy: Crush Tet Attacks on Three Fronts," *Screaming Eagle* 1, no. 4, February 23, 1968.

"Bien Hoa Attackers Repelled: 150 VC Killed in Two Days," *Army Reporter,* February 17, 1968.

Clyde Haberman, "Agent Orange's Long Legacy, for Vietnam and Veterans," *New York Times,* May 11, 2014.

Jack Hurst, "2-506th Patrol Traps Reds in Ambush," *Screaming Eagle* 1, no. 9, March 29, 1968.

"Line Troopers Spend Rare Night at Home," *Screaming Eagle* 1, no. 27, August 5, 1968.

"Commander's Corner," *Screaming Eagle* 1, no. 5, March 1, 1968.

Jon Steinberg, "Recovery Center Puts GIs in the Pink," *Stars & Stripes,* September 27, 1969.

"3rd Brigade Surprises VC: Use New Tactics," *Screaming Eagle* 1, no. 24, July 15, 1968.

Herb Burdett, "Quack Outfit Arrives at Go Dau Bridge," *Tropic Lightning News,* February 1969.

"2 Area Men Cited Posthumously," n.p., n.d.

"Eaton County Inductees Leave for Service," *Eaton Rapids (MI) Journal,* December 15, 1966.

"Soldier Killed in Vietnam," *Vancouver (WA) Columbian,* n.d.

Matthew Burke, "For Those Who Prepared Vietnam's Fallen, a Lasting Dread," *Stars & Stripes,* November 9, 2014.

Lyle Nelson, "R & R—an Island Success Story," *Honolulu Star-Bulletin,* July 1968.

"Viet Cong Use Human Shields in Attack on GIs," *Honolulu Star-Bulletin* 57, no. 250, September 6, 1968.

Currahees Newsletter, 2nd ed., n.d.

Screaming Eagle 1, no. 9, March 29, 1968.

Dennis J. Sauffer, "The Bitter Homecoming," *Grand Rapids (MI) Press,* December 5, 1982.

Herb Burdett, "Quack Outfit Arrives at Go Dau Ha Bridge," *Tropic Lightning News,* February 1969.

PERIODICALS

Bob Bennett, "The 1970 Raid on War Zone D POW Camp," *Delta Troop, 3/17th Air Cavalry,* n.d.

Mike Campbell, "Maisey Building Rededication Honors Fallen Hero of Bien

Hoa," January 31, 2008.

John T. Correll, "Arc Light," *Air Force Magazine*, no. 1, 2009.

Carolyn Later, "The Daisy Cutter Bomb: Largest Conventional Bomb in Existence," *Technical Review*, n.d.

Kip Lindberg, "The Use of Riot Control Agents During the Vietnam War," *Army Chemical Review*, January–June 2007.

Robert N. Neer, "Napalm's Fiery Fall into Infamy," *Vietnam Magazine* (August 2013): 43–52. Compiled by the Memorial Affairs Center (MAC), Fort Lee, VA, March 2000.

"Memorial Affairs Activities – Republic of Vietnam."

EMAILS, LETTERS, TELEPHONE CALLS AND OTHER PERSONAL CORRESPONDENCE

Joe Adams.

Chris Backman.

Dan Bernard.

Harry Brown (Transcripts of letters donated by Mike Scott).

Virginia Brown.

Larry Burton.

John O. Childs.

Everett "Doc" Franks.

Dianne Wheeler Gnass.

David E. Grange.

Chuck Hanson.

Warren Kiilehua.

Chuck Kudla.

Thomas "Biffer" Lamb (Letter extract donated by his sister, Patti Sullivan).

Chuck Limer.

Steve Lyle.

Rebecca McClear.

Mary-Eileen McClear.

Bob Pagano.

Joe Palagyi.

Brandon Plemons.

Basil Rivera.

Jim Roach.

Mike Scott.

Dan Soto.

Dave "Sparrow" Spencer.

Rick St John.

Patti Sullivan.

Mike Tarpley.

Mary Tellis (Copy of letter received in Vietnam by the author).

Ted Tilson.

Terry Van Meter.

Johnny Walker.

Web Sources

Rod Powers, "The Air Assault School," accessed June 16, 2016, http://www.usmilitary.about.com.

Guam Agent Orange, "Summary Document: Agent Orange at Johnston Island," accessed June 15, 2016, htpp://www.guamagentorange.info.

Global Security, "Johnston Atoll," accessed June 15, 2016, htpp://www.globalsecurity.com.

"Agroville Program," accessed June 20, 2016, http://www.Vietnamwar.net.

"Combat Infantry Badge," accessed June 21, 2016, http://www.hrc.army.mil.com.

"Jill St. John," accessed June 24, 2016, http://www.imdb.com.

"DMZ," accessed July 12, 2016, http://www.u-s-history.com.

"Places: Trang Bang and FSB Stuart III," accessed November 10, 2016, www.212warriors.com.

SERTS, accessed November 17, 2016, www.vetshome.com.

"By Courage and Faith: A Love Story Terry '66 and Jacquie Van Meter, accessed May 17, 2017, www.alumni.norwich.edu/vanmeter.

"The Plate of Five Fruits in Vietnamese New Year," accessed June 2, 2017, www.vietnam-culture.com.

Notes

1. www.underground.com.
2. Unit Roster, Headquarters, B Company, 2nd Battalion, 506th Infantry (Airborne) (Fort Campbell, KY), 27 November 1967.
3. Matti Friedman, *Pumpkin Flowers: A Soldier's Story* (Chapel Hill, NC: Algonquin Books of Chapel Hill, 2016), 16.
4. *A Study of the Strategic Lessons Learned in Vietnam: Volume VII, Soldier in Vietnam,* BDM Corporation, McClean, VA, April 11, 1980, EX-4.
5. Doc Franks, email, December 26, 2016.
6. Chris Backman, interview by author, July 11, 2016.
7. "Operational Report for Quarterly Period Ending 31 January 1968, Inclosure 1, After Action Report—Premovement and Movement Phases," Headquarters, 101st Airborne Division, 31 March 1968. 135.
8. Rod Powers, "The Air Assault School," 1, accessed June 16, 2016. http://.www.usmilitary.about.com.
9. "Operational Report for Quarterly Period Ending 31 January 1968," 57.
10. Chuck Hanson, letter, November 2014.
11. Shelby L. Stanton, *The Rise and Fall of an American Army: US Ground Forces in Vietnam 1965–1973* (Presidio, CA: Presidio Press, 1985), 140.
12. "Operational Report for Quarterly Period Ending 31 January 1968," 192–93.
13. Ibid., 55.
14. Sewell Menzel, *Battle Captain: Cold War Campaigning with the US Army in Vietnam, Cambodia, and Laos, 1967–1971* (Bloomington, IN: AuthorHouse, 2007), 30.
15. Steve Lyle, letter, April 21, 2015.
16. *Army Historical Series: American Military History* (Washington, DC: US Army Center of Military History, 1989), 670.
17. "Operational Report for Quarterly Period Ending 31 January 1968," 135.
18. LTG (Ret.) David E. Grange Jr., interview by author, April 23, 2014.
19. Dan Bernard, interview by author, March 3, 2015.
20. "Operational Report for Quarterly Period Ending 31 January 1968," 149.
21. Chuck Hanson, letters, October 8, 10, and 15, 1967.
22. Ibid.
23. "Operational Report for Quarterly Period Ending 31 January 1968," 114.
24. Ibid., 85–92.
25. Ibid., 99.
26. www.history.com.
27. www.infoplease.com.
28. www.biographies.com.
29. www.thepeoplehistory.com.
30. Mike Tarpley, email, May 20, 2015.
31. Grange, interview by author, April 23, 2014.
32. *2/506th Daily Staff Journals,* Headquarters, 2nd Battalion, 506th Infantry, January–December 1968.
33. Steve Lyle, letter, April 21, 2015.
34. Photograph, *Screaming Eagle,* n.d.
35. Dean Phillips, "Vietnamese Sgt Explains Phuoc Vinh Landmark," *Screaming Eagle,* n.d.
36. Michael P. Kelley, *Where We Were in Vietnam: A Comprehensive Guide to the Firebases, Military Installations, and Naval Vessels of the Vietnam War* (Ashland, OR: Hellgate Press, 2002), 877–78.
37. James M. Wright, *History of the 162nd Assault Helicopter Company, 1 April 1968–31 June 1968.*
38. Manzel, *Battle Captain,* 36.
39. Ibid., 39.
40. Rick St John, letter, December 22, 1967.
41. Mike Tarpley, email, July 7, 2016.
42. Rick St John, letter, December 17, 1967.
43. Ibid., December 23, 1967.
44. Chuck Hanson, letter, December 13, 1967.
45. Ibid., January 9, 2017.
46. Steve Lyle, questionnaire, October 2014.
47. Ibid.
48. "Bill Plemons Memorial Service," September 16, 2010.

49. Steve Lyle, questionnaire, October 2014.

50. Ibid.

51. www.theaviationzone.com.

52. *Oxford English Dictionary*, 2nd ed., 1989.

53. Rick St John, letter, December 19, 1967.

54. "Operational Report—Lessons Learned, Headquarters II Field Force Vietnam, Period Ending 31 January 1968," Headquarters, II Field Force Vietnam, 21 February 1968, 39–40.

55. Rick St John, letter, n.d.

56. Ibid., December 24, 1967.

57. Jim Roach, email, May 24, 2017.

58. Ibid., December 30, 1967.

59. Rick Atkinson, *The Long Gray Line: The American Journey of West Point's Class of 1966.* (New York, NY: Holt McDougal, 2009), 1,446.

60. Rick St John, letter, December 17, 1967.

61. "OPORD 1-68," Headquarters, 2nd Battalion, 506th Infantry, Phuoc Vinh, January 3, 1968.

62. "2/506th Daily Staff Journal," Headquarters, 2nd Battalion, 506th Infantry, Phuoc Vinh, January 3, 1968.

63. Ibid.

64. Ibid., January 5–6, 1968.

65. Ibid., January 9, 1968.

66. Dandridge M. Malone, *Small Unit Leadership: A Common Sense Approach* (Novato, CA: Presidio Press, 1983), 89.

67. Spurgeon Neel, MD, *Medical Support of the US Army in Vietnam 1965–1970* (US Army Medical Department, Office of Medical History, Washington, DC), 170.

68. "Operational Report for Quarterly Period Ending 31 January 1968," 4.

69. Chuck Limer, questionnaire, October 2014.

70. Steve Lyle, questionnaire, n.d.

71. Philip Caputo, *A Rumor of War,* (New York, NY: Oel Books, 1977), 174.

72. Warren K. Wilkens, *Grab Their Belts to Fight Them: The Viet Cong's Big Unit War Against the US, 1965–1966* (Annapolis, MD: Naval Institute Press, 2011), 36.

73. Ibid., 33.

74. "2/506th Daily Staff Journal," FSB Dave, 16 January 1968.

75. Everett Franks, email, March 31, 2015.

76. Ibid., January 2016.

77. S. L. A. Marshall and David B. Hackworth, "Vietnam Primer: Lessons Learned May 1966– February 1967, Headquarters, Department of the Army, n.d.

78. James C. Donahue, *Blackjack 33: With Special Forces in the Viet Cong Forbidden Zone* (New York, NY: Ballantine Books, 1999), 8.

79. Wilkens, *Grab Their Belts*, 228.

80. Kelley, *Where We Were in Vietnam*, 1153.

81. "After Action Report on Ranger Operations in Zone D, Phase III, 'Operation Holiday,' 3 January to 9 January 1963," Headquarters, US MAAG Detachment, Phuoc Binh Thanh Special Zone (Song Be), 17 January 1963, 1.

82. Grange, interview by author, April 4, 2014.

83. "Area Analysis Study, 66-16, War Zone D," Combined Intelligence Center, Headquarters, USMACV, Office of the Assistant Chief of Staff J-2 and Headquarters, Armed Forces of RVN, Office of Joint General Staff J-2, 19 May 1966, 5-1.

84. Ibid., 6-1.

85. Bob Bennett, "The 1970 Raid on War Zone D POW Camp," Delta Troop, 3/17th Air Cavalry, n.d.

86. George J. Veith, *Code-Name Bright Light* (New York: Dell, 1998), 397.

87. Grange, interview by author, April 4, 2014.

88. "Operational Report for Quarterly Period Ending 31 January 1968," 2.

89. Ibid., 3.

90. "2/506th Daily Staff Journal," FSB Dave, January 12, 1968.

91. Rick St John, letter, January 13, 1968.

92. Map Tan Uyen (1:50,000), Series L7014, Sheet 6331 II, Headquarters, 29th Engineer Battalion, 1971.

93. "Human Factors Consideration of Undergrounds in Insurgencies" (Washington, DC: Special Operations Research Office, American University, 1966).

94. "Agroville Program," www.Vietnamwar.net, accessed June 20, 2016.

95. "2/506th Daily Staff Journal," FSB Dave, January 13, 1968.

96. Ibid., January 14, 1968.

97. Chuck Hanson, letter, January 18, 1968.

98. Ibid.

99. Ibid.

100. William Treseder, US Marine Corps veteran, n.d.

101. Chuck Hanson, letter, January 18, 1968.

102. Dan Bernard, interview by author, March 3, 2015.

103. Everett Franks, email, March 31, 2015.

104. Basil Rivera, interview by author, March 4, 2015.

105. "Handbook for US Forces in Vietnam," Headquarters, US Army, Vietnam, June 1966, fig. 37.

106. "2/506th Daily Staff Journal," FSB Dave, January 16, 1968.

107. Ibid.

108. Ibid.

109. Rick St John, letter, January 17, 1968.

110. Chris Backman, interview by author, July 11, 2016.

111. Ibid.

112. Chuck Hanson, letter, January 18, 1968.

113. Rick St John, letter, January 17, 1968.

114. Ibid.

115. Chris Backman, interview by author, July 11, 2016.

116. Everett Franks, interview by author, November 12, 2014.

117. Rick St John, letter, January 17, 1968.

118. Everett Franks, interview by author, November 12, 2014.

119. "2/506th Daily Staff Journal," FSB Dave, January 16, 1968.

120. "Operational Report—Lessons Learned," 21 February 1968, 33.

121. The Virtual Wall (Vietnam Veterans Memorial).

122. "2/506th Daily Staff Journal," FSB Dave, January 17, 1968, 7.

123 "101st Airborne Division Operational Report and Lessons Learned," Headquarters, 101st Airborne Division, January 31, 1968, 7.

124. "2/506th Unit History," Headquarters, 2nd Battalion 506th Infantry, n.p., n.d., 3.

125. Rick St John, letter, January 19, 1968.

126. Sebastian Junger, *Tribe* (New York: Machete Book Group, 2016), 67.

127. Ibid.

128. Statistical Information About Fatal Casualties of the Vietnam War (College Park, MD: National Archives), accessed 2014.

129. Rick St John, letter, January 25, 1968.

130. "Combat Infantry Badge," Headquarters, US Army Human Resources Command, Adjutant General Directorate, accessed June 21, 2016, http://www.hrc.army.mil.com.

131. Ibid.

132. "Special Order 33: Combat Infantryman Badge (First Award)," Headquarters, 101st Airborne Division, February 2, 1968.

133. Caputo, *A Rumor of War*, 216.

134. "Special Order 33," February 2, 1968.

135. "VC Lose by a Nose: Airborne Finds Supply Cache," *Screaming Eagle*, n.d.

136. Mike Tarpley, questionnaire, November 6, 2016.

137. Mike Tarpley, email, May 20, 2015.

138. "VC Denied 42 Tons of Rice; Chinooks Lift Out Cache, *Screaming Eagle* 1, no. 5, March 1, 1968.

139. Steve Lyle, questionnaire, October 2014.

140. www.historyguy.com.

141. Steve Lyle, letter, April 21, 2015.

142. Ibid.

143. Ibid.

144. "VC Denied 42 Tons of Rice," March 1, 1968.

145. Chris Backman, email, March 16, 2014.

146. "101st Airborne Division Operational Report and Lessons Learned," January 31, 1968.

147. Dan Soto, questionnaire, March 12, 2017.

148. Steve Lyle, questionnaire, n.d.

149. Rick St John, letter, January 28, 1968.

150. "VC Denied 42 Tons of Rice," March 1, 1968.

151. Lewis Sorley, *A Better War: The Unexamined Victories and Final Tragedies of America's Last Years in Vietnam* (Orlando, FL: Harcourt, 1999), 56–57.

152. "Operational Report for Quarterly Period Ending 31 January 1968," 31 March 1968.

153. Chris Backman, interview by author, July 11, 2016.

154. Matti Friedman, *Pumpkin Flowers: A Soldier's Story* (Chapel Hill, NC: Algonquin Books of Chapel Hill, 2016), 30.

155. Mike Scott, questionnaire, n.d.

156. Ibid.

157. Gordon L. Rottman, *North Vietnamese Army Soldier 1958–1975* (New York: Osprey Publishing, 2009), 101.

158. Rick Atkinson, *An Army at Dawn: The War in North Africa 1942–1943* (New York: Owl Books, 2002), 673.

159. Caputo, 443.

160. "2/506th Daily Staff Journal," Phuoc Vinh, January 28, 1968.

161. Rick St John, letter, January 29, 1968.

162. Rick St John, *Circle of Helmets: Poetry and Letters from the Vietnam War* (Bloomington, IN: AuthorHouse), 2002), 28.

163. "Empty Helmets Tell Full Story," *Nashville Tennessean*, February 15, 1968.

164. Ibid.

165. "Operational Report for Quarterly Period Ending 31 January 1968," 31 March 1968.

166. Richard W. Stewart, ed., *American Military History, Volume II: The United States Army in a Global Era, 1917–2008*, US Army Center for Military History, Washington D.C., 333.

167. "2/506th Daily Staff Journal," Phuoc Vinh, January 31, 1968.

168. Rick St John, letter, February 1, 1968.

169. Keith Nolan, *Battle for Saigon* (New York: Pocket Books, 1996), 178.

170. Neal Sheehan, *A Bright Shining Lie: John Paul Vann and America in Vietnam* (New York: Modern Library, 2009), 1,921.

171. "Project CHECO Southeast Asia Report: The Defense of Saigon," Headquarters, Pacific Air Force, Directorate of Tactical Evaluation, 14 December 1968, 2.

172. Ibid., 3.

173. "Tet Offensive II Field Force, Vietnam: After Action Report 31 January–18 February 1968," Headquarters, II Field Force, Bien Hoa, n.d., 10.

174. Sheehan, *A Bright Shining Lie*, 1,922.

175. "Tet Offensive II Field Force Vietnam," n.d.

176. Mike Campbell, "Maisey Building Rededication Honors Fallen Hero of Bien Hoa," January 31, 2008.

177. James Arnold, *Tet Offensive 1968: Turning Point in Vietnam* (Long Island City, NY: Osprey Publishing, 2012), 51.

178. Nolan, *Battle for Saigon*, 158.

179. Ibid., 159.

180. "145th Combat Aviation Battalion History, Volume II," Headquarters, 145th Combat Aviation Battalion, n.d.

181. "Tet Offensive II Field Force Vietnam," n.d., A-3.

182. Nolan, *Battle for Saigon*, 159.

183. "2/506th Daily Staff Journal," Phuoc Vinh, January 31, 1968.

184. Joe Palagyi, interview by author, February 12, 2015.

185. Steve Lyle, questionnaire, October 2014.

186. Dave Spencer, email, January 26, 2011.

187. Steve Lyle, questionnaire, October 2014.

188. "Screaming Eagles Maul Enemy: Crush Tet Attacks on Three Fronts," *Screaming Eagle* 1, no. 4, February 23, 1968, 1.

189. Everett Franks, letter, undated.

190. Nolan, *Battle for Saigon*, 179.

191. Ibid.

192. Ibid., 173.

193. Basil Rivera, letter, August 13, 2012.

194. Steve Lyle, questionnaire, October 2104.

195. Basil Rivera, letter, August 13, 2012.

196. Ibid.

197. Ibid.

198. Mike Tarpley, interview by author, May 14, 2015.

199. James Arnold, *Tet Offensive 1968: Turning Point in Vietnam* (Long Island City, NY: Osprey Publishing, 2012).

200. "Five Fruits of the Vietnamese New Year," www.vietnam-culture.com.

201. Ted Tilson, interview by author, April 16, 2016.

202. Steve Lyle, questionnaire, October 2014.

203. Basil Rivera, letter, August 13, 2012.

204. Nolan, *Battle for Saigon*, 182–84.

205. Robert N. Neer, "Napalm's Fiery Fall into Infamy," *Vietnam Magazine* (August 2013), 42.

206. www.encyclopedia2@thefree
dictionary.com.

207. Neer, "Napalm's Fiery Fall into Infamy," 42.

208. Nolan, *Battle for Saigon*, 182–84.

209. "2/506th Daily Staff Journal," Bien Hoa,
February 1, 1968.

210. Ibid., January 31–February 1, 1968.

211. Ibid.

212. Don Starry, *Vietnam Studies: Mounted
Combat in Vietnam*, Headquarters, Department of the Army, Washington, DC, 124.

213. Nolan, 187.

214. Nolan, *Battle for Saigon*, 193.

215. Steve Lyle, questionnaire, October 2014.

216. Ibid.

217. Nolan, *Battle for Saigon*, 193.

218. Steve Lyle, questionnaire, October 2014.

219. Nolan, *Battle for Saigon*, 189.

220. Ibid., 192.

221. Steve Lyle, questionnaire, October 2014.

222. Ibid.

223. Ibid.

224. Ibid.

225. Nolan, *Battle for Saigon*, 192.

226. "2/506th Daily Staff Journal," Bien Hoa,
January 31–February 1, 1968.

227. Chuck Hanson, letter, January 27, 2016.

228. Jim Roach, email, April 11, 2015.

229. "2/506th Daily Staff Journal," Bien Hoa,
January 31–February 1, 1968.

230. Nolan, *Battle for Saigon*, 191.

231. Ibid.

232. Nolan, *Battle for Saigon*, 195.

233. "2/506th Daily Staff Journal," Bien Hoa,
February 4, 1968.

234. "Bien Hoa Attackers Repelled: 150 VC
Killed in Two Days," *Army Reporter*, February 17, 1968.

235. Grange, interview by author, April 23, 2014.

236. Ibid.

237. Brandon Gold and Stanley Horowitz,
History of Combat Pay, Institute for Defense Analyses, Alexandria, VA, 7.

238. Ibid.

239. *A System Analysis View of the Vietnam War
1965–1972, Volume 8, Casualties and Losses*,
OASD (SA), Southeast Asia Division,
Washington, DC, 114.

240. Joe Adams, interview by author, January
21, 2015.

241. Harry Brown, letter, February 9, 1968.

242. Tom McClear, letter, February 28, 1968.

243. "2/506 Daily Staff Journal," Phuoc Vinh,
December 1967–January 1968.

244. Atkinson, *An Army at Dawn*, 443.

245. Grange, interview by author, September
11, 2014.

246. Wilkins, *Grab Their Belts to Fight Them*, 48.

247. www.gruntonline.com.

248. Wilkins, *Grab Their Belts to Fight Them*, 49.

249. "2/506th Daily Staff Journal," FSB
Concord, February 28, 1968.

250. Ibid.

251. DA Pam 360-521, "Handbook For US
Forces in Vietnam," June 1966, 93.

252. Thomas B. Bennett, "Night Jungle Operations," School of Advanced Military
Studies, US Army Command and General Staff College, Fort Leavenworth, KS
(1992–93), 23.

253. Dan Bernard, interview by author, March
3, 2015.

254. "2/506th Daily Staff Journal," FSB
Concord, March 1, 1968.

255. "Operational Report for Quarterly
Period Ending 31 January 1968," 31
March 1968, 44.

256. Tom McClear, letter, March 4, 1968.

257. Ibid., March 7, 1968.

258. FM 6-141-1, "Field Artillery Target Analysis
and Weapons Employment," Headquar-
ters, Department of the Army, Washington, DC, February 15, 1978, B-2, fig. B-5.

259. "Report of the Seminar on Night Operations in RVN," Headquarters, US Army,
Vietnam, April 13, 1968, 1.

260. Harry Brown, letter, March 1, 1968.

261. Ibid.

262. "Report of the Seminar on Night Operations in RVN," 2–3.

263. Harry Brown, letter, March 1, 1968.

264. Ibid.

265. Jack Hurst, "2-506 Patrol Traps Reds in
Ambush." *Screaming Eagle* 1, no. 9, March
29, 1968.

266. Ibid.

267. Ibid.

268. Joe Palagyi, interview by author, February 12, 2015.

269. Chuck Hanson, letter, March 3, 1968.

270. Dan Bernard, interview by author, March 3, 2015.

271. Hurst, "2-506 Patrol," March 29, 1968.

272. Harry Brown, letter, March 1, 1968.

273. Dan Bernard, interview by author, March 3, 2015.

274. Chuck Hanson letter, March 3, 1968.

275. Marshall and Hackworth, *Vietnam Primer*, 35.

276. "2/506th Daily Staff Journal," FSB Concord, March 1, 1968.

277. Hurst, "2-506 Patrol," March 29, 1968.

278. Harry Brown, letter, March 1, 1968.

279. Chris Backman, interview by author, July 11, 2016.

280. Chuck Kudla, interview by author, April 4, 2015.

281. "2/506th Daily Staff Journal," FSB Concord, March 1, 1968.

282. Dan Bernard, interview by author, March 3, 2015.

283. Tom McClear, letter, March 7, 1968.

284. Rick St John, letter, February 8, 1968.

285. "Line Troopers Spend Rare Night at Home," *Screaming Eagle* 1, no. 27, August 5, 1968.

286. Ibid.

287. Ibid.

288. Ibid.

289. Ibid.

290. Rick St John, letter, February 18, 1968.

291. Chuck Kudla, email, January 2, 2010.

292. Chuck Limer, questionnaire, October 2014.

293. "2/506th Daily Staff Journal," FSB Concord, March 5, 1968.

294. Ibid.

295. Harry Brown, letter, March 7, 1968.

296. "2/506th Daily Staff Journal," FSB Concord, March 6–11, 1968.

297. Mike Tarpley, email, July 7, 2016.

298. Rick St John, letter, March 16, 1968.

299. "2/506th Daily Staff Journal," FSB Concord, March 12–25, 1968.

300. Chris Backman, interview by author, July 11, 2016.

301. Rick St John, letter, March 24, 1968.

302. Ibid.

303. "Operational Report of 3rd Brigade, 101st Air Cavalry Division, for Period Ending 30 April 1968," Headquarters, 101st Air Cavalry Division, August 12, 1968.

304. "2/506th Daily Staff Journal," FSB Concord, March 26–29, 1968.

305. Everett Franks, email, July 13, 2016.

306. Rick St John, letter, April 13, 1968.

307. St John, *Circle of Helmets*, 32.

308. Ibid., 31.

309. Harry Brown, letter, March 29, 1968.

310. Doc Franks, email, August 6, 2016.

311. Philip Caputo, *A Rumor of War* (New York: Owl Books, 1977).

312. LTC Edwin T. Cook, "AMEDD Study on Combat Stress and Its Effects," Medical Field Service School, Fort Sam Houston, TX., Chapter 6 Combat's Bloodless Casualties.

313. Jim Roach, interview, November 21, 2011.

314. Ibid., 14.

315. Joe Palagyi, interview by author, February 2, 2015.

316. Study 67-080, "NVA Rocket Artillery Units," Headquarters, Armed Forces of RVN, Office of Joint General Staff J-2, September 1, 1967, 10–14.

317. "2/506th Daily Staff Journal," FSB Concord, April–May 1968.

318. Rick St John, letter, May 1, 1968.

319. Ibid., March 19, 1968.

320. Matti Friedman, *Pumpkin Flowers: A Soldier's Story* (Chapel Hill, NC: Algonquin Books of Chapel Hill, 2006), 11.

321. "2/506th Daily Staff Journal," FSB Concord, April 1, 1968.

322. Ibid., April 4, 1968.

323. Harry Brown, letter, April 10, 1968.

324. "2/506th Daily Staff Journal," FSB Concord, April 4, 1968.

325. Harry Brown, letter, April 10, 1968.

326. Ibid.

327. 2/506th Daily Staff Journal", FSB Concord, May 1, 1968..

328. www.historyplace.com/Vietnamtimeline.

329. Chuck Hanson, letter, April 2, 1968.

330. www.historyplace.com/Vietnamtimeline.

331. *Screaming Eagle* 1, no. 9, March 29, 1968.

332. Mike Scott, email, February 7, 2014.

333. Mike Scott, questionnaire, n. d.

334. Ibid.

335. *Medical Support of the US Army in Vietnam, 1965–1970,* Headquarters, Department of the Army, Washington, DC, 1991, 51.

336. Ibid., 65.

337. Mike Tarpley, interview by author, May 2015.

338. Mike Tarpley, questionnaire, November 6, 2016.

339. Harry Brown, letter, April 10, 1968.

340. Ibid, February 27, 1968.

341. Rick St John, letter, August 19, 1968.

342. "25th Infantry Division Operational Report and Lessons Learned, Period Ending 31 July 1968," Headquarters, 25th Infantry Division, August 1, 1968, 33.

343. Ibid.

344. *Medical Support of the US Army in Vietnam,* 36.

345. Ibid.

346. *A Study of the Strategic Lessons Learned in Vietnam: Volume VII, Soldier in RVN,* BDM Corporation, McLean, VA, April 11, 1980, EX-9.

347. Rick St John, letter, May 3, 1968.

348. MG O. M. Barsanti, "Commander's Corner," *Screaming Eagle* 1, no. 5, March 1, 1968.

349. "Handbook for US Forces in Vietnam," sec. XXI.

350. Chuck Limer, email, December 21, 2009.

351. "2/506th Daily Staff Journal," FSB Concord, April 26, 1968.

352. Rick St John, letter, April 26, 1968.

353. Ibid., April 18, 1968.

354. *Study of Strategic Lessons,* EX-10.

355. Ibid.

356. Mike Tarpley, William Beaumont Hospital Escape letter, n.d.

357. Rick St John, letter, April 13, 1968.

358. Chuck Hanson, letter, May 20, 1968.

359. "2/506th Daily Staff Journal," Phuoc Vinh, May 22–26, 1968.

360. Chuck Limer, email, December 17, 2007.

361. James Willbanks, *Vietnam War: The Essential Reference Guide* (Santa Barbara, CA: ABC-CLIO, 2013), 42.

362. Kelley, *Where We Were in Vietnam,* 281.

363. Chris Backman, letter, June 14, 1968.

364. "Handbook for US Forces in Vietnam," sec. IX, para. 36.

365. "Combat Operations After Action Report for Period 24 May–12 June 1968," Headquarters, 4th Infantry Division, June 13, 1968, 2.

366. Ibid., 7.

367. Ibid.

368. "4th Infantry Division Operational Report and Lessons Learned for Period Ending 31 July 1968," Headquarters, 4th Infantry Division, December 23, 1968, 14.

369. Ibid., 43.

370. Ibid., 3.

371. "Combat Operations After Action Report for Period 24 May–12 June 1968," Headquarters, 4th Infantry Division, June 13, 1968, 5.

372. Ibid.

373. Rick St John, letter, May 26, 1968.

374. "Combat Operations After Action Report—Operation Mathews, 4th Infantry Division, Period 24 May–12 June 1968," Headquarters, 4th Infantry Division, 31 January 1969.

375. "3rd Brigade Operational Report and Lessons Learned for Period Ending 31 July 1968," Headquarters, 3rd Brigade, 101st Air Cavalry Division, 12 August 1968.

376. Carolyn Lauer, "The Daisy Cutter Bomb: Largest Conventional Bomb in Existence," *Technical Review,* n.d.

377. Rick St John, letter, May 27, 1968.

378. "Combat Operations After Action Report—Operation Mathews," 13 June 1968, 9.

379. Ibid.

380. Kelley, *Where We Were in Vietnam,* 348.

381. Shelby S. Stanton, *Vietnam Order of Battle: A Complete Illustrated Reference to US Army Combat and Support Forces in Vietnam, 1961–1973* (Mechanicsburg, PA: Stockpile Books, 2003), 246.

382. "Operational Report—Lessons Learned for Quarterly Period Ending 21 July 1968," Headquarters, 4th Infantry Division, 18 August 1968, 4.

383. "2/506th Daily Staff Journal," Dak Pek, May 28, 1968.

384. Map D Go Kram, Vietnam, and Laos (1:50,000) Sheet 6539 III, Series L7014, US Army 1966.

385. Harry Brown, letter, May 29, 1968.

386. Map D Go Kram.

387. Harry Brown, letter, May 29, 1968.

388. "2/506th Daily Staff Journal," Dak Pek, May 30, 1968.

389. Rick St John, letter, May 30, 1968.

390. John T. Correll, "Arc Light," *Air Force Magazine*, no .1 (2009).

391. "4th Infantry Division Operational Report and Lessons Learned, Period Ending 31 July 19- 68," 18 August 1968, 4.

392. Rick St John, letter, May 31, 1968.

393. Letter, incident investigation, Headquarters, 2nd Battalion, 506th Infantry, Dak To, 3 June 1968.

394. Chuck Limer, interview by author, December 11, 2014.

395. Letter, incident investigation.

396. "Chronological Summary of Significant Events, 1 May – 31 July 1968, Headquarters, 4th Infantry Division, n. d., 8.

397. Doc Franks, email, November 21, 2015.

398. Larry Burton, phone call with author, September 29, 2016.

399. Chuck Limer, questionnaire, October 2014.

400. Marshall and Hackworth, *Vietnam Primer*, 7.

401. Chuck Limer, questionnaire, October 2014.

402. Chuck Limer, letter, February 23, 2016.

403. Ibid.

404. Doc Franks, interview by author, November 12, 2014.

405. Chuck Limer, questionnaire, October 2014.

406. Ibid.

407. Chuck Limer, letter, February 23, 2016.

408. www.212warriors.com.

409. Doc Franks, email, n. d.

410. www.1stcavmedic.com.

411. "4th Infantry Operational Report and Lessons Learned for Period Ending 31 July 1968," 23 December 1968, 48.

412. Doc Franks, interview by author, November 12, 2014.

413. Larry Burton, interview by author, June 23, 2015.

414. Rick St John, letter, June 3, 1968.

415. Ibid.

416. Dan Bernard, interview by author, March 3, 2015.

417. Ibid.

418. Doc Franks, email, November 21, 2015.

419. Doc Franks, email, June 10, 2015.

420. Larry Burton, phone call with author, September 29, 2016.

421. Ibid.

422. Doc Franks, interview by author, November 11, 2014.

423. "2/506th Daily Staff Journal," Dak To, June 2–9, 1968.

424. Rick St John, letter, June 19, 1968.

425. "2/506th Daily Staff Journal," Dak To, June 10–14, 1968.

426. Doc Franks, email, November 21, 2015.

427. Jon Steinberg, "Recovery Center Puts GIs in the Pink, *Stars & Stripes*, September 27, 1969.

428. Ibid.

429. General Order 10, Department of the Army, Headquarters, 6th Convalescent Center, June 13, 1968.

430. "Operational Report of the 3rd Brigade, 101st Air Cavalry Division, for the Period Ending 31 July 1968," August 12, 1968, 4.

431. *Vietnam Studies: Tactical and Material Innovations,* Center for Military History Publication 90-21-1, Headquarters, Department of the Army, Washington, DC, n. d., 148.

432. Joe Adams, interview by author, January 21, 2015.

433. Kelley, *Where We Were in Vietnam*, 564.

434. Marshall and Hackworth, *Vietnam Primer*, 7.

435. "3rd Brigade Surprises VC: Use New Tactics," *Screaming Eagle* 1, no. 24, July 15, 1968.

436. Ibid.

437. Dan Bernard, interview by author, March 3, 2015.

438. "2/506th Daily Staff Journal," Cu Chi Base Camp, June 15, 1968.

439. Chuck Kudla, email, October 2, 2012.

440. Ibid.

441. "2/506th Daily Staff Journal," Cu Chi Base Camp, June 15, 1968.

442. Dan Bernard, interview by author, March 3, 2015.

443. Mike Scott, questionnaire, n.d.

444. "2/506th Daily Staff Journal," Cu Chi Base Camp, June 15, 2016.

445. *Infantry in Battle*, 2nd ed. (Washington, DC: *Infantry Journal*, 1939), 16.

446. Joe Adams, interview by author, January 21, 2015.

447. www.vetshome.com

448. "2/506th Daily Staff Journal," Cu Chi Base Camp, June 15, 1968.

449. Ibid., June 16, 1968.

450. Ibid., June 17, 1968.

451. "Operational Report of 101st Air Cavalry Division for Period Ending 31 July 1968," 15 August 1968, 8.

452. Julius Becton, *Autobiography of a Soldier and Public Servant* (Annapolis, MD: Naval Institute Press, 2008), 88.

453. "Airborne All the Way," *Screaming Eagle* 1, no. 31, September 16, 1968, 1.

454. "2/506th Daily Staff Journal," Cu Chi Base Camp, June 18, 1968.

455. Ibid.

456. *Analysis of Communist Vietnamese Special Operations Forces During the Vietnam War* (Naval Postgraduate School, June 2005), v.

457. "2/506th Daily Staff Journal," FSB Lela, June 18, 1968.

458. Ibid., June 19, 1968.

459. Ibid., June 20–24, 1968.

460. "2/506th Daily Staff Journal," Phuoc Vinh Base Camp, June 25, 1968.

461. Ibid., July 2–6, 1968.

462. Ibid., July 5, 1968.

464. Jim Roach, interview, November 21, 2011.

465. "2/506th Daily Staff Journal," Phuoc Vinh Base Camp, July 7–13, 1968.

466. "Operational Report of 3d Brigade, 101st Air Cavalry Division for Period Ending 31 July 1968," 12 August 1968, 18.

467. "3rd Brigade Daily Staff Journal," 15–21 July 1968.

468. Ibid.

469. Philip Caputo, *A Rumor of War* (New York: Owl Books, 1977), 160.

470. Tom Mangold and John Penycate, *The Tunnels of Cu Chi: A Harrowing Account of America's "Tunnel Rats" in the Underground Battlefields of Vietnam* (New York: Random House, 2005), 49.

471. Ibid., 53.

472. www.212warrior.com.

473. Recommendation for Award of the Presidential Unit Citation, Inclosure L, Headquarters, 3rd Brigade, 101st Airborne Division (Airmobile), 25 November 1968, 2.

474. *Currahee Newsletter*, n.d.

475. "3rd Brigade Daily Staff Journal," 15–21 July 1968.

476. Ibid.

477. "Ground Forces Commander's Daily Situation Report (SITREP)," 3rd Brigade, 101st Air Cavalry Division, 22 July 1968.

478. "Award of the Air Medal with "V" Device, 2/506th, 101st Airborne Division, 19 October, 1968.

479. www.tutlemap.net/climate/cu_chi.

480. Larry Burton, interview by author, 2014.

481. "Operational Report of 3rd Brigade, 101st Air Cavalry Division, for Period Ending 31 July 1968," 12 August 1968, 6.

482. Kip Lindberg, "The Use of Riot Control Agents During the Vietnam War," *Army Chemical Review*, January–June 2007, 51.

483. Larry Burton, interview by author, 2014.

484. Ibid., January 2, 2016.

485. "The VC as an Enemy," Office of the Adjutant General, Department of the Army, Washington, DC, 1 July 1966, 4.

486. Larry Burton, interview by author, January 2, 2016.

487. Ibid.

488. Basil Rivera, interview by author, March 4, 2015.

489. www.gruntonline.com/NVAandVC.

490. Friedman, *Pumpkin Flowers*, 48.

491. Dan Bernard, interview by author, March 3, 2015.

492. Basil Rivera, interview by author, March 4, 2015.

493. Ibid.

494. Ibid.

495. General Order 150, Headquarters, U S Army, Vietnam, January 14, 1969.

496. Larry Burton, interviews by author, 2014 and January 2, 2016.

497. Ibid.

498. Joe Adams, interview by author, January 21, 2015.

499. Basil Rivera, interview by author, March 4, 2015.

500. Terry Van Meter, interview by author, May 2, 2017.

501. Rick St John, letter, July 24, 1968.

502. www.virtualwar.org.

503. Ibid.

504. Basil Rivera, interview by author, March 4, 2015.

505. "2 Area Men Cited Posthumously," n.p., n.d.

506. Dianne Wheeler Gnass , email, March 24, 2017.

507. Ibid., January 2011.

508. Ibid., April 14, 2016.

509. "Eaton County Inductees Leave for Service," *Eaton Rapids (MI) Journal*, December 15, 1966.

510. Larry Burton, interview by author, January 2, 2014.

511. Johnny Walker, email, October 30, 2015.

512. Mike Scott, questionnaire, n.d.

513. www.virtualwall.org.

514. Ibid.

515. Rick St John, letter, July 24, 1968.

516. Ibid.

517. Joe Palagyi, interview by author, February 12, 2015.

518. Rick St John, letter, July 24, 1968.

519. Joe Palagyi, interview by author, February 12, 2015.

520. Joe Adams, interview by author, January 21, 2015.

521. Joe Palagyi, interview by author, February 12, 2015.

522. Terry Van Meter, interview by author, May 2, 2017.

523. Ibid.

524. Rick St John, letter, July 25, 1968.

525. Recommendation for the Award of the Presidential Unit Citation, Headquarters, 3rd Brigade, 101st Airborne Division, 25 November 1968, 3.

526. Ibid.

527. "2/506th Daily Staff Journal," Phuoc Vinh, 25 July 1968.

528. Ibid.

529. Ibid.

530. Recommendation for the Award of the Presidential Unit Citation, 3rd Brigade.

531. Major Russell W. Glenn, *Men Against Fire in Vietnam*, School of Advanced Military Studies, US Army Command and General Staff College, Fort Leavenworth, KS, November 8, 1987, 37.

532. Basil Rivera, interview by author, March 4, 2015.

533. Rick St John, letter, July 25, 1968.

534. Basil Rivera, interview by author, March 4, 2015.

535. Dan Soto, interview by author, October 24, 2016.

536. www.virtualwall.com.

537. "Soldier Killed in Vietnam," *Vancouver (WA) Columbian*, n.d.

538. Gordon L Rottman, *US Army Infantryman in Vietnam 1965 – 1973* (Osprey Publishing, 2005), 129.

539. Joe Adams, interview by author, January 21, 2015.

540. www.virtualwall.org.

541. Dan Bernard, interview by author, March 3, 2015.

542. Ibid.

543. Rick St John, letter, August 6, 1968.

544. Dan Bernard, email, May 26, 2017.

545. "Mortuary Affairs Activity—Republic of Vietnam, compiled by Mortuary Activities Center, Fort Lee, VA, March 2000.

546. Matthew M. Burke, "For Those Who Prepared Vietnam's Fallen, a Lasting Dread," *Stars & Stripes*, November 9, 2014.

547. Steve Lyle, questionnaire, n.d.

548. Chris Backman, interview by author, July 11, 2016.
549. Mrs. William J. Tellis, letter, August 9, 1968.
550. Rick St John, letter, August 21, 1968.
551. "2/506th Daily Staff Journal," FSB Patton, 27 July 1968.
552. Ibid.
553. Caputo, *A Rumor of War*, 483.
554. www.vietnam-warfare.tripod.com.
555. "2/506th Staff Journal, FSB Patton, 27 July 1968.
556. Ibid.
557. Ibid.
558. "Mines and Booby Traps," Headquarters, 3rd Brigade Task Force, 25th Infantry Division, 5 August 1967, 1.
559. Rick St John, letter, August 13, 1968.
560. Dan Bernard, interview by author, March 3, 2015.
561. Basil Rivera, interview by author, March 4, 2015.
562. Ibid.
563. Dan Bernard interview by author, March 3, 2015.
564. Gordon L. Rottman, *North Vietnamese Army Soldier 1958–1975* (New York: Osprey Publishing, 2009), 261.
565. Ibid., 260.
566. Dan Bernard, interview by author, March 3, 2015.
567. Ibid.
568. Harry Brown, letter, July 31, 1968.
569. Ibid., July 27, 1968.
570. Rick St John, letter, July 28, 1968.
571. "2/506th Daily Staff Journal," FSB Patton, 2 August 1968.
572. Dan Soto, interview by author, October 24, 2016.
573. Ibid.
574. Rick St John, letter, August 6, 1968.
575. "Operational Report for 3rd Brigade, 10st Airborne Division (Airmobile) for Period Ending 31 October 1968," 15 November 1968.
576. "2/506th Daily Staff Journal," FSBs Patton and Judy, 7–25 August 1968.
577. Tom McClear, letters, 30 July and 6 August 1968.
578. Mike Scott, email, October 28, 2013.
579. Rick St John, letter, August 3, 1968.
580. Mike Scott, email, February 7, 2014.
581. Dandridge M. Malone, *Soldiers*, n.d.
582. Bravo Company Roster, FSB Patton, 12 August 1968.
583. Ted Tilson, questionnaire, November 2016.
584. Mike Tarpley, email, July 7, 2016.
585. Caputo, *A Rumor of War*, 11.
586. *Currahee Newsletter*, 2nd ed., n.d.
587. Rick St John, letter, August 11, 1968.
588. Thomas Lamb, letter, July 28, 1968.
589. Rick St John, letter, July 28, 1968.
590. Intelligence Summary 242-68, 3rd Brigade, 101st Airborne Division (Airmobile), 29 August 1968.
591. Ibid.
592. "2/506th Daily Staff Journal," FSB Judy, 22–26 August 1968.
593. Rick St John, letter, August 26, 1968.
594. Ibid., August 27, 1968.
595. "2/506th Daily Staff Journal," FSB Judy, 27 August 1968.
596. Harry Brown, letter, August 31, 1968.
597. "2/506th Daily Staff Journal," FSB Judy, 27 August 1968.
598. Ibid.
599. Harry Brown, letter, August 31, 1968.
600. www.ugca.org (Utah Gun Club Collectors' Association).
601. Mike Scott, email, April 2015.
602. FM 23-11, "90MM RECOILLESS RIFLE, M-67," Headquarters, Department of the Army, Washington, DC, 6 July 1965.
603. Mike Scott, email, April 2015.
604. Ibid.
605. Ibid.
606. "2/506th Daily Staff Journal," FSB Judy, 28 August 1968.
607. Ibid.
608. Ibid.
609. Ibid.
610. Ibid.
611. Ibid.
612. www.tutiempo.net/en/climate/cu_chi.

326

TIGER BRAVO'S WAR

613. "2/506th Daily Staff Journal," FSB Judy, 28 August 1968.
614. Ibid.
615. Ibid.
616. "By Courage and Faith: A Love Story Terry '66 and Jacquie Van Meter, www.alumni.norwich.edu/vanmeter, 3.

617. Terry Van Meter, interview by author, May 2, 2017.
618. Basil Rivera, interview by author, March 22, 2016.
619. Basil Rivera, email, October 15, 2012.
620. Mike Scott, email, n.d.
621. General Order 8630, Award of the Silver Star, Headquarters, 101st Airborne Division, TCO 320, 30 October 1968.
622. Harry Brown, letter, August 31, 1968.
623. Ibid.
624. Ibid.
625. Mike Scott, email, October 28, 2013.
626. Terry Van Meter, interview by author, May 2, 2017.
627. Terry Van Meter, letter, September 12, 1968.
628. "2/506th Daily Staff Journal," FSB Judy, 28 August 1968.
629. Ibid., 29 August 1968.
630. Recommendation for Award of the Presidential Unit Citation, Headquarters, 3rd Brigade, 101st Airborne Division (Airmobile), 25 November 1968, 11.
631. www.virtualwall.com.
632. Mike Scott, email, February 7, 2014.
633. www.virtualwall.org.
634. "2/506th Daily Staff Journal," FSB Judy, 29 August 1968.
635. Rick St John, letter, August 30, 1968.
636. "2/506th Daily Staff Journal," FSB Pershing, 30 August 1968.
637. John O. Childs, email, May 26, 2017.
638. John O. Childs, questionnaire, November 14, 2014.
639. "2/506th Daily Staff Journal, FSB Pershing, 1–4 September 1968.
640. John O. Childs, email, May 26, 2017.
641. "Operational Report of 3rd Brigade, 101st Airborne Division (Airmobile) for Period Ending 31 October 1968," 3.

642. Ibid.
643. John O. Childs, questionnaire, November 14, 2014.
644. Ibid.
645. "Operational Report of 3rd Brigade, 101st Airborne Division (Airmobile) for Period Ending 31 October 1968," 3.
646. Recommendation for Award of the Presidential Unit Citation, Headquarters, 3rd Brigade, 101st Airborne Division (Airmobile), 25 November 1968.
648. John O. Childs, questionnaire, November 14, 1968.
649. Ibid.
650. Rick St John, letter, September 7, 1968.
651. Ibid.
652. Tom McClear, letter, September 23, 1968.
653. "Viet Cong Use Human Shields," September 6, 1968.
654. Ibid.
655. "2/506th Daily Staff Journal," FSB Pershing, 7 September 1968.
656. Ibid., 8–14 September 1968.
657. Ibid.
658. Ibid.
659. Rick St John, letter, September 15, 1968.
660. "2/506th Daily Staff Journal," FSB Pershing, 15 September 1968.
661. Rick St John, letter, September 16, 1968.
662. "2/506th Daily Staff Journal," FSB Pershing, 15–17 September 1968.
663. Ibid.,19 September 1968.
664. Ibid., 20–22 September 1968.
665. Rick St John, letter, 23 September 1968.
666. Tom McClear, letter, September 23, 1968.
667. "2/506th Daily Staff Journal," Cu Chi Base Camp, 25 September 1968.
668. "Services Herbs Tape: A Record of Helicopter and Ground Spraying Missions, Aborts, Leaks, and Incidents," US Army and Joint Services Environmental Support Group, Washington DC, author compilation of various tables.
669. Greg Knowlton, *Agent Orange: Phuoc Vinh Ground Zero*, 1.
670. John O. Childs, email, July 14, 2016.
671. Steve Lyle, email, June 19, 2016.

672. Jim Roach, email, June 27, 2016.

673. Doc Franks, email, July 13, 2016.

674. Chris Backman, interview by author, July 11, 2016.

675. Doc Franks, email, July 13, 2016.

676. Ted Tilson, email, July 17, 2016.

677. Joe Palagyi, phone call with author, April 10, 2017.

678. *The Air Force and Herbicides in Southeast Asia, 1961–1971*, Headquarters, US Air Force, Office of Air Force History, Washington, DC, 1982, 9.

679. Ibid., iii.

680. Clyde Haberman, "Agent Orange's Long Legacy, for Vietnam and Veterans," *New York Times*, May 11, 2014.

681. William F. Warren, *A Review of the Herbicide Program in South Vietnam*, commander in chief, Pacific, Scientific Advisory Group, August 1968.

682. *The Air Force and Herbicides in Southeast Asia, 1961–1971*, appendix 1, 1.

683. Letter with attached speech, Admiral (Ret.) Elmo Zumwalt, 9 December 1991.

684. Ibid.

685. "Interrogation Report, Subject: Defoliation Operations in War Zone D," Combined Military Interrogation Center, 22 June 1967.

686. www.va.gov.

687. Knowlton, *Phuoc Vinh: Agent Orange Ground Zero*, 2.

688. Harry Brown, letter, September 13, 1968.

689. John Childs, email, July 14, 2016.

690. "Memorial Service, 2nd Battalion, 506th Infantry Airborne, Phuoc Vinh, Republic of Vietnam," n.d., 2.

691. www.vetshome.com.

692. "Operational Report of Headquarters, 3rd Brigade, 101st Airborne Division for Period Ending 31 January 1969," 5 February 1969, 2.

693. *After Action Report for Operation Golden Sword*, Headquarters, 3rd Brigade, 101st Airborne Division, 11 November 1968, 1.

694. "2/506th Daily Staff Journal," Phuoc Vinh, 30 September 1968.

695. www.u-s-history.com, accessed July 12, 2016.

696. Rick St John, letter, 4 October 1968.

697. Joe Adams, interview by author, January 21, 2015.

698. Kelley, *Where We Were in Vietnam*, 172.

699. John O. Childs, letter, November 14, 2014.

700. Joe Adams, interview by author, January 21, 2015.

701. John O. Childs, letter, November 14, 2014.

702. Stewart, ed., *American Military History, Volume II*, 353.

703. *A Study of Strategic Lessons Learned in Vietnam, Volume VII*, 3-2.

704. Ibid.

705. "2/506th Daily Staff Journal," FSB Bastogne, 1–31 October 1968.

706. John O. Childs, email, July 14, 2016.

707. Ibid.

708. "2/506th Daily Staff Journal," FSB Bastogne, 1–31 October 1968.

709. Joe Adams, interview by author, January 21, 2015.

710. "2/506th Daily Staff Journal," FSB Bastogne, 23 October 1968.

711. Joe Adams, interview by author, January 21, 2015.

712. *Project CHECO Southeast Asia Report: Rolling Thunder, January 1967–November 1968*, Headquarters, Pacific Air Force (Hickam AFB, HI), 1 October 1969, 1.

713. Ibid., 38.

714. www.history.com.

715. Project CHECO Southeast Asia Report: Rolling Thunder January 1967–November 1968, 39.

716. Message from the commanding general, 101st Airborne Division, 1 November 1968.

717. "2/506th Daily Staff Journal," Camp Evans, 7 November 1968.

718. www.countrystudies.us/vietnam/33.

719. Bob Pagano, interview by author, July 11, 2016.

720. "Operational Report of Headquarters, 3rd Brigade, 101st Airborne Division for Period Ending 31 January 1968," 2.

721. Bob Pagano, interview by author, July 11, 2016.

722. John Childs, letter, November 14, 2014.

723. *2nd Battalion (Airborne), 506th Infantry Unit History*, appendix A, n.d.

724. Bob Pagano, interview by author, July 11, 2016.

725. Warren Kiilehua, email, May 25, 2017.

726. Ibid.

727. "2/506th Daily Staff Journal," Camp Evans, 12 November 1968.

728. *www.virtualwall.com*.

729. Bob Pagano, interview by author, July 11, 2016.

730. Ibid.

732. "Operational Report of the 3rd Brigade, 101st Airborne Division (Airmobile) for the Period Ending 31 October 1968," Camp Eagle, 15 November 1968, 7.

734. John O. Childs, email, May 26, 2017.

735. Warren Kiilehua, interview by author, May 31, 2017.

736. Ibid.

737. "2/506th Daily Staff Journal," Camp Evans, 24–31 November 1968.

738. Rick St John, letter, 21 November 1968.

739. Ibid., 25 November 1968.

740. "Operational Report and Lessons Learned, 101st Airborne Division, for Period Ending 31 January 1969," 24 February 1969, 16.

741. Mike Tarpley, email, n.d.

742. Steve Lyle, questionnaire, November 30, 2016.

743. Chuck Limer, email, December 8, 2009.

744. Doc Franks, email, December 25, 2016.

745. Chuck Hanson, letter, November 19, 1968.

746. Basil Rivera, questionnaire, February 2, 2017.

747. Dan Soto, questionnaire, March 12, 2017.

748. Stephen E. Ambrose, *Band of Brothers: E Company, 506th Regiment, 101st Airborne Division from Normandy to Hitler's Eagle's Nest* (New York: Simon & Schuster, 2001), 819.

749. Harry Brown, letter. July 27, 1968.

750. "Personnel Pending Awards," Headquarters, 2/506 Infantry, 16 November 1968, 3.

751. Chuck Limer, questionnaire, October 2014.

752. www.506infantry.org.

753. Robert H. Scales, "Life Is like a Box of C Rations," speech at the Truman Library, September 12, 2009.

754. Dennis J. Stauffer, "The Bitter Homecoming," *Grand Rapids (MI) Press*, December 5, 1982.

755. Chuck Hanson, letter, January 9, 2017.

759. Caputo, *A Rumor of War*, 2.

760. www.virtualwall.org

Acknowledgments

While putting pen to paper is a solitary exercise, researching and writing a history book is certainly not one. *Tiger Bravo's War* could not have been written without the unstinting support, words of encouragement and extensive contributions of my Tiger Bravo brothers-in-arms and their families. Each, in their own way, opened themselves up to me and shared a piece of their life that was not always easy to revisit. All too often, the memories were gut-wrenching and highly personal. For that level of commitment and raw honesty, I am eternally grateful.

During this journey, I have discovered a wonderful community of professionals who graciously offered their time and talents to help me take a dream and shape it into literature. First among those has to be Michael Lund, Professor Emeritus of English at Longwood University and Director of Home & Abroad, a free writing program for military, veterans and family. He has been my much needed and often called upon mentor from the beginning. I also received invaluable assistance from Pete Molin, Mentoring Program Coordinator, of the Veterans Writing Project — another organization dedicated to helping veterans tell their stories. Also, my deep appreciation goes to Ron Toelke, of Toelke Associates, for his marvelous graphic design skills and collaborative nature — he brought a level of expertise to the project which was well beyond my meager knowledge of book design and publishing.

I would also like to thank my wife, Susan, for understanding my deep connection with the Vietnam War, and for giving me the space to process my all-to-often consuming memories of the war and put them into book form.

About the Author

Rick St John — storyteller, poet and historian — has authored a wide range of creative works, from light-hearted children's stories to poetry of the Vietnam War. In his latest book, *Tiger Bravo's War*, he returns to Vietnam to chronicle the exploits of a company of young paratroopers in the 101st Airborne Division during some of the war's heaviest fighting. A story he knows only so well, as he was one of those paratroopers. With two successful careers behind him, retiring as a US Army Colonel in 1993 and as a Group Executive of a global, financial transactions processing company in 2012, Rick is now working on his "bucket list" — teaching and writing. He lives and writes in the woods, by a small lake, in South Georgia.

Left: Rick St John, outside Tiger Bravo headquarters, Phuoc Vinh Base Camp, date unknown. Right: Rick St. John today.